T0252684

The Welfare of Cattle

Edited by
Terry Engle
Donald J. Klingborg
Bernard E. Rollin

CRC Press
Taylor & Francis Group
Boca Raton London New York

CRC Press is an imprint of the
Taylor & Francis Group, an **informa** business

CRC Press
Taylor & Francis Group
6000 Broken Sound Parkway NW, Suite 300
Boca Raton, FL 33487-2742

First issued in paperback 2020

© 2019 by Taylor & Francis Group, LLC
CRC Press is an imprint of Taylor & Francis Group, an Informa business

No claim to original U.S. Government works

ISBN-13: 978-1-138-19706-0 (hbk)
ISBN-13: 978-0-367-89349-1 (pbk)

This book contains information obtained from authentic and highly regarded sources. Reasonable efforts have been made to publish reliable data and information, but the author and publisher cannot assume responsibility for the validity of all materials or the consequences of their use. The authors and publishers have attempted to trace the copyright holders of all material reproduced in this publication and apologize to copyright holders if permission to publish in this form has not been obtained. If any copyright material has not been acknowledged please write and let us know so we may rectify in any future reprint.

Except as permitted under U.S. Copyright Law, no part of this book may be reprinted, reproduced, transmitted, or utilized in any form by any electronic, mechanical, or other means, now known or hereafter invented, including photocopying, microfilming, and recording, or in any information storage or retrieval system, without written permission from the publishers.

For permission to photocopy or use material electronically from this work, please access www.copyright.com (http://www.copyright.com/) or contact the Copyright Clearance Center, Inc. (CCC), 222 Rosewood Drive, Danvers, MA 01923, 978-750-8400. CCC is a not-for-profit organization that provides licenses and registration for a variety of users. For organizations that have been granted a photocopy license by the CCC, a separate system of payment has been arranged.

Trademark Notice: Product or corporate names may be trademarks or registered trademarks, and are used only for identification and explanation without intent to infringe.

Library of Congress Cataloging-in-Publication Data

Names: Engle, Terry (Terry Eugene), 1971- editor.
Title: The welfare of cattle / editors: Terry Engle,
Donald J. Klingborg, and Bernard E. Rollin.
Description: Boca Raton : Taylor & Francis, 2018.
Identifiers: LCCN 2018019673 | ISBN 9781138197060 (hardback : alk. paper)
Subjects: LCSH: Cattle. | Animal welfare.
Classification: LCC SF197 .W45 2018 | DDC 636.2—dc23
LC record available at https://lccn.loc.gov/2018019673

Visit the Taylor & Francis Web site at
http://www.taylorandfrancis.com

and the CRC Press Web site at
http://www.crcpress.com

The Welfare of Cattle

*This book is dedicated to our wives: Shannon Archibeque
Engle, Beverly Klingborg, and Linda Rollin*

Contents

Donald J. Klingborg

Munashe Chigerwe

Jan Shearer and James Reynolds

Ivette Noami Roman-Muniz

Jesse Robbins and Alex Beck

Preface

As is the case in all areas of animal agriculture, the cattle industry has undergone unprecedented changes over the last century. With the exception of extensive ranching, which remains structurally very close to its historical roots, all aspects of the cattle industry have been significantly restructured. This includes cattle feeding in feedlots, highly industrialized dairies milking huge numbers of cows, and slaughterhouses killing unimaginable numbers of animals who have been transported great distances. Most importantly, the core values of traditional cattle agriculture—husbandry, stewardship, and way of life—have been supplanted to a significant extent by industrial values of efficiency and productivity. This in turn necessitates a new reevaluation of cattle welfare.

The purpose of this book is to provide the reader with a basic understanding of contemporary cattle production and how one achieves good welfare for the animals.

Pursuant to this goal, the book is divided into four segments. The first segment provides an account of major factors that have led to, and been occasioned by, the industrialization of the cattle industry. An introductory chapter sets the stage for the seven White Papers that follow, and are designed to provide the reader with relevant empirical information setting the stage for the remainder of the book.

The second segment focuses on considerations relevant to all aspects of cattle production, both beef and dairy. The third segment is specifically devoted to the production of beef cattle, with the final segment focusing on dairy. The reader should be aware that there is a significant amount of overlap between the industries since many dairy cattle end their lives as beef.

Veterinarian and historian Calvin Schwabe once remarked that the cow is "the mother of the human race," and we need to be especially careful of how that mother is treated. It is in the spirit of that injunction that we have produced this book.

Bernard E. Rollin

Introduction to the White Papers

Welfare does not exist in a vacuum as a single issue, but rather is a complex set of values and beliefs that are strongly held and that "color" the conversations that revolve around the subject.

To better inform our exploration of cattle welfare, we have commissioned a number of White Papers, prepared by notable experts in their field, to help provide factual context around selected topics that impact cattle welfare and production systems. The detailed papers follow this short introduction of the experts and summary of their comments.

Whether you agree with the comments of these experts or not, we hope you find the dialogue useful and informative as you challenge your beliefs and evolve your values while reading this book.

Donald J. Klingborg, DVM

Can Food Production Keep Pace with Population Growth, Changing Dietary Habits, and a Desire for Higher Quality Protein?

Daniel M. Dooley
Sacramento, CA

As a former Vice President of the University of California's Division of Agriculture and Natural Resources, and former Chief Deputy Director of the California Department of Food and Agriculture, Mr. Dooley is well positioned to speak to the issue of feeding the world.

Current projections suggest there will be insufficient food for the world's human population by 2050. Today 800 million people on Earth exist on less than 1,800 calories per day, 2 billion people are considered as malnourished, and 900 million people earn less than $1.90 per day. Over 50% of the world's current population are less than adequately nourished and suffer all the limitations in life associated with their condition.

The demand for adequate nutrition and the demand for food that satisfies one's preferences both impact what and how much we eat. As personal income goes up globally so does the demand for higher value protein from meat, milk, and eggs and the economic benefits to society associated with those markets.

Challenges and opportunities:

- Global population growth, demographic change, and increasing affluence and urbanization
- The appropriate role of science and technology in meeting demand and reducing impacts, how will society restrict the use of tools and methods
- Global climate and other environmental changes, including changing weather patterns, extreme weather events, and expanding pest ranges
- Environmental impacts, minimizing the impact of agriculture production to meet productivity objectives
- Key resources, including limitations on land, air, and water
- Social and cultural change, including land tenure, international security, dietary preferences, etc.
- Economic factors, including competitiveness, food markets, volatility, supply, distribution, regulation, affordability, etc.

Arable land is finite, as are fresh water resources. Production per unit of land, gallon of water, etc., will have to increase to meet today's demands and those projected for 2050.

Economics of Global Cattle Industries

Daniel A. Sumner, PhD
Agriculture Issues Center
University of California, Davis, CA

Daniel Sumner is a professor of agricultural economics and has been director for the UC Davis Agriculture Issues Center since 1997. His contribution provides context about the global economic role of beef and dairy cattle production.

Beef and dairy production contributes a little more than a third of the global value of all agricultural output in the world. In the USA, about 28% of the total value of agricultural output is associated with the sale of cattle for meat and by-products, and the sale of milk. Beef and dairy production combined make up significant portions of the agricultural contributions to state economies including employment across the nation.

North America is the most efficient beef producer in the world, producing 18% of the world's meat from only 10% of the world's slaughter numbers. As a comparison, East and Southeast Asia produce about 10% of the world's meat from about 19% of the world's slaughter numbers, and Africa produces about 9% of the world's beef from 12% of the world's slaughter animals.

South America, Oceania, and North America are big exporters of meat, while East and Southeast Asia, North America, Europe, and Africa are big importers. In North America, most meat imports are ground beef and exports are higher value muscle cuts. The major exporters of beef are Australia, New Zealand, USA, Uruguay, Brazil, and India and major importers are China, Japan, Korea, Egypt, USA, and Russia.

Milk production is widely distributed across the globe with about 30% produced in Europe/ Oceana from 13% of milked bovines. South Asia produces about 25% of the world's milk from 37% of the world's milked bovines. South America produces about 11% of the world's milk from 13% of the world's milked bovines, and Africa produces 5% of the world's milk supply from 21% of the world's milked bovines. North America produces more than four times the world average of milk per cow contributing 13% of the world's milk supply from 3% of the world's milked bovines. Developed countries account for about 85% of the world's dairy exports and developing countries account for about 80% of dairy imports.

Livestock and Climate Change: Facts and Fiction

Frank Mitloehner, PhD
Department of Animal Science
University of California, Davis, CA

Professor Mitloehner is an internationally recognized air quality specialist researching impacts of livestock on air quality. He is a professor and air quality Extension Specialist with a background in agricultural engineering (University of Leipzig, Germany) and animal science (Texas Technical University). He is chaired committees of the global United Nations Food and Agriculture Organization partnership project to benchmark the environmental impacts of livestock production, served as a workgroup member on the President's Council of Advisors on Science and Technology, and as a member on the National Academies of Science Institute of Medicine committee on assessing the health, environmental and social effects of the food system.

Leading scientists throughout the USA and with the US Environmental Protection Agency have quantified the impacts of livestock production at 4.2% of greenhouse gasses. The transportation sector contributes 27%, and the energy sector contributes 31% of greenhouse gasses. Both globally and in North America industries associated with energy production and the transportation sector are the largest anthropogenic contributors of greenhouse gasses.

Within the 4.2% contribution from the livestock sector, beef cattle were responsible for 2.2%, dairy cattle 1.37%, swine 0.47%, poultry 0.08%, sheep 0.03%, goats 0.01%, and other (including horses) 0.04%.

Today, with higher production per animal for both milk and meat operations, dairy cattle numbers have decreased 59% from 1,950 levels and the amount of milk produced has gone up 79%. Beef cattle numbers have decreased by 36% and meat production has remained the same. The carbon footprint for both industries has declined, and today dairy cattle's footprint is now one third of the 1,950 level.

The United States has the lowest carbon footprint per unit of product (meat, milk, eggs) of all countries because fewer animals are needed to produce the same amount of product. The average cow in the US produces ~22,248 pounds of milk/year while the average cow in Mexico produces ~10,500 pounds of milk/year. India's average cow produces ~2,500 pounds of milk/year. India produces about nine times the methane and manure production for the same amount of milk as the average US cow.

More production per cow decreases the environmental impacts of livestock. To feed a growing world population in 2050 requires intensification of livestock production that will also reduce deforestation and environmental pollution.

Environmental Impacts—Dairy and Beef Sustainability: Nutrient Loss and Use Efficiency

Joe Harrison, PhD
Washington State University
Alan Rotz, PhD
USDA—ARS

Professor Harrison is the nutrient management specialist in the Department of Animal Sciences at Washington State University's Puyallup Research and Extension Center. Dr. Harrison is a nutritionist by training and has focused his research and extension activity on whole-farm nutrient management. He has been an active researcher and proponent of managing nitrogen and phosphorus waste to protect the environment. Dr. Rotz is an agricultural engineer working for USDA's Agriculture Research Service in areas associated with managing farms for environmental stewardship, including evaluating alternative technologies and management strategies on integrated farming systems for dairy and beef production.

Sustainability as used in this discussion reflects environmental, economic, and social systems that meet present societal needs without compromising the needs of future generations. Environmental issues include impacts on biodiversity, air quality, energy consumption, water emissions, water consumption, land management, and waste and by-product impacts on the environment. The potential pollution of water from livestock production includes concerns about nitrogen and phosphorus nutrient losses, pathogens, toxic compounds, hormones, and others products.

Cattle feed includes nitrogenous proteins, and up to 80% of nitrogen intake may be excreted in their urine and feces. Ammonia volatilization to the atmosphere can represent up to half the nitrogen in these cattle wastes. Runoff in water also contributes to loss from the farm.

Much of the conversation today is about the sustainability of beef and dairy operations in extensive systems, characterized by pasture, low tech, and animals foraging for much or most of their own feed, versus intensive systems characterized by housing with higher density of animals and animals are provided with most of what they eat. Young stock of both beef and dairy cattle are raised in more extensive systems until they reach a point where they are close to starting their first lactation if a dairy cow, or ready to go to a feedlot to finish their growth and fatten up prior to slaughter for beef cattle.

The driving factor toward intensification has to do with shrinking profit margins based on low prices to the producer for their products and higher costs of inputs. Increasing efficiency is often as simple as increasing the number of animals per unit of land. In this setting, nutrients are imported, while milk and meat are exported and the wastes remain on the farm. This is a parallel to people moving from farms to cities so more wastes are managed in a smaller area.

Genetic and management improvements including nutritional support have resulted in fewer animals making more milk and/or meat. Today's dairy cattle, for example, captures 35% of the consumed protein as meat or milk where a few decades ago that figure was ~20%. Beef cattle have also improved and compared to 1977 and now excrete 12% less nitrogen and 10% less phosphorus. Comparing 1977 and 2007, beef produced the same amount of meat with 30% fewer cattle, 19% less feed, 12% less water and 33% less land, and had a 16% decrease in their carbon footprint.

Costs of reducing environmental impacts are largely borne by the producer. Consumers need to share in these costs, including assuming some of the costs of production and by decreasing wastage of food produced. Estimates are about 30% of food produced is wasted today, and that represents the largest single component of the life cycle costs associated with beef and dairy products.

Microbial Pathogens in Extensive and Intensive Animal Agriculture Systems

Pramod Pandey, MS, PhD and E. R. Atwill, DVM, MPVM, PhD
School of Veterinary Medicine
University of California, Davis, CA

Professor Pandey is an agricultural engineer and Professor Atwill is an epidemiologist who works on understanding and managing microbial threats to the environment and food. Dr. Atwill is the director of the Western Institute of Food Safety and Security.

Microbes dominate the life forms in our world, and include viruses, bacteria, protozoa, and fungi. They make critical contributions to the health of soils, plants, animals, and humans. They also can threaten the health of plants and animals. Sewage waste from animals and people may contain pathogenic microbes that can cause disease in susceptible species.

Today's animal agriculture systems in the USA produce more than a billion tons of manure. How manure is handled on-farm can influence the types and number of microbes found in it.

Confinement operations, so called intensive systems, are able to implement methods to collect and treat manure that lower or eliminate disease-causing microbes. Composting, anaerobic digesters, and liquid/solid separation lower both the number of microbes and the number of pathogens.

Extensive systems such as pastures tend to leave the manure where it falls and rely on Mother Nature (via sunlight, temperature, and time) to lower the number or eliminate pathogens. About 8% of manure from beef cattle's extensive systems is recoverable for treatment or transportation compared to ~75% of dairy cattle's intensive system.

Animal manure is seen as a commodity with some economic value especially to improve crop production, but untreated manure may be a threat to human and animal health. Limiting environmental exposure to these organisms is a worthy goal. Microbes are not very mobile, relying instead on hitching a ride on something else that's moving to get from one point to another. The leading vehicle for spread is water runoff and this can be a problem in both intensive and extensive production systems.

Intensive systems produce more food per unit of resource input (water, land, feed) and a lower level of environmental contamination per unit of food produced. About 29% of the world is land.

Of that 29%, about 36% is used for agriculture, 30% are in forests, and 3% in urban use. The remaining 31% is represents various landscapes including mountains, permanent snow, rivers, lakes, and reservoirs that are not considered suitable for food production. Most of agriculture's 36% of land are already in productive use so increased production of food will require more intensive production systems.

There is insufficient land to move the current meat, by-product, milk, and egg production operations from intensive to extensive systems. Accepting the current published stocking density standards: Wisconsin would require about 60% of all the land in the state to house their current inventory of dairy cattle; Iowa would require 20 times all of the land in the state to house their swine population; and Georgia would require 477 times all of the land in the state to house their boiler chicken population.

Cattle Ectoparasites in Extensive and Intensive Cattle Systems

Alec C. Gerry, PhD
College of Natural and Agricultural Sciences
University of California, Riverside, CA

Professor Gerry is an entomologist and Cooperate Extension Specialist who focuses on arthropods and disease vectors that impact the growth or productivity of animals. A portion of his work is to understand the impacts on human populations working on farms or living adjacent to farming areas. His focus is on integrated pest management approaches to control these pests.

As the world becomes more populated, there are more and more people living in proximity to animal production operations and they are often concerned about insect pests that impact their neighborhoods. Concern also exists about the role intensive production systems have on ectoparasites (flies, mites, ticks, grubs mosquitoes) as compared to extensive systems.

The seasonal timing of peak ectoparasite activity is predictable by temperature and rainfall during the preceding months, but the abundance of ectoparasites and the severity of their impact depend on the operational characteristics of the facility.

The differences between extensive and intensive systems are primarily: (a) presence or absence of pasture; (b) sharing habitat with wildlife; (c) density of cattle in the pen or on the pasture; (d) storage of cattle feed and supplemental rations on site; (f) collection and storage of manure on site; and (g) ease of implementing ectoparasite control including applying insecticides to animals.

Neither extensive nor intensive production systems are the "best" for managing all ectoparasites. The presence of ectoparasites in herds without a control program is dependent on the availability of immature development habitat, survival of ectoparasites when off the host, and the opportunity for ectoparasites to acquire a new host whenever needed for feeding.

The diversity of cattle ectoparasites will be greater in extensive production systems compared to intensive systems. This includes the presence of most but not all ticks, horn and face flies. Stable and house flies are expected to be more frequently associated with intensive systems unless cattle on pasture are being supplemented or bedded with hay or other organic material. Lice, mites, and grubs could all be common in either extensive or intensive systems. Aquatic habitats are common in and around pasture operations and may be found around intensive operations with water "leaks" creating areas of standing water. Aquatic habitats support the growth of midges, mosquitoes, horse, deer, and black flies.

Intensive systems enjoy a higher level of human contact, ready facilities for handling cattle, and more opportunities for early recognition and treatment and therefore should result in lower abundance levels and lower impacts.

The Use of Biotechnology to Improve Animal Welfare

Alison L. Van Eenennaam, MS, PhD
Department of Animal Science
University of California, Davis, CA

Professor Van Eenennaam comes from Melbourne, Australia and received her Masters and PhD degrees from UC Davis. She's been a Cooperative Extension Advisor working on dairy farms in California and since 2002 has been a Cooperative Extension Specialist in Animal Genomics and Biotechnology. Her research interests include the use of biotechnology to address some of the animal welfare concerns and environmental impacts of animal production.

Genetic engineering is a process where scientists use recombinant DNA technology to introduce desirable traits into an organism. All organisms share the same four-nucleotide building blocks in their DNA, meaning that the same gene in different organisms will encode the exact same protein whether it is being manufactured in an animal, plant, or microbe. Recombinant DNA is DNA with fragments from two or more sources that have been joined together in the laboratory. A recombinant animal carries a known sequence of recombinant DNA in its cells and will pass that on to its offspring.

Genetic engineering accelerates the change that is associated with selective breeding. Gene editing can precisely add, delete, or replace the genetic code to introduce a new gene or turn off an existing gene. It can be used to correct diseases or disorders that have a genetic basis by altering the error, change a less desirable trait, or insert a new desirable trait.

In cattle, it is possible to use gene editing to prevent the growth of horns, thereby eliminating the pain and costs associated with dehorning, or increase resistance to mastitis, a major reason for cattle death and removal from a herd. It can also prevent a protein from being made that is a major cause of milk allergies, and increase resistance in cattle to tuberculosis and respiratory disease. Similar benefits exist in other food animal species. Beyond health, but of great importance for animal welfare, gene editing could produce only single gender of offspring (female for egg layers or dairy production) or eliminating testes development so castration is no longer necessary.

Gene editing does not necessarily add any foreign DNA to the animal being edited and the intended changes are similar to those already found naturally in the species. Editing allows a more rapid change than natural breeding and it is expected to be used in combination with ongoing genetic selection of the animals from natural breeding and artificial insemination programs. As an example, the hornless Holstein carries the polled gene sequence that came from Angus cattle via a breeding program, but could be more easily and quickly achieved via gene editing.

The following genetic traits are now recommended for consideration in a breeding program to improve the welfare of dairy cattle: fertility, disease resistance, feet and leg conformation, calving ability, udder shape and support, body size and weight, productive live and longevity, mastitis susceptibility, and daughter pregnancy rates. This is a positive step expanding the goals beyond the previous emphasis on milk yield, milk fat, and milk protein. These newer additions are identified as important traits associated with cattle welfare and may be more quickly accomplished by allowing gene editing in combination with selective breeding.

List of Contributors

Jason K. Ahola is a professor of animal sciences at Colorado State University. His research program focuses on beef cattle management and pain mitigation strategies for castration, branding, and dehorning of beef cattle.

Shawn L. Archibeque is a professor of animal sciences at Colorado State University. He is a ruminant nutritionist focused on environmental dimensions of animal agriculture.

E. R. Atwill is a professor of veterinary medicine at the School of Veterinary Medicine, University of California, and Davis. He is an epidemiologist focused on microbial impacts on food and the environment. He serves as director of the Western Institute of Food Safety and Security and Director of Veterinary Medicine Extension.

Fuller W. Bazer is a regent fellow, distinguished professor, and O.D. Butler Chair in animal science at Texas A&M University.

Alex Beck is a veterinary practitioner specializing in dairy in Banks, Oregon.

Frank H. Buck Jr, UC Davis, Davis, California.

Munashe Chigerwe is an associate professor of medicine and epidemiology in the School of Veterinary Medicine at the University of California, Davis.

Johann F. Coetzee is a professor and the department head of anatomy and physiology at Kansas State University College of Veterinary Medicine, where he also serves as interim director of nanotechnology at the Innovation Center of Kansas State (Nicks) and Institute of Computational Comparative Medicine (ICCM).

Candace Croney is the director, Center for Animal Welfare Science and a professor of Animal Behavior and Well-Being at the Purdue University College of Veterinary Medicine, Department of Pathobiology.

Courtney Daigle is an assistant professor in the Department of Animal Science at Agriculture and Life Sciences Texas A&M University. Dr. Daigle specializes in evaluating management practices to optimize animal health, productivity, and welfare. Her laboratory quantifies behavior to develop and validate technologies designed to measure species specific behaviors important to health, welfare, and productivity.

Daniel M. Dooley is an attorney and former Vice President of the Division of Agriculture and Natural Resources, University of California and former Chief Deputy Director of the California Department of Food and Agriculture.

Lily N. Edwards-Callaway is currently an assistant professor of Livestock Behavior and Welfare at Colorado State University in the Animal Science Department. Her research interests include investigating management strategies to reduce animal stress.

Terry Engle is a professor of animal sciences at Colorado State University. His research interests include the influence of trace minerals on animal health and production.

Tom Field is the director of the Engler Agribusiness Entrepreneurship program and the Paul Engler Chair of Agribusiness Entrepreneurship at the University of Nebraska-Lincoln's Institute of Agriculture and Natural Resources.

Franklyn Garry is a professor and veterinary extension specialist of the College of Veterinary Medicine and Biomedical Sciences, Department of Clinical Sciences at Colorado State University. His research interests include Johnes's disease control, causes of mortality in adult cattle, livestock worker education, and calf health management.

Désirée Gellatly, Agriculture and Agri-Food Canada, Lethbridge, AB, Canada

Alec C. Gerry is a professor of entomology and cooperative extension specialist at The College of Natural and Agricultural Sciences, University of California, Riverside. His specialty is pests that impact animals.

Temple Grandin is professor of animal science at Colorado State University. Temple's research interests include understanding animal behavior, design of livestock handling facilities, and improving animal handling.

Joe Harrison is a nutrient management specialist in the Department of Animal Sciences, Washington State University and works from the Puyallup Research and Extension Center. His interest is in whole-farm nutrient management with an emphasis on protecting the environment from nitrogen and phosphorus waste.

Andy D. Herring is a professor in the Department of Animal Science at Agriculture and Life Sciences Texas A&M University. He has researched genetic and environmental influences on milk production in beef cows, breed differences for feedlot and carcass characteristics, and genetic influences on beef cow reproduction and productivity, cattle temperament, and immune responses.

Eugene Janzen, University of Calgary, Calgary, AB, Canada

Michael D. Kleinhenz, Kansas State University, Manhattan, Kansas

Donald J. Klingborg is an associate dean emeritus at The School of Veterinary Medicine, University of California, Davis. He spent the majority of his career as a practicing veterinarian with an emphasis on cattle health and wellbeing.

Sonia Marti, Agriculture and Agri-Food Canada, Lethbridge, AB, Canada

Daniela Mélendez Suarez, Agriculture and Agri-Food Canada, Lethbridge, AB, Canada

Frank Mitloehner is a professor and air quality extension specialist in the College of Agriculture and Environmental Sciences, University of California, Davis. His interest includes air quality impacts of animal agriculture.

Jerry D. Olson is a consulting dairy veterinarian focused on understanding the influence of nutrition on dairy cow productivity and health.

Pramod Pandey is an assistant professor at the School of Veterinary Medicine, University of California, Davis. His interest is in how to control microbial pathogens in animal manure, especially cattle, and poultry manure.

Jesse Robbins is post-doctoral research associate, College of Veterinary Medicine, Iowa State University, Ames, Iowa.

Bernard E. Rollin is University Distinguished Professor at Colorado State University with appointments in the departments of Philosophy, Animal Sciences, and Biomedical Sciences.

Ivette Noami Roman-Muniz is an associate professor and extension dairy specialist in the Department of Animal Science at Colorado State University. Her research interests include management strategies to improve worker safety and overall dairy cattle health.

Alan Rotz is an agricultural engineer with USDA's Agriculture Research Service in Pennsylvania. His interest is in managing farms for environmental stewardship as it relates to dairy and beef production.

Karen Schwartzkopf-Genswein is a Canadian federal scientist with expertise in farm animal behavior, health and welfare. She works for Agriculture and Agri-Food Canada at the Lethbridge Research and Development Centre.

Daniel A. Sumner is a professor of agriculture economics and director of the Agriculture Issues Center at the College of Agriculture and Environmental Sciences, University of California, Davis.

Alison L. Van Eenennaam is a cooperative extension specialist in Animal Genomics and Biotechnology, College of Agriculture and Environmental Sciences, University of California, Davis. Her research interests include using biotechnology to address animal welfare and environmental impacts of animal production.

Beth Ventura is a teaching assistant professor—companion animal and equine behavior and animal welfare faculty member in the Department of Animal Science at the University of Minnesota. Dr. Ventura is developing a teaching program that will equip students with the foundational skills to navigate the issues facing the animal industries in a rapidly changing society. She aims to engender her students with the knowledge of both the science and values that affect the practice of raising and keeping animals for companionship, food, entertainment, and science.

Kurl D. Vogel is an associate professor at the University of Wisconsin—River Falls Campus. His research interests include food animal welfare and behavior, impacts of management strategies on animal welfare, physiology, and product quality, and the ethics of animal use.

Marina A. G. von Keyserlingk is a professor in Animal Welfare at the University of British Columbia. Dr. Von Ketserlingk research interests include understanding the links between behavior and nutrition, particularly in welfare related issues.

John J. Wagner is a professor of animal science at Colorado State University. His research interests include understanding the impacts of nutrition and management strategies on feedlot cattle performance and stress.

Jennifer Walker is the director of Dairy Stewardship for Dean Foods. In her role she is working with customers, suppliers, and dairy farmers to develop an industry wide standard that promotes the good welfare of dairy cattle.

PART **I**

White Papers

Introduction

We humans tend to carefully select the materials we read, watch, and listen to based on whether the material agrees with what we already believe. Those that present alternative positions don't receive the same effort from us to try to understand and to thoughtfully consider the point the author is trying to make. You're encouraged and we believe you'll find it useful to read all of the book. You'll find some materials you like, and some you don't. The goal is to have us all broadly think about the issues and for each of us to apply those concepts we can to improve the welfare of cattle.

It's my belief we as a society can have whatever kind of animal production system we want. We can't collectively seem to decide what that is, however, because we have not yet been able to achieve the purposeful dialogue in order to find the common ground required to foster change. This book is one step in the effort to achieve that purposeful dialogue.

As background know that I love cattle, especially dairy cattle. But more important than my emotional attachment to them is the respect I have for what they are and what they do. "My girls" produce one of the world's highest quality and most nutritious foods, milk, in the extremes of Nordic and desert climates, at high altitudes and at sea level, in high tech production systems and in low and no tech grazing systems. Much of the food they consume is non-competitive with simple stomached animals, including we humans, because the cow's rumen harvests nutrients from feedstuffs we can't digest. From 20% to 40% of the average dairy cow's diet captures nutritional value from recycled waste streams associated with the production of human food, fuel, or fiber products. These products would otherwise be burned or buried, wasting their value, and contributing to environmental degradation.

The nutrients provided by these by-products, along with nutrients coming from hay, silage, and grain, allow cows to provide about 17% of the protein consumed in America today. They've also made milk a major economic force in the USA. The number one agriculture commodity in California, the USA's largest ag economy, is milk. In many other states milk production is critical to the state's economy as a major provider of employment.

While milk and eggs provide the highest quality proteins relative to digestibility and in providing human's essential amino acid requirements, meat from cattle is also a high quality and valuable protein. In North America about 20% of the beef consumed comes from dairy cattle. Cattle butchered for meat also provide many important and economically beneficial by-products including leather, felt, fats, lubricants, soaps, creams, buttons, glues, gelatin, and much more. It's reported that 99% of slaughtered cattle today are utilized commercially to make useful and salable products.

Beef production continues to operate under similar "norms" as those that applied decades ago, but over the last 50+ year's milk pricing policies and dairy operational costs of production has created a situation where higher turnover rates and shorter cow longevity are economically possible and, in some situations, are financially rewarded.

In the 1996 publication *The Lost Art of Healing* physician Bernard Lown, emeritus professor of cardiology at Harvard and previously a senior physician at Brigham and Women's Hospital in Boston, laments *the industrialization* of human health care and makes a passionate appeal to restore the *"3,000-year tradition that bonded doctor and patient in a special affinity of trust."* He notes the biomedical sciences had begun to dominate our conception of health care, and he warned that as a consequence *"healing is replaced with treating, caring is supplanted by managing, and the art of listening is taken over by technological procedures."* The parallels between his remarks and our discussion about animal welfare strike me, and Dr. Lown reminds us all to guard against "managing" at the expense of "caring". Please also note that the modern germ theory of disease causation, and the resulting attention to hygiene that resulted in improved surgery survival rates, only existed in the last ~150 of those 3,000 years. The goal is to care AND pay attention to science-based facts.

"Caring", for me in the context of food animals, includes a commitment first to the animal, then to "managing" the animal's performance. The word "husbandry" reflects a bond between the animal and the farm owner, and implies a moral responsibility on the part of the owner to care and provide for the animals under their care. The animal is and must be more than a "commodity" or "profit center", and attention to the cow's needs and wants beyond those required to simply produce milk or meat is more than expected – it's promised in the husbandry contract we entered into when we domesticated animals. This contract is founded on the provision of benefits to both the involved animals and humans. At the same time the producer that invests more than the value received for their product isn't in business for long.

Stewardship is another term that will appear often in this book, reflecting the belief that it's not about what the animal does for us, but what we do for them to compensate for their contributions to our well-being. We rightfully expect more consideration for the animal from people who raise animals for food, use animals as power in their work, or keep animals as pets. Animals are sentient beings and beef and dairy producers as their custodian have a solemn responsibility for their care and keeping.

The commitment to food animals is not, however, one of longevity. Food animals go to market at various ages based on market preferences, carcass yields, and cultural influences. Veal calves and broiler chickens go to slaughter at weeks of age, while lambs and market pigs go at months of age. Beef go to market between 1 and 2 years of age for prime cuts, and younger or older for less valuable cuts. Dairy cattle go to slaughter at whatever age they are no longer competitive with their herd mates.

Rather than longevity, our commitment is to how these animals are cared for during the time they are in our production systems. The contract may include a responsibility for providing some level of protection from extreme weather, predation, starvation, malnutrition, bullying, premature pregnancy as well as avoidable pain, disease, and injury. If injury or disease occurs our commitment is to treat it rapidly and appropriately. At the same time, we should be committed to provide housing designed and maintained to prevent injury or disease and that allows for many normal behaviors.

Dairy cattle have value as meat in the slaughter market, with that option triggered when a cow's milk production falls to a level that does not provide the farm sufficient income over expense. At that point, they are removed from the milking herd to enter the slaughter beef market and replaced with a milking animal that promises to be more economically competitive. An increase in income per unit of milk, or a decrease in operational expenses, or a combination of the two allows for lower turnover rates and longer herd longevity.

The current discussion about welfare in the dairy industry seems to be distracted by a focus on the business entity that owns the dairy (pejoratively referred to by many as "corporate farms"). The truth is today's dairy industry is dominated by the family farm, usually with multiple family generations living and working on the property. Dairy businesses are organized into three legal entities: owners as individuals, including sole proprietorships and partnerships (most often family partnerships including parents and kids, or brothers, sisters, and in-laws); family corporations wholly owned by family members; and non-family related corporations. Family corporations are still family run businesses, just legally organized to provide: tax benefits; the ability to more easily transfer assets between individuals and generations; improved liability protection and easier access to borrowing.

According to USDA's Census of Agriculture there were 5,223 non-family corporation dairy farms in the US in 1997 (3.1% of 168,473 dairies); 3,355 in 2007 (3.9% of 86,022 dairies); and 3,220 in 2012 (7% of 46,005 dairies). Of all farms (including dairy and all other crops) in the US, only 3% were not family owned in 2012.

The change toward larger industrialized dairy operations is a direct result of economic forces that have shaped many businesses since the early 1950's, lowering costs while improving product quality and consistency. Today, successful dairy operations have "right-sized" their herds to balance operational costs with the price of milk they receive. In general, those areas with higher milk prices and lower operational costs enjoy economically successful herds that are smaller and less industrialized, while in areas with lower milk prices and higher operational costs the need is to be larger and more process oriented.

Between 1992 and 2012 USDA reports that the number of US dairy herds decreased by 61% while the average number of cows per herd increased from 74 to 142. While most of the cows in the USA remain in what we consider to be smaller herds, much of our milk comes from dairies that are larger in size, provide higher density housing, and maintain cattle on solid flooring rather than predominantly on dirt or grass. Some individuals refer to these operations as "factory farms"; a label intended to evoke a negative opinion among many. These farms are definitely industrialized (meaning process and outcomes oriented) and they are confinement operations (meaning more milking cows per unit of land). Milk quality is generally better, and environmental impacts are lower when production per cow is higher.

The issue before us isn't whether farms should be industrialized, that decision has been made by the earth having a limited supply of arable land (with that acreage shrinking every year due to urban expansion), limited supplies of fresh water with insufficient reservoir storage capacity to support agriculture, and a rapidly growing population of hungry human mouths to feed. Intensive agriculture is here to stay.

The size of the herd isn't necessarily related to the welfare of the animals in it. Smaller herds often have fewer laborers sharing a wide variety of chores (generalists), while larger herds tend to have laborers with specific training and accountability in a narrow area (specialists – such as dedicated milkers, calf raisers, breeders, ration mixers, and feeders, etc.). Record keeping in smaller herds tends to be via paper, making analysis more difficult and small changes a challenge to recognize and manage. Larger herds use on-farm computers that can provide timely reports or alerts, and can invest in improving husbandry to capture a small outcome (such as 1% less illness or death rate) where that would not be visible in the small herd.

I moved from Colorado in the early 1970's to practice in California because Colorado had a relatively high milk price to the farm, lower land, labor, feed, tax, and environmental costs, and the farms could survive without paying as much attention to detail. They tended to be more reactive that proactive. By comparison, California had lower milk prices, higher costs for land, feed, labor, taxes, and environmental costs and to be successful the producer had to pay more attention to "the little stuff". The impact of health programs implemented to prevent disease and injury could be visible and measured in larger herds, and paid for in a shorter period of time.

In my experience, the middle-sized dairy may be the most challenging relative to promoting animal welfare. They may not have right-sized such that it remains dependent on general laborers without specialized training often in crowded facilities, a lack of resources to invest in technology and have too many time-demands to manage all aspects of the enterprise. Other strategic threats that may negatively impact animal welfare include those dairies that are expanding, most often due to economic pressures, and add more cattle to the same facilities thereby creating overcrowding; or grow the facilities and slowly add more cattle, but not able to grow the labor force due to the investment made to expand the facilities that won't bring in income until it's filled with milking cows.

High- producing herds have invested in better genetics and provide carefully formulated rations to support their dairy cattle athletes. Yes, athletes. High producing cows have a metabolism while lactating that is similar to that of a marathon runner in the middle of a race. They may look placid and calm standing or lying around, but they are metabolically working very hard. They naturally lose weight after birthing and into the first quarter of their lactation, during the time their ability to consume feed increases but doesn't yet equal the demands associated with making milk. These cows require diets carefully formulated and balanced to provide the high-density nutrients required to support their milk production and get them ready to become pregnant. Like all athletes, they are in a delicate balance between health/well-being and performance.

A historical policy to provide USA citizens with inexpensive food has resulted in very low margins of profit for farmers in good years, and losses in bad years. As a result, those farms producing products with highly regulated prices and "on the fiscal edge" have had to apply the same process and outcomes controls to their production systems as we've experienced in manufacturing of durable goods, thereby lowering the costs of production, keeping retail prices low, and improving quality control.

The farmer assumes the highest risk relative to return-on-investment in our food system that spans from production to consumption, including assuming the risk of market fluctuations and adverse weather. At the same time the farmer receives a very small portion of the money society spends on food. In 2015 every dollar we spent on all food items included only 8.6 cents to the farm of origin. Compare that to 15.6 cents for food processing, 15.3 cents for packaging, transportation, and wholesale trade, 34.4 cents for food services (eating out or buying prepared foods), and 13.5 cents for energy, finance, Insurance, advertising, and other food-associated expenses (source USDA ERS Food Dollar Series). Comparing prices in the 60 years between 1946 and 2006, tuition at Harvard University increased 70+ times; average home prices increased over 50 times; the cost of rental housing or purchasing a new car went up ~25 times; gas per gallon, postage, a loaf of bread or a movie ticket went up ~10 times; the price of milk went up ~5 times and the price of eggs went up ~2 times (*Signal Magazine* 09/06 from US Government, professional, and trade association sources).

Today consumers spend just less than half their food dollar on food prepared outside of the home. As a society, we're spending the lowest percentage of our income on food in history, much lower than the citizens of Europe and third world countries. Restaurant prices have increased from about double retail (grocery store) prices in 1990 to about four times retail prices in 2014 and the amount of our diet that comes from foods we don't prepare is at an all-time high.

I think we all agree, producers and consumers alike, that we want the animal to be the major point of concern in our food production systems. To assure that reality the producer needs help. While it is not possible to feed the world's population by using organic production methods, and there are serious concerns about the welfare status of animals in organic production systems, the increased income to the farm from consumers willing to pay more for organic products is a model that may help change role of animals used for producing food. The retail costs of organic products are significantly higher than similar foods produced in conventional systems because it takes more land, inputs, and in some cases, time, to produce the same amount of conventionally produced food. Without consumers willing to pay those higher costs, organic producers could not economically survive.

Surely consumers of domesticated animal products also share with the producer the solemn responsibility for the animal's care. If consumers will pay more for their dairy and beef food products producers can do more to promote animal welfare, environmental stewardship, and other important elements of sustainable agricultural production. Basing one's food purchasing decisions on the lowest retail price can only result in a less animal-centered food production systems.

An unresolved question exists about how a "preferred food production system" sustains itself in the face of global competition from foreign producers who are often free to use banned chemicals/drugs, may have no or smaller environmental costs of production, may pay non-living employee wages without housing or health care, and often escape the welfare considerations we're discussing. Welfare expectations are very different around the world where laws, cultures, and ethics differ.

As an example of the problem, I live in a college town with an extraordinary farmer's market. I can purchase vegetables, fruits, meats, and baked goods from independent vendors all of whom claim to be providing fresh and wholesome food. I note that almost all of the items avoid USDA inspection for pesticides, or USDA slaughter inspection, rarely have labels with the date they were packaged, and almost never have a country of origin label. Many are certified organic but that's different than an inspection system. Food safety is a real concern and farm-to-table and farm-to-school programs that by-pass USDA inspections remove one of the important safety nets associated with assuring a safe food supply. As we create more stringent requirements for our food systems we need to financially reward that system in order for producers to survive.

So, we as a society can have whatever kind of animal production system we want; we just have to be willing to share in the costs required to pay for it. Larger herds are able to provide some benefits to their cattle that are not feasible in smaller herds just as smaller herds are able to provide some benefits that are not feasible in larger herds. Consumers need to decide what benefits they will support with higher prices to the farm.

Every cattle producer I know would love to have a viable alternative to castrating and dehorning calves. Thankfully Angus cattle carry a dominant allele that codes for polled offspring, and the beef industry has made major strides in using breeding cattle that pass on this allele so their calves do not require dehorning. It's a problem in dairy breeds, however, because using Angus semen will produce offspring without horns AND with less milk. There is an answer, gene editing, which can insert the Angus polled allele into a dairy breed's genetics and produce dairy calves that will not require dehorning. The technology is precise and makes use of a natural allele without significant alteration to other genes. It's also possible to control the sex of calves using genetic technology thereby avoiding the need for castration, but at this point society has not demonstrated any willingness to use these kinds of technologies to improve animal welfare. More thoughtful dialogue about the appropriate use of technologies like gene editing is needed if society wants to seriously address animal welfare issues.

Do you like cheese? More than 80% of our cheese is produced using a product produced by a GMO microbe that has been modified to manufacture an enzyme that clots and solidifies cheese. Before this GMO product became available naturally occurring rennet was used to accomplish the task. Rennet is an extract from the stomach lining of nursing calves. Much of this rennet came from the slaughter of veal calves and as that industry largely disappeared based on societal concerns, the GMO product became the dominant method for today's cheese production. Similarly, many effective and safe medical therapeutics are produced using GMO technology today, including: most of the insulin in use to treat diabetes; blood products and clotting factors; allergens; monoclonal antibodies; immunoglobulin's; antivenoms; some vaccines, and more. If you've had a serious illness and been hospitalized, or had surgery, you've probably received products that have come from a GMO. An exploration of the acceptable use of GMO and associated technologies would be useful and could play an important role in animal welfare tomorrow, but only if thoughtfully contemplated today by an informed society carefully considering benefits, disadvantages, and consequences associated with their use.

For more than 20 years, six or seven days a week, you'd find me in rubber boots and coveralls, called to a dairy or beef operation because the producer cared enough about his or her animals to provide regular and preventive medical services. In my personal experience, I found the overwhelming majority of owners were proud and respectful of their animals and did all they could to care for them. To most of these producers their animals were family. Differences in the quality of care provided were primarily associated with the economic limitations of the enterprise and not the lack of concern of the producer. Regardless of herd size, dairy and beef producers were proud of the genetics they've invested in their herds, often representing a family legacy that spans multiple generations of effort to produce the best animals possible. Dairy cattle require care seven days and nights a week, 365 days a year, and want to be milked every day, preferably at the same time, regardless of holidays, weather, or special events. It is a demanding responsibility. I've found the producer's personal status, identity, and self-image are all tightly connected to their cattle.

My hope is that this book presents information in a manner that allows the reader with strong opinions on all sides of the issues to pause, to invest in the required struggle to truly understand the points that are being made by the authors, to ruminate thoughtfully about what the reader's personal truth is, and then to strive to take the first step, or the next step, in improving the welfare of cattle under their stewardship (and I include consumers as sharing in the responsibility of stewardship through their purchasing decisions). This is a journey we humans have been on with cattle for thousands of years, and that will continue as long as humans and animals depend on each other.

Donald J. Klingborg
School of Veterinary Medicine Emeritus
University of California, Davis, CA

CHAPTER 1

Can Food Production Keep Pace with Population Growth, Changing Dietary Habits, and a Desire for Higher Quality Protein?

Daniel M. Dooley
University of California

CONTENTS

An accelerating discussion about population growth, changing dietary habits, and the ability of the global food system to keep pace with demand has occupied the minds of governments, producers, processors, and nonprofit stakeholder communities. The conversations center around how productivity can be enhanced to keep pace while minimizing impacts to the global environment and respecting local customs and cultures. Current trajectories indicate that global food production systems are perilously close to losing the race to meet the need expected by 2050.

Pressures to curb expansion of arable farmland to preserve forestlands, rainforests, watershed, and other natural habitats are limiting options to expand productivity by creating more farmland. These pressures have focused more attention on intensification of existing farmland to achieve even greater productivity.

While estimates vary, food production will have to increase dramatically to meet demand in 2050.* This issue must be considered in context of current estimates that nearly 800 million people in the world exist on less than 1800 calories a day. Two billion people today are not getting sufficient nutrients and approximately 900 million people today earn less than $1.90 a day.

This white paper will briefly look at population trends, projected food demand and dietary trends, productivity growth rates and requirements, the impact of climate change, and challenges to implementation of concerted global actions necessary to achievement of productivity required to meet demand.

---

* The future of food demand: understanding differences in global economic models, Hugo Valin, et al., December 10, 2013.

Table 1.1 Population for the World and Major Areas, 2015, 2030, 2050, and 2100 According to the
Medium Variant Projection

Major Area	Populations (Millions)			
	2015	2030	2050	2100
World	7.349	8.501	9.725	11.213
Africa	1.186	1.679	2.478	4.387
Asia	4.393	4.923	5.267	4.889
Europe	738	734	707	646
Latin America and the Caribbean	634	721	784	721
North America	358	396	433	500
Oceania	39	47	57	71

Source: United Nations, Department of Economic and Social Affairs, Population Division (2015). World Population
Prospects: The 2015 Revision. New York: United Nations.

POPULATION GROWTH

The United Nations Department of Economic and Social Affairs/Populations Division estimated world population in 2015 at 7.349 billion people. They estimate the number to increase to 9.725 billion by 2050 and 11.213 billion by 2100.

The projections shown in Table 1.1 suggest modest growth in Asia, Latin America, and the Caribbean. They show substantial growth in Africa, Oceania, and North America. Interestingly, Europe's population is expected to fall.

World population did not reach one billion people until 1804. It took another 123 years to reach two billion, 33 additional years to reach three billion, and 14 additional years to reach five billion. While the rate of growth has slowed somewhat, growth is occurring nonetheless. Global population is not expected to stabilize until the year 2100.

CHANGING DIETARY DEMAND

In addition to population growth, several other factors impact demand for food. Notably, the increase in personal income worldwide is driving change in dietary demand. In many regions of the world, low-income populations are moving into middle-income categories. Furthermore, there is a general view that this trend will continue.

Generally, rising incomes lead to higher demand for meats and other higher value foods. As this trend continues, further strains on the food production systems will occur in as much as this transition is more resource intensive to support. Per capita consumption of meat and milk has increased considerably in both developed and developing countries in recent decades, but the increase in developing countries has been more dramatic. If developing countries reach the levels seen in the developed countries, there will be dramatic overall increases in the coming decades, which will further stress resources.[*]

These trends in dietary changes continue notwithstanding the argument of some suggesting that the food system would be far more efficient and have less adverse impacts if global diets were plant based.

[*] Don Hofstrand, AgMRC Renewable Energy &Climate Change Newsletter, February 2014.

COUNTERVAILING TRENDS

Notwithstanding the increasing demand for higher value proteins resulting from increasing personal income, there is ample evidence that malnourishment and undernourishment exists at alarming numbers. While the aggregate numbers are declining, the regional distribution is shifting also.

The changing distribution of hunger in the world.

The changing distribution of hunger in the world.

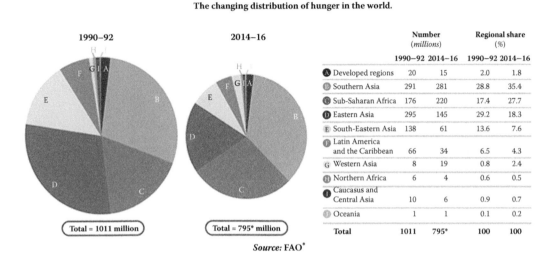

		Number (millions)		Regional share (%)	
		1990–92	2014–16	1990–92	2014–16
A	Developed regions	20	15	2.0	1.8
B	Southern Asia	291	281	28.8	35.4
C	Sub-Saharan Africa	176	220	17.4	27.7
D	Eastern Asia	295	145	29.2	18.3
E	South-Eastern Asia	138	61	13.6	7.6
F	Latin America and the Caribbean	66	34	6.5	4.3
G	Western Asia	8	19	0.8	2.4
H	Northern Africa	6	4	0.6	0.5
I	Caucasus and Central Asia	10	6	0.9	0.7
J	Oceania	1	1	0.1	0.2
	Total	1011	795*	100	100

Source: FAO*

1990–92   Total = 1011 million

2014–16   Total = 795* million

In many countries that have failed to meet international hunger targets, vulnerabilities have increased because of natural and human-induced emergencies and political instability. On the present course, it is easy to assume some of the uncertainties are intractable and not easily resolved without concerted international attentions.

AGRICULTURAL PRODUCTIVITY

Over the past five decades, global agricultural output grew by about 2.24% per year. Over this period, the primary source of productivity growth changed from input-based growth to mainly Total Factor Productivity (TFP)-based growth.[†] Historical productivity was measured in total production. For example, what has the growth rate been in total production of milk, pounds of beef produced or acres farmed. Conversely, TFP is a measure of output compared to total inputs. For example, TFP growth might be a measure of the unit of milk produced per unit of feed provided, or the amount of corn produced per unit of nitrogen or water applied. Unfortunately, producing more per unit of input may not increase overall food production. Rather, it tends to be a measure of input management and efficiency (land, water, nutrients, management, labor, and capital).

---

[*] FAO, the State of Food Insecurity in the World 2015.
[†] Growth in Global Agricultural Productivity: An Update, Keith Fuglie and Nicholas Rada, International Agricultural Productivity, USDA, Economic Research Service, January 2017.

Total factor productivity has replaced resource intensification as the primary source of growth in world agriculture

Total output growth (percent/year)

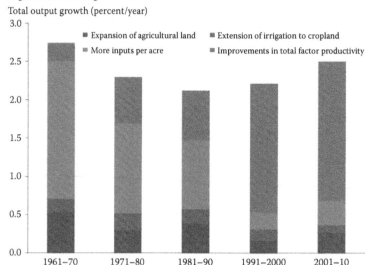

Source: USDA, Economic Research Service, derived from Food and Agriculture Organization of the United Nations and other agricultural data using methods described in Fuglie et al. (2012).

The United States Department of Agriculture, Economic Research Service estimates that the recent uptick in productivity in the last two decades is due to increases in the developing countries and the transition of the economies of the former Soviet Union and Eastern Europe. A larger number of countries in sub-Saharan Africa and South Asia have yet to see such increased productivity. In developed countries, total inputs to agriculture have fallen while output has continued to grow.

The uncertain potential impacts on productivity of climate change must also be considered. These potential impacts present a significant challenge for the world. While previous modeling summarized climate change and potential adaption as a function of temperature, they have not examined the uncertainty, the timing of impacts, or the relative effectiveness of various adaptation strategies.[*] Additionally, figuring out how to enhance productivity while reducing agricultural contributions to the emission of greenhouse gases continues to be a vexing challenge.

These statistics raise the principal question about whether productivity is growing at a rate essential to meet demand. Several recent studies suggest that yields on 24%–39% of the globe's most important croplands are no longer increasing.

Some suggest that the easiest way to accelerate productivity increases is to work with the least productive by aiding them to adopt best practices. Others say we must intensify further the most highly productive lands to increase output and minimize the consumption of resources. Perhaps ramped up investments in research and development is the answer. More likely, it will require an integrated strategy using all levers on the control panel.

GLOBAL FOOD SECURITY DRIVERS AND CHALLENGES[†]

Success in meeting the increasing demand for food is a function of how the global community meets the many challenges and opportunities facing it. The paper has previously mentioned the

[*] A meta-analysis of crop yield under climate change and adaptation, A. J. Challinor et al., Nature Climate Change (2014).
[†] Adapted in part from the Global Food Security Strategic Plan (2017), Global Food Security, a multiagency program bringing together the main UK public sector funders of research and training related to food.

implications of population growth, shifting dietary expectations, and present rates of malnourishment. Following is a summary of these and other challenges and opportunities:

- *Global population growth, demographic change, and increasing affluence and urbanization*. In addition to the factors discussed above, it is critical to evaluate the production, processing, and distribution challenges created by a rapidly urbanizing population.
- *The appropriate role of science and technology in meeting demand and reducing impacts*. Will values of the developed countries limit the applicable science and technology available to meet the challenge of feeding the world? Will we take some production systems, science and technologies off the table in confronting this challenge? Or will be recognize that we must employ all known systems, science and technology to meet the test before us.
- *Global climate and other environmental changes*. Dealing with warmer temperatures, rising sea levels, changing rainfall patterns, and increasing extreme weather events will create great challenges about how we manage and utilize our infrastructure and protect public and private investments in the food production enterprise. These climate change impacts will generate many challenging social economic impacts. New pests and expanding ranges of old pests will be confronted.
- *Environmental impacts*. Determining how to minimize or avoid the adverse impacts of producing food on land, air, and water will escalate as a challenge. Increasing scarcity and competition for clean water and healthy land will challenge our productivity objectives. At its core, the food production system is dependent upon other ecosystems.
- *Key resources*. The core resources upon which agriculture relies are limited. Land, air, and water resources are all under siege to some extent. The challenge to use them sustainably is and will always be at the core of the effort to meet food demand. Many questions arise in this context. For example, will we accept localized impacts associated with intensifying production in one area to benefit the global good?
- *Social and cultural*. There are many challenges associated with urbanization, demographic change, land tenure, governance and international security, and changing patterns of consumer needs, preferences, habits and practices, affecting the demand for and consumption of different foods, and patterns of waste. There are also many challenges associated with local custom, culture and religion that must be navigated, including, but not limited to, dietary preferences.
- *Economic*. We must understand how factors of trade, land tenure, trends in production and demand and potential for shocks, competitiveness of food and farming businesses profitability, food markets and volatility, supply and distribution, regulation, affordability and availability (particularly in developing countries) assist or interfere with achieving the goal of meeting food demand.
- *Political*. How do changes in government policy, development of new multilateral or bilateral agreements and/or political instability affect strategies to meet escalating food demand. The impact of rising nationalism in Europe (the UK, in particular) and the United States raises many challenges and difficulties moving forward with a progressive plan to meet the global food challenges.

THE PATH FORWARD

In view of the complexity of the issues to be confronted, it would be easy to throw up one's hands and concede defeat. Frankly, that is not an option. All international bodies, governments, producers, processors, nonprofit organizations (NGOs), foundations, and universities have an obligation to align resources and actions to meet the demand for food.

Alignment means an integrated international effort led by the developed countries of the world and the international organizations they support. It means the developed countries, including the United States, must strategically deploy capital and resources to build institutions and capacity in the developing world to address these critical challenges.

Alignment means producers, processors, NGOs, foundations, and universities must rise above their narrow view and accept that all systems, practices, and technologies must be appropriately deployed and integrated to meet the challenge of feeding the world population.

It will require creativity, diligence, and persistence. It is the Manhattan project for food.

CHAPTER 2

Economics of Global Cattle Industries, with Implications for On-Farm Practices

Daniel A. Sumner
University of California Agricultural Issues Center

CONTENTS

CONTRIBUTION TO ECONOMIC ACTIVITY

Cattle produce a substantial share of the value of agricultural output globally and in many regions. Of course, shares vary from year to year as market prices rise and fall, but in higher income countries and regions, cattle represent almost one-third of the value of farm output. For example, in the United States, sale of cattle comprises about 18 percent of US farm sales and milk adds another ten percent (USDA ERS, 2017). In the European Union, cattle output comprises about nine percent and milk about 20 percent of the value of production (FAO). Developing countries typically have lower shares of livestock output. But, in India, which is bovine-intensive for its per capita income, bovine meat (cattle and buffalo) is about seven percent and milk is about nine percent of the value of agricultural output (FAO). In India, bovines still provide substantial on-farm power for cultivation and hauling, but I have found no systematic measure of the value of this economic contribution.

ECONOMIC OUTPUTS AND PRELIMINARY ISSUES

In higher income regions, cattle are used primarily as food-producing animals for milk and meat. Of course, substantial economic value is also generated from many by-products, primarily leather from hides, including fertilizer from manure; household goods from fats, hair and collagen; medicine ingredients from internal organs; and nutritional supplements from bones. In poor regions where mechanization still lags, in addition to the uses listed above, cattle and buffalo remain important as power sources for transport and cultivation, competing with human labor and tractors, and manure is used as a fuel.

We include buffalo along with cattle in our discussion and data. Buffalo are unimportant in much of the world. However, buffalo comprise a significant share of meat and milk production in several places, especially in South Asia, which is the largest milk-producing region in the world, much of it from buffalo (FAO). Where there is no ambiguity, we will use the term "cattle" to include buffalo.

In most places in the world, the same cattle breed and farms produce both milk and meat. Main exceptions are North America and some parts of Europe and Oceana where different breeds are specialized for meat or milk. In North America, for cows of the familiar breeds used for beef, calves are the only consumers of the milk produced. In all markets, specialized dairy milk breeds do produce substantial marketable beef from male and cull female calves that enter the beef stream and from cull cows at the end of their economic life as milk producers.

Specialization between milk and meat is dominant in North America. For example, in the United States the primary beef cattle industry has no commercial milk production and the dairy industry receives more than 90 percent of revenue from milk sales. However, even in the United States, where specialization is the dominant pattern, dairy steers and heifers comprise about 15 percent of the steers and heifers marketed as fed cattle for meat consumption (USDA ERS). Because milk cows are culled and replaced when they are no longer economically viable and dairy bulls are replaced as younger bulls come along with higher genetic potential, the two groups comprise about 40 percent of the total cull cattle marketed (USDA NASS).

DISTRIBUTION OF FARMED BOVINES ACROSS REGIONS

Figure 2.1 shows the distribution of cattle and buffalo across regions of the world. About one-quarter of the total number of animals are found in South Asia, which roughly equals its share of the world's human population, and another quarter are found in the area including South and Central

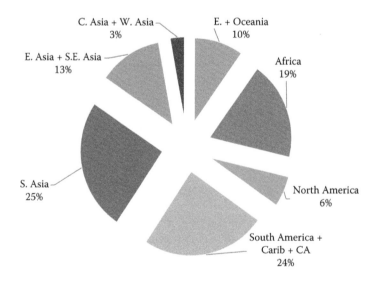

Figure 2.1   Live bovines, distribution across regions 2014.

Source: http://www.fao.org/faostat/en/#data/QL.

America and Caribbean. Africa has a large cattle herd comprising almost 20 percent of the world total. East and South East Asia, with about one-third the global human population, have about 13 percent of bovines on farms. Europe has about eight percent and Oceania (with a tiny human population) has about two percent of bovines. (Despite geographic separation, in our data discussion, we put the small region of Oceania together with Europe since both have highly developed cattle industries and Oceania is quite small separately.) North America has another six percent of bovines, which is less than its nine percent share of human population. Thus, cattle tend to be where people are, except that South America and Oceania have a much larger share of cattle than people.

DISTRIBUTION OF MILK PRODUCTION AND TRADE

Figure 2.2 documents the global distribution of milk production from milked bovines across the world. Milk production is widely distributed with about one-third of milk production in Europe and Oceania, which has only 11 percent of world human population. These regions have high per capita milk consumption and are major world exporters, especially, Oceania, which exports most of its milk production. Europe and Oceania have only 13 percent of milked bovines. Milk production per milked animal is about 5.6 tons per year, well above the world average of 2.26 tons per cow per year. South Asia produces one-quarter of the world's milk with about 37 percent of the milked bovines, about half of which are buffalo. India, the world's largest milk-producing country, is also a major exporter. But other large nations, such as Bangladesh, are importers. North America produces 13 percent of the world's milk with three percent of the milked cows. Milk per cow in North America, more than 10 tons per year, is more than four times the world average. High milk per cow in North America is derived from cows kept in barns or open area pens, fed intensively on rations that include hay, silage, grains, and oilseeds and kept for fewer lactations.

South American milk production is about 11 percent of the world total with about 13 percent of the cows. East and South East Asia are not major milk producers or major milk consumers per capita. Productivity per cow roughly equals the world average. Central and Western Asia have about

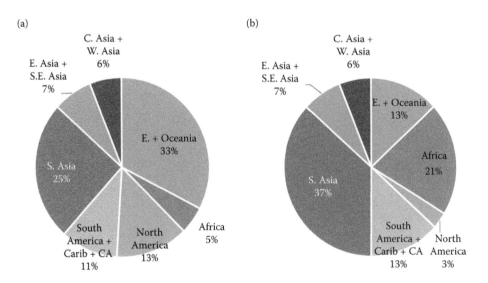

Figure 2.2   Milk production and milk bovines. (a) Production of buffalo or cow milk by regions in 2014. (b) Number of milked buffalos or cows by regions in 2014.

Source: http://www.fao.org/faostat/en/#data/QL.

six percent of both the milk and the cows. Finally, Africa has a significant 21 percent of milked cows in the world but only five percent of the milk production. Density varies widely across the continent with much of the milking herd in East Africa. Nonetheless, low productivity in terms of milk production means that per capita consumption is low despite significant imports.

International trade in milk is mainly in dry products. About ten percent of milk production is traded. Big exporting regions include Europe, Oceania, and North America. The big importing regions are East and Southeast Asia and Africa. Europe and North America have importing and exporting nations and trade high-priced dairy products among themselves even though both regions are major net exporters. Africa is a major dairy import region. Overall, Europe, North America and Oceania account for about 85 percent of dairy exports and African and other low income regions account for about 80 percent of dairy imports. But, of course, the poorest countries participate in little of this trade accounting for only about six percent of world dairy imports. Trade barriers, production subsidies, and production quotas, which used to severely distort world dairy trade, still have impacts, but are much less influential in recent years.

DISTRIBUTION OF BOVINE MEAT PRODUCTION AND TRADE

Figure 2.3 documents the global distribution of bovine meat production and numbers of animals slaughtered (including cattle and buffalo). Meat production is more evenly distributed geographically than is milk production, with about 27 percent of production in the area including South American, Central America, and the Caribbean. This region slaughters about 25 percent of the animals, so meat per animal (226 kg) is about the world average of 210 kg. Europe and Oceania together produce 20 percent of world beef from about 17 percent of world slaughter again with only 11 percent of world human population. Both South America and the Europe and Oceania regions have high per capita beef consumption and, as with milk, are major world exporters, especially Oceania, which exports most of its beef production. South Asia produces only seven percent of the world's bovine meat, much of it buffalo with about 12 percent of the slaughter, so the slaughtered

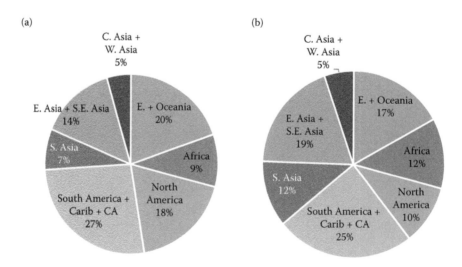

Figure 2.3   Shares of bovine meat production and numbers slaughtered, by region 2014. (a) Shares of meat production. (b) Shares of number slaughtered.

Source: http://www.fao.org/faostat/en/#data/QL.

animals have relatively little meat. Because beef consumption is low in India, it is a significant exporter of bovine meat to other low-income countries.

North America produces 18 percent of the world's beef with about ten percent of the slaughter numbers. Meat per animal averages 373 kg. The cattle slaughtered come from two sources: cull dairy and beef cows and bulls that are used primarily for ground and processed beef products (and represent less meat per animal than the regional average); and young steers and heifers, slaughtered at less than 2 years of age, that have been fed grains and oilseed rations for the last 6 months of their lives.

East and South East Asia produce about ten percent of the world's meat with about 19 percent of the bovines. Buffalo are still important for on-farm work in South East Asia and animals slaughtered have relatively little meat compared to the world average. Finally, Africa has about nine percent of beef production and 12 percent of slaughter.

International trade in bovine meat accounts for about 14 percent of production. Big exporting regions include South America, Oceania, and North America. The big importing regions are East and Southeast Asia, North America, Europe, and Africa. The United States is a major importer of ground beef and a major exporter of muscle cuts of beef. Africa is a major importer of lower-priced bovine meat, some of it from India, which is a major exporter. Overall, high income regions account for about 40 percent of bovine meat production and 45 percent of exports. Australia, New Zealand, the United States, Uruguay, Brazil, and India are major exporters. Individual countries that are major importers of bovine meat include China, Japan, Korea, Egypt, United States, and Russia.

REFERENCES

Food and Agriculture Organization of the United Nations. 2017. http://www.fao.org/faostat/en/#data/QV.

United States Department of Agriculture, Economic Research Service. 2017. https://data.ers.usda.gov/reports.aspx?ID=17830.

United States Department of Agriculture, National Agricultural Statistics Service. 2017. http://nass.usda.gov/.

http://www.fao.org/fileadmin/templates/est/COMM_MARKETS_MONITORING/Dairy/Documents/FO_Dairy_June_2016.pdf.

CHAPTER 3

Dairy and Beef Sustainability: Nutrient Loss and Use Efficiency

Joe Harrison
Washington State University

Alan Rotz
USDA—Agricultural Research Service

CONTENTS

SUSTAINABILITY DEFINED

Sustainability likely has as many definitions as individuals who wish to define it, and the definition is normally biased toward the perception or goal of the one defining. A general definition is normally stated something like "Meeting the needs of the present without compromising the ability of future generations to meet their needs." Developing a more specific definition and quantifying sustainability becomes difficult. Sustainability is generally viewed to consist of three major categories of issues: environmental, economic, and social. Each of these major areas consists of many smaller categories (Figure 3.1). For example, environmental issues include air quality, water quality, water use, energy use, resources use, etc. Within the category of water quality, there are pollution issues related to nutrient losses, pathogens, toxic compounds, hormone levels, etc. Quantifying these individual environmental impacts is possible, but often requires much effort. Quantifying social issues such as animal welfare is even more difficult, often relying on qualitative measures.

Life cycle assessment (LCA) has become a common tool used to evaluate and quantify sustainability. This accounting tool is used to integrate all impacts over a full life cycle and express them per unit of product or service received. For dairy, the intensity of the impact or footprint is normally expressed per unit of fat and protein corrected milk consumed. For beef, it is the unit of meat consumed. Greenhouse gas emission intensity or carbon footprint has received much attention as a measure of the sustainability of beef and dairy cattle products (Steinfeld et al., 2006), but a true assessment of sustainability must be much more comprehensive. Greenhouse gas emissions are only

Figure 3.1 Sustainability can be defined by many factors or metrics making up environmental, economic, and social issues (obtained from the National Cattlemen's Beef Association).

one of many metrics that should be considered in assessing the sustainability of cattle or any other product or service.

Nutrient cycling and loss to the environment is an important issue in cattle production systems with the most important nutrients being nitrogen and phosphorus. Large amounts of nitrogen in the form of protein are consumed by cattle with 70 to 85% excreted in urine and feces. During the handling of the manure, up to 50% of the nitrogen can be lost through ammonia volatilization to the atmosphere (Rotz, 2004). Most of the manure is returned to crop and pasture land where the remaining nitrogen is susceptible to runoff in surface water and leaching to ground water. Depending upon soil, crop, and manure management, leaching losses can be substantial. Runoff losses are normally small compared to other pathways, but these small losses can contribute to eutrophication of waterways. Most of the phosphorus consumed by cattle is also excreted. Volatilization of phosphorus does not occur so nearly all of that excreted is normally applied to crop and pasture land. Runoff losses of phosphorus, although small compared to that applied, are a major contributor to eutrophication of surface waters in many regions.

EXTENSIVE VS INTENSIVE SYSTEMS

The discussion of dairy and beef sustainability often focuses on resource use and systems can be defined as extensive or intensive. The most common characteristics differentiating these two systems relate to the access to pasture, amount of grain fed, and the type of housing provided for the animals. Stage of life of the animal can also have an impact on the type of system utilized. For instance, extensive pasture systems are often used for beef cow-calf and stocker cattle in order to raise the animals on low-cost forages. The cattle are then finished during the last 4–6 months of the production cycle on intensively managed feedlots. A debate continues on whether extensive or intensive production systems are more sustainable (Capper 2012).

The primary driving factor in the global movement toward more intensive livestock production is a declining profit margin. Input costs have increased greater than the wholesale/retail price paid for the marketable meat, milk, and eggs. However, consumers with ample dispensable income have facilitated the growth of organic and non-GMO production of beef and dairy products which generally use more extensive systems.

One of the most effective ways to increase efficiency is to increase the number of cows per unit of land (Rotz et al., 1999). This approach is not without consequences as nutrient import and accumulation in soils often occurs on farms with increased animal density (Harrison et al., 2007). Imported nutrients come in purchased feed, which may be transported long distances making it impractical and uneconomical to return the manure nutrients to the land producing the crops. An accumulation of nutrients on-farm can lead to greater movement of nutrients off farm via air and water resulting in degradation of water and air (Harrison et al., 2007).

A county-by-county estimate was published in 2000 of the manure nutrients relative to the capacity of cropland and pastureland to assimilate nutrients (Kellogg et al., 2000). The comparison encompassed the time period of 1982 to 1997, and clearly showed a trend for more concentration of animals per operation and to be more spatially concentrated in high-production areas. A result of this spatial concentration is that feed (nutrients) are imported from areas of grain and forage production to areas of livestock production. Without the movement of nutrients in manure back to the areas of feed production, soils at livestock production sites become nutrient sinks (see maps 24 and 25).

Map 24   Capacity of cropland and pastureland to assimilate manure nitrogen, 1997

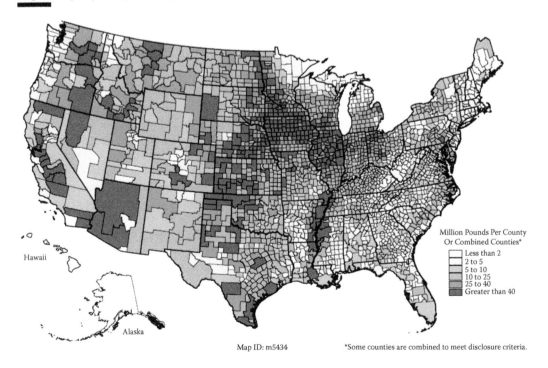

Million Pounds Per County
Or Combined Counties*

☐ Less than 2
☐ 2 to 5
☐ 5 to 10
☐ 10 to 25
☐ 25 to 40
☐ Greater than 40

Hawaii

Alaska

Map ID: m5434   *Some counties are combined to meet disclosure criteria.

Map 25 Capacity of cropland and pastureland to assimilate manure phosphorus, 1997

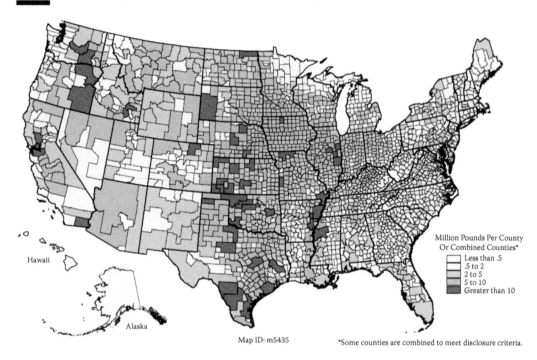

Million Pounds Per County
Or Combined Counties*
- Less than .5
- .5 to 2
- 2 to 5
- 5 to 10
- Greater than 10

Hawaii

Alaska

Map ID: m5435

*Some counties are combined to meet disclosure criteria.

The application of nitrogen in excess of crop needs can result in elevated concentrations of nitrate-nitrogen in soil, a form of nitrogen that can be lost to groundwater (Carey et al., 2017; Harter et al., 2001). In areas where groundwater is a sole source of drinking water, elevated concentrations of nitrates can be a health concern (WHO, 2011).

EFFICIENCY OF PRODUCTION

Coincident with the concentration of nutrients per operation, the genetic selection of animals for improved efficiency, along with the adoption of management technologies, has afforded the dairy industry to produce 59% more milk in 2007 as compared with 1944. In addition, this was accomplished while reducing the carbon footprint by 41% (Capper et al., 2009). When considering protein efficiency, the current-day precision feeding of the dairy cow provides the opportunity to capture ~35% of consumed protein by the cow as human edible protein, an increase from ~20% over a few decades ago (Harrison et al., 2007).

While improvements in efficiency have also been obtained in the beef industry, those improvements have generally been less than that obtained in dairy. Capper (2011) reported that compared to 1977 an equivalent amount of beef was produced in the United States in 2007 using 30% fewer cattle, 19% less feed, 12% less water, and 33% less land with a 16% reduction in carbon footprint. Along with these benefits, 12% less nitrogen and 10% less phosphorus were excreted per unit of beef produced in 2007. With less excreted, less nitrogen and phosphorus should be lost to the environment. These improvements came through producing more beef with fewer cattle and through greater use of culled dairy animals. In an analysis of a beef production system in the Midwestern United States, Rotz et al. (2013) found that the carbon footprint of the cattle produced decreased by 6.5% from the practices used in 1970 to those in 2005. During this period, reactive nitrogen losses were reduced 11%. Following the use of distiller's grain for cattle feed in 2005, the carbon footprint had a slight increase by 2011 and reactive nitrogen losses increased 10%. Feeding distiller's grain caused less-efficient protein utilization and thus greater nitrogen excretion and greater emissions to air and water.

IMPROVING SUSTAINABILITY

To meet the challenge of feeding the world while reducing environmental impacts and improving economic viability, progress must continue to be made in improving sustainability. There are solutions to many of our environmental issues, but these solutions often come with some cost to the producer and this cost may be considerable. The trade-offs between environmental, social, and economic issues must be considered to develop more sustainable systems.

In improving the sustainability of beef and dairy products (along with all other foods), consumers have a role as well. It has been estimated that about 30% of food produced is not consumed and this includes dairy and beef products (Buzby et al., 2014). This waste has a major impact on all metrics for measuring sustainability where a 30% loss effectively increases all measures by about 43%. It is unlikely that any other single component of the life cycle of beef and dairy products can have this much impact on the quantification of sustainability.

REFERENCES

Buzby, J C., H F Wells, and J Hyman. 2014. The estimated amount, value, and calories of postharvest food losses at the retail and consumer levels in the United States. Bulletin number 121, Economic Research Service, USDA. https://www.ers.usda.gov/webdocs/publications/43833/43680_eib121.pdf. Accessed 13 June, 2018.

Capper, J L, R A Cady, and D E Bauman. 2009. The environmental impact of dairy production: 1944 compared with 2007. J. Anim. Sci. 87:2160–2167.

Capper, J L. 2011. The environmental impact of beef production in the United States: 1977 compared with 2007. J. Anim. Sci. 89:4249–4261.

Capper, J L. 2012. Is the grass always greener? Comparing the environmental impact of conventional, natural, and grass-fed beef production systems. Animals. 2:127–143.

Carey, B, C F Pitz, and J H Harrison. 2017. Field nitrogen budgets and post-harvest soil nitrate as indicators of N leaching to groundwater in a Pacific Northwest dairy grass field. Nutr. Cycl. Agroecosyst. doi:10.1007/s10705-016-9819-5.

Harrison, J H, T D Nennich, and R White. 2007. Nutrient management and dairy cattle production. CAB Reviews: Perspectives in Agriculture, Veterinary Science, Nutrition and Natural Resources. 2(020). www.cababstractsplus.org/cabreviews/reviews.asp. Accessed June 13, 2018.

Harter, T, M C Mathews, and R D Meyer. 2001. Effects of dairy manure nutrient management on shallow groundwater nitrate: A case study. 2001 ASAE Annual International Meeting. Sacramento, CA. July.

Kellogg, R L, C H Ladner, D C Moffitt, and N Gollehon. 2000. Manure nutrients relative to the capacity of cropland and pastureland to assimilate nutrients: Spatial and temporal trends for the United States. USDA NRCS ERS. Publication No. nps00–0579.

Rotz, C A. 2004. Management to reduce nitrogen losses in animal production. J. Anim. Sci. 82(E. Suppl.):E119–E137.

Rotz C A, B J Isenberg, K R Stackhouse-Lawson, and E J Pollak. 2013. A simulation-based approach for evaluating and comparing the environmental footprints of beef production systems. J. Anim. Sci. 91(11):5427–5437.

Rotz C A, L D Satter, D R Mertens, R E Muck. 1999. Feeding strategy, nitrogen cycling, and profitability on dairy farms. J. Dairy Sci. 82:2841–2855.

Steinfeld, H, P Gerber, T Wassenaar, V Castel, M Rosales, and C de Haan. 2006. Livestock's long shadow: Environmental issues and options. FAO, Rome. www.europarl.europa.eu/climatechange/doc/FAO%20 report%20executive%20summary.pdf. Accessed 13 June, 2018.

World Health Organization. 2011. Nitrate and nitrite in drinking-water: Background for development of guidelines for drinking-water quality (accessed May 23, 2017; www.who.int/water_sanitation_health/ dwq/chemicals/nitratenitrite2ndadd.pdf).

CHAPTER 4

Livestock and Climate Change: Facts and Fiction

Frank Mitloehner
University of California

CONTENTS

As the November 2015 Global Climate Change Conference COP21 concluded in Paris, 196 countries reached agreement on the reduction of fossil fuel use and emissions in the production and consumption of energy, even to the extent of potentially phasing out fossil fuels out entirely. Both globally and in the U.S., energy production and use, as well as the transportation sectors, are the largest anthropogenic contributors of greenhouse gasses (GHG), which are believed to drive climate change. While there is scientific consensus regarding the relative importance of fossil fuel use, anti-animal agriculture advocates portray the idea that livestock is to blame for a lion share of the contributions to total GHG emissions.

One argument often made is U.S. livestock GHG emissions from cows, pigs, sheep, and chickens are comparable to all transportation sectors from sources such as cars, trucks, planes, trains, etc. The argument suggests the solution of limiting meat consumption, starting with "Meatless Mondays," which will show a significant impact on total emissions.

When divorcing political fiction from scientific facts around the quantification of GHG from all sectors of society, one finds a different picture. Leading scientists throughout the U.S., as well as the U.S. Environmental Protection Agency (EPA)[*] have quantified the impacts of livestock production in the U.S., which accounts for 4.2%[†] of all GHG emissions, very far from the 18% to 51% range that advocates often cite. Comparing the 4.2% GHG contribution from livestock to the 27% from the transportation sector or 31% from the energy sector in the U.S. brings all contributions to GHG into perspective. Rightfully so, the attention at COP21 was focused on the combined sectors consuming fossil fuels, as they contribute more than half of all GHG in the U.S.

Breaking down the 4.2% EPA figure for livestock by animal species shows the following contributors: beef cattle 2.2%, dairy cattle 1.37%, swine 0.47%, poultry 0.08%, sheep 0.03%, goats 0.01%, and other (horses, etc.) 0.04%. It is sometimes difficult to put these percentages in perspective, however; if all U.S. Americans practiced Meatless Mondays, we would reduce the U.S. national GHG emissions by 0.6%. A beefless Monday per week would cut total emissions by 0.3% annually.

---

[*] www3.epa.gov/climatechange/ghgemissions/sources/agriculture.html.
[†] www3.epa.gov/climatechange/Downloads/ghgemissions/US-GHG-Inventory-2015-Main-Text.pdf.

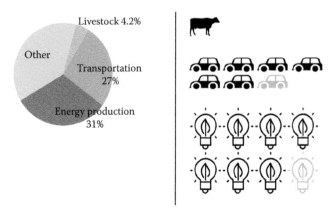

One certainly cannot neglect emissions from the livestock sector but to compare them to the main emission sources would put us on a wrong path to solutions, namely to significantly reduce our anthropogenic carbon footprint to reduce climate change.

U.S. Population replace incadescent   U.S. Population "Meatless Monday"
with energy star bulbs – 1.2%   = GHG emission – 0.6%

In spite of the relatively low contributions to total GHG emissions, the U.S. livestock sector has shown considerable progress during the last six plus decades, and commitment into the future, to continually reduce its environmental footprint, while providing food security at home and abroad. These environmental advances have been the result of continued research and advances in animal genetics, precision nutrition, as well as animal care and health.

U.S. Dairy & Beef Production Continuous Improvement

	1950	2015
Total dairy cows	22 million dairy cows	9 million dairy cows (–59%)
Milk production	117 billion lbs	209 billion lbs (+79%)
Carbon footprint		1/3 that of 1950
	1970	**2015**
Total beef cattle	140 million head	90 million head (–36%)
Beef production	24 billion lbs	24 billion lbs

Globally, the U.S. livestock sector is the country with the relatively lowest carbon footprint per unit of livestock product produced (i.e., meat, milk, or egg). The reason for this achievement largely lies in the production efficiencies of these commodities, whereby fewer animals are needed to produce a given quantity of animal protein food, as the following milk production example demonstrates: the average dairy cow in the U.S. produces 22,248 lbs milk/cow/year. In comparison, the average dairy cow in Mexico produces 10,500 lbs milk/cow/year; thus, it requires 2-plus cows in Mexico to produce the same amount of milk as one cow in the U.S. India's average milk production per cow is 2,500 lbs milk/cow/year, increasing the methane and manure production by a factor of 9 times compared to the U.S. cow. As a result, the GHG production for that same amount of milk is much lower for the U.S. versus the Mexican or Indian cow. Production efficiency is a critical factor in sustainable animal protein production and it varies drastically by region.

More milk produced per cow = Less methane and waste

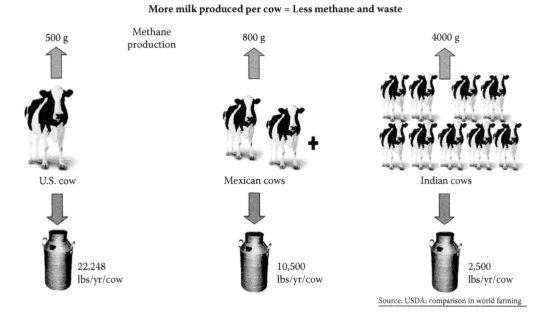

Source: USDA; comparison in world farming

Improvements in livestock production efficiencies are directly related to reductions of the environmental impact. Production efficiencies and GHG emissions are inversely related—when the one rises, the other falls.

The 2050 challenge to feeding the globe is real: throughout our lifetime, the global human population will have tripled from three to more than nine billion people without concurrent increases of natural resources to produce more food. Our natural resources of land, water, and minerals (fertilizer) necessary for agricultural production have not grown but in fact decreased. As a result, agriculture will have to become much more efficient worldwide and engage in an efficient path similar to the one it has traveled down in U.S. livestock production in recent decades.

HOW CAN EMISSIONS ACCURATELY AND FAIRLY BE ASSESSED TO LAY GROUND FOR A PATH FOR SOLUTIONS?

In its quest to identify a sustainable, scientific path toward fulfilling the future global food demand, the Food and Agriculture Organization of the United Nations (FAO) has formed an international partnership project to develop and adopt a "gold standard" life cycle assessment (LCA) methodology for each livestock specie and the feed sector. The "Livestock Environmental Assessment and Performance Partnership" (LEAP) engaged with more than 300 scientists from the world's most prestigious academic institutions in developing this unprecedented effort in developing a global benchmarking methodology. The first 3-year phase project was finalized in December 2015 with six publically available LCA guidelines.* This globally harmonized quantification methodology will not only allow the accurate measurement by livestock species and production regions across the globe today but also will identify opportunities for improvement and the ability to measure that progress in each region going forward.

SUMMARY

Addressing the 2050 challenge of supplying food to a drastically growing human population can sustainably be achieved through intensification of livestock production. Indeed, intensification provides large opportunities for climate change mitigation and can reduce associated land-use changes such as deforestation. Production efficiencies reduce environmental pollution per unit of product.

The U.S. livestock, poultry, and feed industries are one of the most efficient and lowest environmental impact systems in the world. The research, technologies, and best practices that have been developed and implemented over time in the U.S. can also be shared with other production regions around the world. It is important to understand that all regions have unique demands and abilities, and thus require regional solutions. However, the advances in the U.S. agriculture and food system can be adapted within these regional solutions. These significant environmental advances and benefits are in addition to the well-documented human health and developmental value of incorporating animal protein in the diets of the growing population.

The livestock sector is committed to continuous improvement of their environmental impact in North America, and to doing its part in transferring knowledge, technologies, and best practices to enhance global environmental livestock impact by region. Now is the time to end the rhetoric and separate facts from fiction around the numerous sectors that contribute emissions and to identify solutions for the global food supply that allow us to reduce our impact on the planet and its resources.

---

* www.fao.org/partnerships/leap/en/.

CHAPTER 5

Microbial Pathogens in Extensive and Intensive Animal Agriculture Systems

Pramod Pandey and E. R. Atwill
University of California

CONTENTS

INTRODUCTION

We live in a world filled with microorganisms. Viruses, bacteria, protozoa, and fungi are commonly found throughout our world and most play important roles in the Earth's ecosystems, including making many critical contributions to the health of soils, plants, animals, and humans.

Huge numbers of microorganisms live on both the outside (skin/hair/scales) and inside (primarily in the gastrointestinal tract) of animals and humans. Today, we have a much better understanding of the importance of these microorganisms, collectively referred to as the microbiome, in supporting animal and human health. Some of these microorganisms, referred to as pathogens, also cause illness in plant and animal species. Animal manure and human sewage both contain high numbers of microbes, some of which may be pathogenic when they gain entry into susceptible animals and humans.

MANURE PRODUCTION AND MICROBIAL NUMBERS

Over the last 60 years, changes in economic realities and advancing science have moved animal agriculture toward more intensive production systems—those that concentrate animals in small areas where humans provide the animal's feed (CAFO's—Concentrated Animal Feeding Operations) and away from extensive systems that require animals to forage on their own to meet their nutritional needs on larger areas of land.

Table 5.1   Descriptions of Animal Type, Body Weight, Manure Production, and Recoverable Manure

Animal Type	Body Weight kg	Manure Production kg/day	Manure kg/day/kg-Body Weight)	Manure kg/year/500 kg-Body Weight	Recoverable Manure %
Beef (grazing)	634.2	35.1	0.055	10,101	7.8
Beef (brood cow)	498.3	27.6	0.055	10,108	80
Beef (service bull)	770.1	45.5	0.059	10,783	80
Dairy (milking)	588.9	42.1	0.071	13,047	75
Hogs and pigs (brood sow)	158.5	4.9	0.031	5,642	75.8
Hogs and pigs (market hog)	61.2	4.9	0.080	14,612	75.8
Chicken layers	3.2	0.14	0.044	7,984	92.5
Chicken broilers	3.2	0.18	0.056	10,266	93.3
Sheep (ewe)	81.5	2.8	0.034	6,270	33.3
Sheep (market lamb)	31.7	1.1	0.035	6,333	33.3

Data source: United States Department of Agriculture (USDA)-Natural Resources Conservation Service (NRCS) (2016); Taylor and Ricker (1995).
Note: Body weight and daily manure production was extracted from USDA-NRCS (2016) and Taylor and Ricker (1995), and calculations were made for annual manure estimation.

Today, the animals producing milk, meat, eggs, and important by-products produce more than a billion tons of animal manure being produced in the USA per year (US EPA, 2013). How manure is handled on-farm can influence the number and types of microbes found in the manure on-farm and in the environment.

Table 5.1 shows animal type, body weight, and manure production in various livestock operations. Confinement operations are able to implement systems to collect and treat manure while extensive systems allow most of the manure to sit where it was deposited by the animal. Confinement operations use manure treatment methods such as anaerobic digestion, composting, and separation of liquid and solid fractions to lower total microbial and pathogen levels prior to using it as a soil amendment. Extensive systems rely primarily on time, sunlight, and temperature to reduce microbial levels.

Anaerobic digesters and lagoons used for treatment have a potential to reduce pathogens such as *Escherichia coli* and *Salmonella*. Previous studies showed that 5–6 log reductions of pathogens can be achieved, when manure is stored in lagoons for about 2–3 months (a 1-log reduction means the number of bacteria left would be 10% of those that were there before the treatment. A 5-log reduction would be 10% of 10% of 10% of 10% of 10 of 10%, or a reduction from 100,000 to 1, and a 6-log reduction would be from 1,000,000 to 1). Similarly, anaerobic digestion experiments showed that *E. coli* and *Salmonella* can be reduced by 5 logs (Pandey et al., 2015; Pandey et al., 2016a). Under composting environment, when temperature was increased to 55 °C, the *E. coli* and *Salmonella* levels were reduced to undetectable levels in less than 5 days (Pandey et al., 2015; Pandey et al., 2016b). In addition to pathogen inactivation, both anaerobic digestion and composting processes produces a digestate that is used as soil amendment, and anaerobic digestion produces biogas, a source of renewable energy (Pandey et al., 2015). In general, animal manure is seen as a valuable commodity, which provides precious recycled nutrients required for improved crop production. This natural fertilizer reduces the necessity of chemical fertilizers.

Both anaerobic digestion and composting are well-established methods for treating animal waste and widely used all over the world. The recovery of manure under grazing conditions is relatively low compared to confined systems. For example, approximately 7.8% manure is recovered from beef operation under grazing conditions, while around 80% manure can be recovered under confined operations (Table 5.1).

Table 5.2   Pathogen Prevalence and Loads in Fresh Animal Manure

Animal Type/ Description	Pathogen Prevalence and Load in Animal Manure				
	Escherichia coli O157	Salmonella	Campylobacter	Giardia intestinalis	Cryptosporidium parvum
Cattle					
Prevalence (%)	13.2	7.7	12.8	3.6	5.4
Load (CFU/g)	2.9×10^6	3.9×10^4	7.6×10^3	2.2×10^2	2.7×10^2
Pig					
Prevalence (%)	11.9	7.9	13.5	2.4	13.5
Load (CFU/g)	6.9×10^4	9.6×10^3	1.9×10^3	5.3×10^4	3.0×10^2
Poultry					
Prevalence (%)	0	17.9	19.4	0	0
Load (CFU/g)	0	5.0×10^3	4.2×10^3	0	0
Sheep					
Prevalence (%)	20.8	8.3	20.8	20.8	29.2
Load (CFU/g)	1.1×10^4	1.1×10^3	8.6×10^2	3.8×10^2	5.3×10^1

Data source: Hutchison et al. (2004). Note: Number of manure samples for cattle, pig, poultry, and sheep were 810, 126, 67, and 24, respectively. CFU/g is the number of colony-forming units/gram of manure.

Note: Different pathogens have different effective doses required to cause disease. Some need a high dose of CFU, some a small dose.

PATHOGEN PREVALENCE

A pathogen's prevalence (how often they are found) and load (how many are found in the manure tested) in fresh manure of cattle, pig, poultry, and sheep shown in Table 5.2. Untreated manure does represent a potential risk to animal and human health (Atwill et al., 2006; Atwill et al., 2012; Pandey et al., 2014), and minimizing or eliminating their impact on the environment is an important goal.

PATHOGEN TRANSPORT

Microbes do not have the ability to travel any significant distance on their own, and must rely on hitching a ride on or in something that is moving such as water, dust, boots, tires, other animals, etc. The leading method of transporting pathogenic microorganisms from manure is in ambient water that is runoff from agricultural land (agricultural nonpoint source (NPS) pollution) (Atwill et al., 2002; EPA, 2013; EPA, 2017a; EPA, 2017b). Both intensive and extensive animal production systems that inadequately manage their manure, including that used as fertilizer on crop lands, often result in an increase of microbial pathogens in ambient water (Chapin et al., 2005; Hollenbeck, 2016; Rogers and Haines, 2005).

The pathogen type and load originating from animal manure and found in cropland and soil depends on various factors such as amount of sunlight, temperature, moisture, rainfall, pH, presence or absence of specific soil nutrients, and the source, timing, and amount of manure applied. Unintended runoff associated with rainfall is a major cause of contamination. The major factors that control pathogen levels in a natural environment are time and higher temperatures and radiation of sunlight (Baumgardner, 2012; Hipsey et al., 2008).

In North America and the European Union, there are major efforts ongoing to minimize or eliminate the transport of manure pathogens from confined feeding operation to natural water resources such as surface and ground water. In the EU, member states are required to design a "program of measures" to ensure compliance with the existing microbiological standards of bathing waters (CEC, 2006; Kay et al., 2007; WHO, 2003). In the USA, a series of Best Management Practices (BMPs) and technologies such as using vegetative filter strips (small areas of grass that surface runoff must transverse before leaving the farm) adjacent to confined animal feeding operations, and the use of anaerobic digesters, animal lagoons, and composting for treating animal manure before applying it as fertilizers, have been suggested and widely implemented for treating animal waste and controlling runoff from landscapes receiving animal manure (Atwill et al., 2002; Kearney et al., 1993; Krieger et al., 1975; Lund et al., 1983; Nicholson et al., 2005; Oeschner and Doll, 2000). Research is ongoing to identify and implement methods that will further reduce or eliminate pathogens in manure.

IMPORTANCE OF CONFINED ANIMAL FEEDING OPERATIONS IN FOOD SUPPLY AND MANURE MANAGEMENT

The debate about the future of confined animal feeding operations is ongoing with opponents and proponents on both sides of the confinement issue. What is considered as fact is that confined/intensive operations produce more food per unit input of resources such as land and water, and also provide an option for lower environmental contamination of pathogens per unit of food produced. The low recovery of manure in extensive systems with large areas results in more environmental contamination on a per animal basis.

NECESSITY OF INTENSIVE ANIMAL PRODUCTION SYSTEMS

Approximately 71% of earth surface area is under water, and only 29% of earth surface is land area. Of this remaining 29%, we are already using more than 36% of land for agriculture, with 30% in forests and 3% in urban use (World Bank, 2016). The remaining 31% is covered by various landscapes such as mountains, permanent snow, river, lakes, reservoirs, and is unlikely to be suitable for food production.

Table 5.3 represents a scenario analysis of the land area required to support the numbers of beef, dairy, swine & poultry currently residing in the top five states in the USA for each animal type if they were moved from confinement to extensive systems (using standard stocking density numbers currently in use). It shows that there is insufficient land to support the current animal production numbers and the shortfall will be much greater as the world's human population continues to grow.

There is a finite amount of land suitable for food production, and there is a recognized need to preserve rain forests and other important ecosystems to maintain the Earth's life-sustaining environment. There is a foreseeable increased demand for food to meet the needs of the Earth's growing population and a market economically rewarding those who produce meat, milk, eggs, and animal-associated by-products. The markets for more food and higher quality protein cannot be met in extensive animal production systems because there is insufficient land to do so.

SUMMARY

- Microbial organisms are everywhere, including on the outside and inside of animals and humans.
- Most of these microbes are critical to our health; however, some can cause disease (pathogens) and some can cause disease in both animals and humans (zoonotic pathogens).
- Animals and humans produce a significant amount of manure and sewage that contain high numbers of microbes; some of which are pathogens.

Table 5.3 Scenarios of Potential Land Acreages Required for Transforming Existing CAFOs into Pasture Operations

States	Inventory in CAFOs (Heads)	Average Acreage Required for Pastures[a]	Total Land Acreage of State[b]	% of Total State Land Area Needed for Pastures
Beef Production Scenarios (for Top Five Beef Producing States)				
Texas	3,056,260	2,521,415	171,079,000	1.47%
Nebraska	2,736,201	2,257,366	49,198,000	4.59%
Kansas	2,673,400	2,205,555	52,657,000	4.19%
Iowa	1,738,545	1,434,300	36,000,000	3.98%
Colorado	1,130,652	932,788	66,700,000	1.40%
Dairy Production Scenarios (for Top Five Dairy States)				
California	1,840,730	30,372,045	101,000,000	30.07%
Wisconsin	1,249,309	20,613,599	34,761,000	59.30%
New York	626,455	10,336,508	30,200,000	34.23%
Pennsylvania	553,321	9,129,797	28,605,000	31.92%
Idaho	536,463	8,851,640	57,100,000	15.50%
Hog and Pig Production Scenarios (for Top Five Hog and Pig Producing States)				
Iowa	19,295,092	723,565,950	36,000,000	2009.91%
N. Carolina	10,134,004	380,025,150	31,175,000	1219.01%
Minnesota	7,652,284	286,960,650	54,009,000	531.32%
Illinois	4,298,716	161,201,850	37,065,000	434.92%
Indiana	3,669,057	137,589,638	23,307,000	590.34%
Broiler Chickens Production Scenarios (for Top Five Broiler Producing States)				
Georgia	235,400,227	17,655,017,025	37,000,000	47716.26%
Arkansas	202,397,626	15,179,821,950	34,030,000	44607.18%
Alabama	178,338,741	13,375,405,575	32,480,000	41180.44%
Mississippi	150,596,764	11,294,757,300	30,030,000	37611.58%
N. Carolina	149,921,809	11,244,135,675	31,175,000	36067.80%
Laying Hens Production Scenarios (for Top Five Egg Producing States)				
Iowa	53,793,712	4,034,528,400	36,000,000	11207.02%
Ohio	27,070,109	2,030,258,175	26,400,000	7690.37%
Indiana	24,238,513	1,817,888,475	23,157,000	7850.28%
Pennsylvania	21,982,408	1,648,680,600	28,600,000	5764.62%
California	21,091,629	1,581,872,175	101,000,000	1566.21%

[a] Pasture beef cow acreage calculation were based on 11–22 animals/20 acres (USDA-NRCS, 2016); hogs and pigs density in pasture was 25–50 head/acre; dairy cows density in pasture was 11–22 head/acres (Horner, 2011); chicken density in pasture was 50–100 chickens/acres (Plamondon, 2017).
[b] Total acreage of each state were retrieved from Geography Statistics (2016) (www.statemaster.com).

- Confinement animal production systems are about 10 times more effective in capturing manure and any associated microbes for further treatment compared to extensive systems.
- Treating manure using lagoons, digesters, and composting significantly reduce microbial levels in the environment, including the levels of pathogens.
- There are many groups working to develop methods to completely eliminate microbes from manure in a way that does not significantly harm the environment.
- Insufficient land exists to meet the world's demand for meat, milk, and eggs without the use of confinement animal production systems.

REFERENCES

Atwill, E.R., Hou, L., Karle, B.M., Harter, T., Tate, K.W., Dahlgren, R.A. (2002). Transport of Cryptosporidium parvum oocysts through vegetated buffer strips and estimated filtration efficiency. Applied and Environmental Microbiology 68(11): 5517–27.

Atwill, E., Li, X., Grace, D., Gannon, V. (2012). Zoonotic waterborne pathogen loads in livestock. London, UK: IWA Publishing.

Atwill, E.R., Tate, K.W., Pereira, Md.G.C., Bartolome, J., Nader, G. (2006). Efficacy of natural grassland buffers for removal of Cryptosporidium parvum in rangeland runoff. Journal of Food Protection 69: 177–84.

Baumgardner, D.J. (2012). Soil-related bacterial and fungal infections. The Journal of the American Board of Family Medicine 25: 734–44.

CEC (2006). Directive 2006/7/EC of The European Parliament and of The Council of 15th February 2006 concerning the management of bathing water quality and repealing Directive 76/160/EEC. Council of the European Communities. Official Journal of the European Communities L64: 37–51.

Chapin, A., Rule, A., Gibson, K., Buckley, T., Schwab, K. (2005). Airborne multidrug-resistant bacteria isolated from a concentrated swine feeding operation. Environmental Health Perspectives 113(2): 137–42.

Geography Statistics (2016). StateMaster. www.statemaster.com/graph/geo_lan_acr_tot-geography-land-acreage-total (accessed on 5/19/2017).

Hipsey, M.R., Antenucci, J.P., Brookes, J.D. (2008). A generic, process-based model of microbial pollution in aquatic systems. Water Resources Research 44, 1–26.

Hollenbeck, J.E. (2016). Interaction of the role of Concentrated Animal Feeding Operations (CAFOs) in Emerging Infectious Diseases (EIDS). Infection, Genetics and Evolution 38: 44–6.

Horner, J.L. (2011). Starting a 75-cow intensive rotation grazing dairy. University of Missouri Extension. http://extension.missouri.edu/p/G3052 (accessed on 5/19/2017).

Hutchison, M.L., Walters, L.D., Moore, A., Crookes, K.M., Avery, S.M. (2004). Effect of length of time before incorporation on survival of pathogenic bacteria present in livestock wastes applied to agricultural soil. Applied and Environmental Microbiology 70: 5111–8.

Kay, D., Aitken, M., Crowther, J., Dickson, I., Edwards, A.C., Francis, C., Hopkins, M., Jeffrey, W., Kay, C., McDonald, D., Stapleton, C.M. (2007). Reducing fluxes of faecal indicator compliance parameters to bathing waters from diffuse agricultural sources: The Brighouse Bay study, Scotland. Environmental Pollution 147(1): 138–49.

Kearney, T.E., Larkin, M.J., Levett, P.N. (1993). The effect of slurry storage and anaerobic digestion on survival of pathogenic bacteria. Journal of Applied Microbiology 74: 86–93.

Krieger, D.J., Bond, J.H., Barth, C.L. (1975). Survival of Salmonella, total coliforms, and fecal coliforms in swine waste lagoon effluents. Page 11–14 in Proc. 3rd Int. Symp. Addressing Animal Production and Environmental Issues. Urbana-Champaign, IL.

Lund, E., Nissen, B. (1983). The survival of enteroviruses in aerated and non-aerated cattle and pig slurry. Agricultural Wastes 7: 221–30, 33.

Nicholson, F.A., Groves, S.J., Chambers, B.J. (2005). Pathogen survival during livestock manure storage and following land application. Bioresource Technology 96: 135–43.

Oeschner, H., Doll, L. (2000). Inactivation of pathogens by using the aerobic-thermophilic stabilization process. Pages 522–31 in Proc. 8th. Int. Symp. on Animal, Agricultural and Food Processing Wastes. Des Moines, IA.

Pandey, P.K., Kass, P.H., Soupir, M.L., Biswas, S., Singh, V.P. (2014). Contamination of water resources by pathogenic bacteria. AMB Express 4: 51.

Pandey, P., Biswas, S., Vaddella, V., Soupir, M. (2015). Escherichia coli persistence kinetics in dairy manure at moderate, mesophilic, and thermophilic temperatures under aerobic and anaerobic environments. Bioprocess and Biosystems Engineering 38(3): 457–67.

Pandey, P.K., Soupir, M.L., Ikenberry, C.D., Rehmann, C.R. (2016a). Predicting streambed sediment and water column Escherichia coli levels at watershed scale. JAWRA Journal of the American Water Resources Association 52: 184–97.

Pandey, P., Vaddella, V., Wenlong, C., Biswas, S., Colleen, C., Hunter, S. (2016b). In-vessel composting systems for converting food and green waste into pathogen free soil amendment for sustainable agriculture. Journal of Cleaner Production 139: 407–15.

Plamondon, R. (2017). Practical Poultry Tips. www.plamondon.com/wp/how-many-chickens-per-acre/ (accessed on 5/19/2017).

Rogers, S., Haines, J. (2005). Detecting and mitigating the environmental impact of fecal pathogens originating from confined animal feeding operations: review.

Taylor, D.C., Ricker, D.H. (1995). Livestock manure: a nonpoint source environmental hazard in South Dakota. Cattle 95-15. www.sdstate.edu/sites/default/files/ars/species/beef/beef-reports/upload/CATTLE_95-15_Taylor.pdf (accessed on 5/15/2017).

United States Department of Agriculture (USDA)-Natural resources Conservation Service (NRCS). (2016). Animal Manure Management. www.nrcs.usda.gov/wps/portal/nrcs/detail/null/?cid=nrcs143_014211 (accessed on 5/18/2017).

United States Environmental Protection Agency (US EPA). (2013). Literature review of contaminants in livestock and poultry manure and implications for water quality. Office of Water (4304T), EPA 820-R-13-002. https://nepis.epa.gov/Exe/ZyPDF.cgi/P100H2NI.PDF?Dockey=P100H2NI.PDF (accessed on 5/20/2017).

US EPA. (2017a). Nonpoint source success story. Controlling nonpoint source pollution from agricultural areas restores Abbott's Mill Pond. www.epa.gov/sites/production/files/2016-02/documents/de_abbotts.pdf (accessed on 5/18/2017).

US EPA. (2017b). Polluted Runoff. Non-point Source Pollution. www.epa.gov/nps/what-nonpoint-source (accessed on 5/20/2017).

USDA-NRCS. (2016). Balancing your animals with your forage. www.nrcs.usda.gov/Internet/FSE_DOCUMENTS/stelprdb1097070.pdf (accessed on 5/18/2017).

WHO. (2003). Guidelines for safe recreational waters, Volume 1: Coastal and fresh waters. Geneva: World Health Organization.

World Bank. (2016). Center for international earth science information network (CIESIN). http://data.worldbank.org/indicator/AG.LND.TOTL.UR.K2 (accessed on 5/19/2017).

CHAPTER 6

Cattle Ectoparasites in Extensive and Intensive Cattle Systems

Alec C. Gerry
University of California

CONTENTS

University of California, Riverside Ectoparasites can negatively affect cattle health, welfare, and productivity in many ways, ranging from reductions in cattle weight gain or milk production, to severe health consequences and even death of parasitized animals. These negative impacts can be categorized as (1) physical damage to cattle caused by the feeding of ectoparasites on blood, skin, or hair, (2) irritation and disturbance of cattle resulting in unproductive pest avoidance behaviors in response to the painful or irritating bites of ectoparasites, (3) transmission of disease agents to cattle by ectoparasites, and (4) nuisance to facility employees and neighbors by the activity of some ectoparasites even when these ectoparasites have no measureable impact on cattle (Table 6.1). Furthermore, ectoparasites can reduce feed conversion efficiency thereby impacting production even when no obvious damage to cattle has occurred.

The presence and abundance of ectoparasites at any individual cattle facility is driven primarily by the local environment and by the operational characteristics of the facility. Most ectoparasites exhibit seasonal activity that is relatively consistent among years and across cattle production systems within a geographic area, with the timing of peak activity often predictable by temperature and rainfall during the preceding months. But while environmental characteristics determine the timing of peak ectoparasite activity, it is the operational characteristics of the facility, particularly facility design and herd management, that often determine the abundance of ectoparasites and the severity of their impacts to cattle health and production.

Table 6.1 Cattle Ectoparasites in North America and Their Impacts to Cattle Production

Cattle Ectoparasites	Contact with Cattle	Life Stage on Cattle	Food Source	Impact to Cattle Production[a]			
				Damage	Disturbance	Disease	Nuisance
Lice	Permanent	All	Blood or skin	x			
Scabies or "mange" mites	Permanent	All	Skin, lymph	x			
New World screwworm fly (*Cochliomyia hominivorax*)	Intermittent	Immature	Body tissues	x			
Cattle grub (*Hypoderma* spp.)	Intermittent	Immature	Body tissues	x	x		
Spinose ear tick (*Otobius megnini*)	Intermittent	Immature	Blood	x	x		
Cattle fever ticks (*Rhipicephalus* spp.)	Intermittent	All	Blood	x		x	
Pajaroello tick (*Ornithodoros coriaceus*)	Intermittent	All	Blood	x	x	x	
3-host ticks (e.g., *Dermacentor, Amblyomma* ticks)	Intermittent	Mainly adult	Blood	x		x	
Horn fly (*Haematobia irritans*)	Temporary	Adult	Blood	x	x		
Stable fly (*Stomoxys calcitrans*)	Temporary	Adult	Blood	x	x		x
Horse and deer flies	Temporary	Adult	Blood	x	x	x	x
Biting midges (*Culicoides* spp.)	Temporary	Adult	Blood	x	x	x	
Black flies	Temporary	Adult	Blood	x	x		
Mosquitoes	Temporary	Adult	Blood	x		x	x
Face fly (*Musca autumnalis*)	Temporary	Adult	Exudates			x	x
House fly (*Musca domestica*)	Environmental pest	Adult	Exudates			x	x

[a] Negative impacts can be categorized as (1) physical damage to cattle caused by the feeding of ectoparasites on blood, skin, or hair, (2) irritation and disturbance of cattle resulting in unproductive pest avoidance behaviors in response to the painful or irritating bites of ectoparasites, (3) transmission of disease agents to cattle by ectoparasites, and (4) nuisance to facility employees and neighbors by the activity of some ectoparasites even when these ectoparasites have no measureable impact on cattle.

EXTENSIVE VS. INTENSIVE SYSTEMS: ECTOPARASITE PRESENCE AND ABUNDANCE

Cattle production systems are loosely categorized as extensive or intensive depending upon the level of mechanization, human control over cattle nutrition, and amount of labor required for cattle care, as described in greater complexity elsewhere in this book. Of importance to this review of ectoparasite impacts to cattle production, extensive cattle systems are primarily pasture-based with cattle feeding at will on a mix of pasture grasses available to them and with little to no supplemental feed, while intensive cattle systems are those where cattle are held predominantly off pasture in dry lot pens, free stall barns, or feedlots where they receive carefully mixed feed rations balanced to meet the production goals of the herd. From the perspective of ectoparasite impacts to cattle operations, extensive and intensive systems differ primarily in (a) presence or absence of pasture,

Table 6.2   Production System Characteristics affecting Ectoparasite Impacts

System Characteristics	Extensive	Intensive
Presence of pasture habitat	Yes	No
Habitat shared with wildlife	Yes	No
Density of cattle	Low	High
Cattle feed stored on site	No	Yes
Cattle manure collected and stored on site	No	Yes
Herd manager interaction with cattle	Low	High
Easy application of insecticides to cattle	No	Yes

(b) sharing of habitat with wildlife, (c) density of cattle in a pasture or pen, (d) storage of cattle feed and supplemental rations on site, (e) collection and storage of cattle manure on site, (f) frequency of interaction between herd managers and cattle, and (g) ease of implementing ectoparasite control including the application of insecticides to animals (Table 6.2).

While the decision to design and operate a cattle facility as an extensive or intensive production system is not likely to be driven solely by concern over ectoparasite effects on cattle, it is nevertheless important to consider ectoparasite impacts and management when considering design and operational parameters. Neither production system is the "best" one for management of all ectoparasites. Each production system provides advantages to some ectoparasite species while disadvantaging other species. *Absent any pest control measures applied by herd managers, the presence and abundance of ectoparasites on a cattle facility will be related to the availability of immature development habitat, survival of ectoparasites when off the host, and the opportunity for ectoparasites to acquire a new host whenever needed for feeding.* In addition, a herd manager trained to recognize ectoparasites and effectively apply control measures can reduce or even eliminate some ectoparasites even when production system characteristics might seem ideal for the ectoparasite.

The presence of suitable immature development habitat is perhaps the most important cattle system characteristic to determine both presence and abundance of cattle ectoparasites (Table 6.3). It should be no surprise that a pasture-based cattle system provides suitable immature habitat for most cattle ectoparasites, as cattle and their ectoparasites co-evolved in similar natural habitats. As cattle move across a pasture, they leave behind fresh, intact fecal pats that serve as the required development site for horn flies and face flies. These flies are rarely abundant in intensive cattle systems, because cattle move back and forth across their pens disturbing and breaking apart the fecal pats. In contrast, stable flies and house flies can be far more numerous in intensive production systems where cattle feces is often collected and stored on site providing a substantial amount of the moist and fermenting feces that these two species prefer. Intensive systems also must keep quantities of animal feed on site, much of which is fermented either deliberately or as a result of piling or stacking moist feed; or due to placement of dry feed in a location where it is wetted by rainfall, sprinklers, or runoff from pens. Stored feed may include hay, silage, grains, and fruit or nut waste among many other possible plant materials. Stable flies can be particularly numerous where intensively managed cattle are provided plant-based bedding (typically hay or straw) resulting in a fermenting mixture of plant material, feces, and urine. On pastures, stable flies can be numerous when cattle are provided supplemental hay placed at fixed locations for days or weeks allowing for the mixture of hay, feces, and urine at these locations.

Many biting flies develop in aquatic or semiaquatic habitats that are probably not an intentional design component of a cattle operation, but are instead simply natural features of the surrounding habitat. Aquatic habitats are not unusual in or near pasture-based systems. However, even intensive systems may be constructed near wetlands, rivers, streams, or other aquatic features that might result in large numbers of biting flies. It is really distance between animals and aquatic habitat, rather than the design of the facility or the choice of an extensive or intensive production system that

Table 6.3 Immature Development Habitat for Cattle Ectoparasites

Cattle Ectoparasites	Cattle Body	Cattle Pen or Pasture	Fresh Cattle Feces	Manure & Fermenting Feed	Manure-Polluted Ponds	Other Aquatic
Lice	x					
Scabies or "mange" mites	x					
New World screwworm fly (*Cochliomyia hominivorax*)	x	x				
Cattle grub (*Hypoderma spp.*)	x	x				
Spinose ear tick (*Otobius megnini*)	x	x				
Cattle fever ticks (*Rhipicephalus* spp.)	x					
3-host ticks (e.g., *Dermacentor, Amblyomma* ticks)		x				
Pajaroello tick (*Ornithodoros coriaceus*)		x				
Horn fly (*Haematobia irritans*)			x			
Stable fly (*Stomoxys calcitrans*)				x		
Horse and deer Flies						x
Biting midges (*Culicoides* spp.)					x	x
Black flies						x
Mosquitoes					x	x
Face fly (*Musca autumnalis*)			x			
House fly (*Musca domestica*)				x		

determines the abundance and likely impact of these biting flies. However, the greater cattle density in intensive operations may actually decrease damage and disturbance caused by these biting flies as the number of bites received by any individual animal would be reduced.

Some biting midges (e.g., *Culicoides sonorensis*) and some mosquito species are more abundant in aquatic habitats with high concentrations of animal feces or other organic pollutants. These species can be quite abundant in intensive systems, particularly where manure-polluted wastewater is accumulated in poorly designed storage systems. Increasing abundance of *Culicoides* is associated with increasing transmission of bluetongue virus to cattle. Pasture systems are expected to have relatively low numbers of the biting midges and mosquitoes that develop in polluted aquatic habitats, but only if animals and their fecal waste are excluded from any aquatic habitats on or near pastures.

For ectoparasite species that must spend at least part of their life off the host, survival is greatly influenced by habitat characteristics associated with the production system. While ectoparasites are off the host and on the ground, they are subject to mortality from unsuitable environmental conditions. Pasture vegetation provides refuge from direct sunlight, high temperatures, low humidity, and dusty conditions. In contrast, dry dirt pens typical of intensive systems offer little refuge. Ectoparasites that drop from animals in an intensive system may have to crawl or wriggle very far indeed to reach a shaded area. Perhaps of equal importance for survival is the likelihood of being squashed beneath the hooves of your host. In pasture systems, ectoparasites have little risk of being stepped on as cattle density in these systems is relatively low. While in intensive systems where cattle density is high, it must be quite a wild race across the busy pen for an unfortunate ectoparasite that drops from its host near the middle of the pen! One might consider the risk to ectoparasites from predators (e.g., ants, spiders, lizards, rodents) to be higher in pastures than in dry pens, but this risk is probably minor relative to the other factors that affect off-host survival.

An additional aspect of ectoparasite survival must be considered for ticks that require blood meals from non-cattle hosts during their immature stages. Most 3-host ticks commonly require blood meals on smaller mammals or birds during their immature life stages. Pastures may provide the necessary range of alternate hosts needed for these ticks, while dry pen systems provide only cattle leaving these ticks unable to feed during their immature stages. In contrast, 1-host ticks that feed only on cattle (e.g., spinose ear tick, cattle fever tick) may be able to persist in a dry pen environment if other aspects of the habitat are suitable. The spinose ear tick in particular seems to be increasingly abundant in some types of intensive systems that include animal bedding where adult ticks can survive to lay eggs and larval ticks can readily acquire a new host.

The opportunity for ectoparasites to acquire a new host is also critically important. Some ectoparasites can be managed simply by removing their hosts from pens or pastures for a suitable period of time to kill the ectoparasite from lack of a blood meal. Once the pen or pasture is ectoparasite free, uninfested animals can be safely moved in. This technique works especially well for the ectoparasites that cannot survive off their host for more than a few days (e.g., lice and mites). In an intensive system, leaving pens vacant for a few days before rotating the next age-group of animals into the pen will prevent transfer of lice and mites from one herd to the next. In pasture-based systems, cattle are often held in mixed-age herds with contact among animals allowing for transfer of lice and mites from infested to uninfested animals. Ticks have also been managed by taking cattle off a pasture ("pasture spelling"), but the time required to eliminate ticks can be very long as some tick species can survive for months or even years without a blood meal, and deer or other wild animals may serve as substitute hosts in the absence of cattle. To acquire hosts, ticks climb grasses or other vegetation and wave their front legs about until a suitable host brushes past allowing the tick to grasp a few animal hairs. Where vegetation is lacking, ticks cannot easily acquire a host; unless ticks and cattle encounter one another in animal bedding areas, as is the case for both the spinose ear tick and the Pajaroello tick.

Increasing cattle density often also increases the opportunity for ectoparasites to acquire a new host. When cattle density is high, contact among animals is greater, increasing transfer of lice and mites among animals. Of course, ticks would also benefit from higher cattle density, decreasing the time to acquire a new host and therefore increasing tick survival while they wait in the habitat for a suitable host to wander past. Higher cattle density also increases the quantity of feces, feed, and wastewater, so that biting flies that develop in these materials will be more abundant and cattle will experience an increase in bites by these ectoparasites, perhaps resulting in increased opportunity for the transmission of disease agents (e.g., bovine pinkeye, bluetongue) among animals. Paradoxically, increasing cattle density may actually decrease the impact of biting flies that develop in non-manure-polluted aquatic habitats, since ectoparasite abundance would remain unchanged leading to fewer bites per individual animal, thereby lowering production losses as bite avoidance behaviors are reduced.

ECTOPARASITE LIFE HISTORY CHARACTERISTICS

Insects, ticks, and mites that harm cattle generally feed on blood, skin, hair, or exudates (tears or mucus) at the external body surface of cattle, resulting in their common description as external parasites or "ectoparasites." Exceptions include the cattle grubs (*Hypoderma lineatum* and *H. bovis*) and the New World screwworm fly (*Cochliomyia hominivorax*) that invade cattle tissues to feed during their immature life stages, and are therefore more accurately described as internal parasites or "endoparasites." For simplicity, the term ectoparasite is used loosely in this review to include all of the insects, ticks, and mites that negatively impact cattle in some way.

Ectoparasites often cause negative impacts to cattle production related to the type and duration of their contact with a single host animal (Figure 6.1 and Table 6.1). Ectoparasites can complete all life stages living and feeding on a single host animal (*permanent ectoparasites*), they can maintain long-term contact with a single host animal during some portion of their life while also requiring

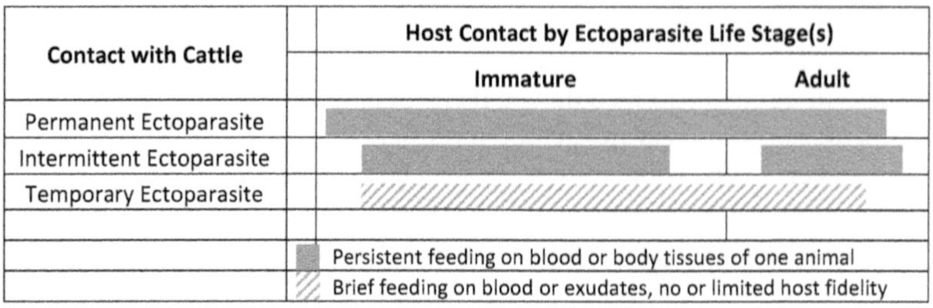

Contact with Cattle	Host Contact by Ectoparasite Life Stage(s)	
	Immature	Adult
Permanent Ectoparasite	██	
Intermittent Ectoparasite	████████████████████	████████
Temporary Ectoparasite	//	
	Persistent feeding on blood or body tissues of one animal	
	Brief feeding on blood or exudates, no or limited host fidelity	

Figure 6.1 Category of cattle ectoparasites and host association by life stage.

time spent off the host (*intermittent ectoparasites*), or they may contact hosts only briefly during one or more life stages to feed on blood or body exudates (*temporary ectoparasites*). In addition, some insect species may not feed on cattle, but can impact cattle production due to nuisance or transmission of pathogens acquired from animal feces (*environmental pests*).

Permanent Ectoparasites

Permanent ectoparasites of cattle in North America include five species of lice and four species of mites. The more damaging blood feeding lice are the longnosed cattle louse (*Linognathus vituli*), shortnosed cattle louse (*Haematopinus eurysternus*), cattle tail louse (*H. quadripertusus*), and little blue louse (*Solenopotes capillatus*). A single species of chewing louse, the cattle biting louse (*Bovicola bovis*), feeds on skin rather than blood. Cattle mites feed on skin debris or lymph within the dermal tissues and include the important scabies or "mange" mites *Psoroptes ovis, Sarcoptes scabiei*, and *Chorioptes bovis*, as well as the cattle follicle mite (*Demodex bovis*). Feeding by lice and mites can be quite irritating to the host, and may result in considerable physical damage due to dermatitis, tissue destruction, and hair loss. Lice and mites can also cause damage to hides, particularly as animals rub and scratch against objects in their environment to alleviate the itching caused by lice and mite feeding. Heavy infestations of lice and/or mites can reduce weight gain and milk yield. Additionally, poor physical condition of heavily infested animals, often coupled with substantial hair loss, can result in death of young calves and older cattle when exposed to severe weather conditions or low nutritional levels.

Management of lice and mites is commonly achieved by treating cattle with topically applied insecticides and acaricides, and by limiting contact among infested and uninfested animals or herds. Injection of ivermectin or related parasiticides may also provide control of lice and mites.

Intermittent Ectoparasites

The New World screwworm fly (*C. hominivorax*) is intimately associated with the cattle on which they live and feed during their immature life stages. The adult fly lays eggs in wounds of cattle (and other animals) where the immature larvae (maggots) consume living tissue, a condition called myiasis. Infested wounds often encourage additional egg deposition as wounds are expanded by the feeding maggots. Damage to cattle caused by these flies can be severe, often resulting in death of the animal when infestation is not promptly treated. The New World screwworm fly was eradicated from North America by 1966 following years of mass releasing sterile male flies to mate with wild female flies. This eradication effort is one of the greatest success stories of insect management by the U.S. Department of Agriculture! Unfortunately, New World screwworm flies persist in South America and on some Caribbean islands, and a reintroduction of these flies to the Florida Keys in 2016–2017 demonstrates that the cattle industry must remain vigilant. While release of sterile males is the primary tool to eradicate these flies from a region, immediate control of screwworm infestation of cattle or other animals is achieved by treating with topical insecticides the infested and uninfested wounds of all animals in the area. Injection of cattle with the parasiticide doramectin is also effective.

Like the New World screwworm fly, the common cattle grub (*H. lineatum*) and the northern cattle grub (*H. bovis*) live and feed on cattle during their immature life stages. Adult cattle grubs are also called "heel flies" as adults of both species land on the legs and lower body of cattle where they deposit eggs onto the hairs at these locations. While adult cattle grubs do not bite cattle, the presence of an adult fly may cause cattle to run madly with their tail raised in the air in an apparent effort to avoid these flies. This panicked running is called "gadding" and can result in cattle injuries as cattle run into objects in their environment. However, it is the immature flies that cause the most significant damage. Newly hatched fly larvae burrow into the skin and migrate through internal body tissues until reaching the back, where larvae cut a breathing hole in the hide resulting in a swelling ("warble") within which the larvae feed on exudates to complete immature development before dropping to the ground to become adult flies. Cattle grubs are effectively controlled using systemic insecticides applied once each year in late summer when the younger larvae are just beginning to migrate through cattle tissues.

Ticks are intermittent ectoparasites that typically remain on their animal host for days to weeks during each blood feeding period. Ticks generally require a bloodmeal for each of their three active life stages (larva, nymph, and adult), though there are some exceptions to this feeding pattern. The majority of tick species (3-host ticks) feed on different individual animals during each life stage, dropping off the host between bloodmeals to molt to the next stage or for adult ticks to lay eggs. Many of these 3-host ticks will feed on cattle mainly during the adult stage, feeding on smaller mammals or even birds during their immature life stages. In contrast, a few tick species attach as larvae to cattle and then remain on the same host animal through all feeding stages, dropping off only when feeding is no longer required (1-host ticks). Adult ticks will deposit up to thousands of eggs on the ground where the tick dropped from its last host. Unfed ticks can survive off the host animal for months or even years depending upon the tick species, making these pests very difficult to control.

Cattle fever ticks (*Rhipicephalus annulatus* and *R. microplus*) are 1-host ticks that feed on the same animal during all life stages, dropping off the host after a final bloodmeal during the adult stage. Cattle fever ticks commonly feed on cattle, antelope, and related bovids, in addition to several species of deer, making control of these ticks difficult when alternate hosts are available. These ticks can be infected with *Babesia* parasites passed from female ticks to their offspring and then to cattle during feeding, resulting in bovine babesiosis or "Texas cattle fever" which presents as anemia, wasting, and eventually death of cattle. Due to their significant impacts to the cattle industry, the U.S. Department of Agriculture in 1906 initiated a cattle fever tick eradication program that coupled an aggressive tick surveillance program with mandatory animal treatments using topically applied acaricides. This highly successful program resulted in eradication of these ticks from the U.S. in 1943. Cattle fever ticks remain common in northern Mexico, and the U.S. maintains an active quarantine program at the southern U.S. border to prevent the reintroduction of these ticks.

Another 1-host tick, the spinose ear tick (*Otobius megnini*), feeds on a single host animal only during the immature stages, then drops off the host to complete development to a non-feeding adult. These ticks attach and feed within the folds of the cattle ear. Spines on the tick body help hold these ticks in place within the ear. Spinose ear ticks are known to cause restlessness and head shaking behavior in infested animals, leading to possible decreases in weight gain or milk yield when infestation is heavy. Spinose ear ticks are common to the southwestern U.S., particularly in pasture settings and increasingly in freestall barns where animal bedding may provide a refuge for adult ticks and host-seeking larvae.

Other ticks of concern to cattle are 3-host ticks. For these ticks, cattle are simply one of several (or many!) suitable hosts from which they can obtain a bloodmeal. Common 3-host ticks that are known to feed on cattle include *Dermacentor* and *Amblyomma* ticks that primarily feed on cattle during their adult stage. Heavy infestations of these ticks can reduce cattle grazing, resulting in reduced weight gains. Cattle infested with adult *Dermacentor* ticks can also suffer paralysis due to the introduction of toxins during tick feeding.

Due to their prolonged association with host animals while feeding, most ticks are primarily controlled by application of topical acaricides to cattle and other hosts during seasons when adult ticks are common. Leaving pastures or pens vacant of cattle and other suitable hosts has been suggested for control of ticks, but this is challenging as most ticks can survive for months to years off the host.

Temporary Ectoparasites

Most of the temporary ectoparasites of cattle are blood-feeding or "biting" flies. These flies include the horn fly (*Haematobia irritans*), stable fly (*Stomoxys calcitrans*), horse and deer flies, biting midges (*Culicoides* spp.), black flies, and mosquitoes. Biting flies complete their immature development off cattle, seeking cattle as hosts only as adults to acquire protein from animal blood during short feeding periods typically lasting just a few minutes. Biting flies often take several bloodmeals from hosts during their adult life, with each bloodmeal usually acquired from a different individual animal. After acquiring a bloodmeal, most biting flies leave the host to rest and digest the bloodmeal in the surrounding habitat. The horn fly is a notable exception to this general life history, as these flies take many small bloodmeals from cattle each day, and rest on their cattle hosts even when not feeding. However, horn flies disturbed by host movement and defensive behaviors will readily fly to a nearby animal, so that bloodmeals are often taken from different animals.

Biting flies often give sharply painful bites, using blade-like mouthparts to cut and tear through host skin to pool blood at the wound site. This is particularly true for horn flies, stable flies, and horse/deer flies. These painful bites often result in considerable cattle disturbance, with cattle exhibiting distinct bite avoidance behaviors including tail flicks, leg stamps, kicks, head throws, and bunching. The type and frequency of bite avoidance behavior is related to the biting fly species and the number of bites that an animal is receiving (biting rate). These avoidance behaviors increase metabolic activity of cattle and reduce feed/water consumption, negatively impacting cattle productivity. Some biting flies can also transmit blood-borne pathogens among cattle as they feed on multiple animals.

The immature development site for biting flies varies by species. Of the species that develop in cattle feces, horn flies develop only in undisturbed fecal pats, while stable flies develop in aged cattle feces and fermenting animal feeds, particularly when animal manure is mixed with feed or plant waste. The remaining biting flies develop in aquatic or semiaquatic habitats. A few species of biting midges and mosquitoes can be quite numerous on cattle facilities that have wastewater ponds or other aquatic habitats polluted with cattle feces.

Two non-biting fly species of importance to cattle production are the face fly (*Musca autumnalis*) and the house fly (*Musca domestica*). These fly species affect cattle production by transmitting pathogens or parasites among cattle, or by the nuisance they cause to humans who work on or live near cattle facilities. While not a blood-feeding fly, the face fly does feed on cattle; feeding on mucus and eye exudates. With this feeding behavior, the face fly is of particular concern as the vector of a bacterium (*Moraxella bovis*) causing bovine pinkeye. The face fly, like the horn fly, develops only in fresh cattle feces, while the house fly commonly develops in many fermenting materials including animal feces, animal feed, and even household kitchen waste. House flies are particularly numerous where cattle feces are collected and stored wet for more than a few days on an animal facility, or where animal feces is allowed to accumulate within pens or barns, especially where manure is wetted by animal urine, sprinkler systems, or spill from water troughs.

With adult flies typically on cattle for only a very short period of time, the application of insecticides to cattle is generally of limited effectiveness for control of these pests. Horn flies are the exception to this rule, as they spend most of their adult life on cattle and are therefore readily managed by application of insecticides to animals. Additionally, recent research using insect repellents applied to cattle to provide short-term protection against horn flies and perhaps other biting flies seems promising. Overall, these flies are best controlled by reducing available immature development sites or by killing ectoparasites while they are concentrated in development sites.

A somewhat unusual tick, the Pajaroello tick (*Ornithodoros coriaceus*), is found in animal bedding sites throughout the California coastal mountain range and the foothills of the Sierra Nevada Mountains. Unlike most other ticks of concern to cattle, this tick feeds multiple times as an adult, with each feeding period lasting only 10–20 minutes. Since the 1950s, the Pajaroello tick was associated with a condition of increased abortions in beef cattle called epizootic bovine abortion or sometimes "foothill abortion." Recently, researchers at UC Davis identified a bacterium (*Pajaroellobacter abortibovis*) from these ticks that appears to be responsible for causing the increase in cattle abortions.

EXTENSIVE VS. INTENSIVE SYSTEMS: ECTOPARASITE MANAGEMENT

As mentioned earlier, a herd manager trained to recognize and control cattle ectoparasites can reduce or even eliminate some ectoparasites, even when these ectoparasites are expected based upon the type of production system. With respect to ectoparasite management, extensive and intensive systems differ most importantly in the frequency and duration of interaction by the herd manager and other facility employees with each animal, and in the ease by which animals are gathered and treated for ectoparasites by topical application or injection of insecticides and acaricides (see Table 6.2).

In intensive systems, the herd manager is expected to have much more frequent contact with each animal as cattle are given daily care, to include providing feed and checking water systems. With the higher frequency of contact, herd managers should readily identify the presence of ectoparasites while their abundance and impacts are low. Some extensive systems, such as pasture-based dairies, may also require frequent contact between herd managers and cattle, thus providing this same management benefit. But in many pasture systems, contact between the herd manager and cattle is infrequent, limiting the ability of herd managers to recognize ectoparasite problems until impacts to cattle production are noticeable. The importance of quickly recognizing ectoparasite presence in the herd varies by ectoparasite. For example, early recognition that cattle are infested with chewing lice may not be so important given their relatively low impact to cattle, but recognizing that cattle are infested with New World screwworm when the number of infested animals is small is incredibly important to avoid culling a large number of animals with gaping wounds that are untreatable.

Cattle in intensive production systems are typically housed in pens allowing the herd manager to easily quarantine parasitized animals or to efficiently administer insecticide treatments to all or part of a herd. Cattle head gates, squeeze chutes, and alley stops built into intensive system facilities make it relatively easy for herd managers to ensure treatment of all animals in a herd, whether the treatment is an injectable, a topical pour-on, or a spray. In many extensive systems, cattle are difficult to gather in one location and can be even more difficult to move through portable squeeze chutes for treatment. The difficulty of treating animals in pasture systems is one reason that insecticide-treated cattle ear tags have been widely used for many years for season-long control of horn flies and face flies. Another recent improvement in treating pasture animals is the delivery of insecticide to pasture cattle by firing an insecticide-filled gel capsule from a CO_2-powered gun

Table 6.4   Ectoparasite Impacts by Cattle Production System and Level of Ectoparasite Management

Cattle Ectoparasites	Extensive Systems		Intensive Systems	
	Active Management	No Management	Active Management	No Management
Lice	xx	xxx		xxx
Scabies or "mange" mites	xx	xxx		xxx
New World screwworm fly (*Cochliomyia hominivorax*)	x	xxx		xxx
Cattle grub (*Hypoderma spp.*)		xxx		xx
Spinose ear tick (*Otobius megnini*)	x	xx	x	xxx
Cattle fever ticks (*Rhipicephalus spp.*)	x	xxx		
3-host ticks (*e.g. Dermacentor, Amblyomma ticks*)	x	xx		
Pajaroello tick (*Ornithodoros coriaceus*)	x	x		
Horn fly (*Haematobia irritans*)	x	xxx		
Stable fly (*Stomoxys calcitrans*)	x	xx	x	xxx
Horse and deer flies	x	x		
Biting midges (*Culicoides spp.*)	x	x	xx	xx
Black flies	x	x		
Mosquitoes	x	x		
Face fly (*Musca autumnalis*)	x	xxx		
House fly (*Musca domestica*)	x	x	x	xxx

x, xx, and xxx indicate an expected increasing abundance and impact of the ectoparasite(s).

(essentially a paintball gun firing insecticide loaded balls). However, the efficacy of treatment using this method is still unclear given the low number of control trials performed to date.

Overall, the diversity of cattle ectoparasites will be greater in extensive, pasture-based systems relative to intensive dry pen systems (Table 6.4). Most ticks, including 3-host ticks and the cattle fever tick, will be more abundant in pasture systems than in intensive systems. However, the spinose ear tick has proven to be an important pest in intensive systems where animal bedding provides these ticks with refuge. For the most part, ticks are difficult to control in pasture settings and with the exception of the cattle fever tick, herd managers do not put great effort into their control. Also more abundant in pasture systems are horn fly and face fly which require undisturbed fecal pats for development. These pests can be controlled on pasture using insecticide treated ear tags, feed-through insecticides, or topical application of insecticidal dusts and sprays. They will not be present in high numbers on intensive operations as manure pats do not remain intact. In contrast, the stable fly and house fly develop in a wider range of fermenting organic materials, particularly in feces and animal feed which are abundant in intensive systems. Control of these flies is difficult, with manure management being the most important means to control them. Lice, mites, screwworm fly (if reintroduced), and cattle grub could all be common in either extensive or intensive systems. However for intensive systems, the high level of contact between herd managers and cattle along with the ease of applying insecticides, acaricides, and parasiticides for control of these pests should result in low abundance and low impacts from these pests. In extensive systems, while treatments for these pests can be applied, the difficulty in quickly recognizing pest outbreaks and in applying treatments to infested animals is likely to keep pest abundance and impacts higher than most herd managers would like.

REFERENCES

Alexander, J. L. 2006. Screwworms. J. Am. Vet. Med. Assoc. 3:357–367.

Broce, A. B. 2005. Winter feeding sites of hay in round bales as major developmental sites of *Stomoxys calcitrans* (Diptera: Muscidae) in pastures in spring and summer. J. Econ. Entomol. 98:2307–2312.

Campbell, J. B. 1985. Arthropod pests of confined beef. In R. E. Williams, R. D. Hall, A. B. Broce, and P. J. Scholl (Eds.), *Livestock Entomology* (pp 207–221), New York: John Wiley & Sons.

Gerry, A. C., A. C. Murillo. 2017. Promoting biosecurity through insect management at animal facilities. In J. Dewulf, and F. Van Immerseel (Eds.), *Biosecurity in Animal Production and Veterinary Medicine* (pp 243-281), Leuven, ACCO.

Gerry, A. C., N. G. Peterson, and B. A. Mullens. 2007. *Predicting and Controlling Stable Flies on California Dairies*. University of California, Division of Agriculture and Natural Resources, Oakland, ANR Publication 8258.

Gerry, A. C., B. A. Mullens, N. J. Maclachlan, and J. O. Mecham. 2001. Seasonal transmission of bluetongue virus by *Culicoides sonorensis* (Diptera: Ceratopogonidae) at a southern California dairy and evaluation of vectorial capacity as a predictor of bluetongue virus transmission. J. Med. Entomol. 38:197–209.

Mullen, G. R. and L. A. Durden. 2009. *Medical and Veterinary Entomology* (2nd Ed.). Oxford: Elsevier.

Mullens, B. A. 1989. A quantitative survey of *Culicoides variipennis* (Diptera: Ceratopogonidae) in dairy wastewater ponds in southern California. J. Med. Entomol. 26:559–565.

Schmidtmann, E. T. 1985. Arthropod pests of dairy cattle. In R. E. Williams, R. D. Hall, A. B. Broce, and P. J. Scholl (Eds.), *Livestock Entomology* (pp 223–238), New York: John Wiley & Sons.

Taylor, D. B. 2012. Economic impact of stable flies (Diptera: Muscidae) on dairy and beef cattle production. J. Med. Entomol. 49:198–209.

Wright, R. E. 1985. Arthropod pests of beef cattle on pasture or range land. In R. E. Williams, R. D. Hall, A. B. Broce, and P. J. Scholl (Eds.), *Livestock Entomology* (pp 191–206), New York: John Wiley & Sons.

CHAPTER 7

The Uses of Biotechnology to Improve Animal Welfare

Alison L. Van Eenennaam
University of California

CONTENTS

Biotechnology is defined in the Cartagena protocol as "any technological application that uses biological systems, living organisms, or derivatives thereof, to make or modify products or processes for specific use." From this definition, it is clear that some applications of biotechnology have been used in animal agriculture for many years. Biotechnologies have directly benefitted the three core scientific disciplines of animal science: genetics, nutrition, and health, as summarized in Table 7.1.

ANIMAL WELFARE ASPECTS OF BIOTECHNOLOGY

Some biotechnologies such as vaccinations clearly benefit animal health and welfare, whereas others such as ionophores and recombinant proteins improve production efficiency. Some people are opposed to the use of certain biotechnologies to improve production efficiency, and argue that their use is associated with decreased animal welfare. In fact, the very use of a subset of specific biotechnologies is prohibited in certain production systems. This brings up the very real tensions that frequently exist between the three components of sustainability: environmental, economic, and social. Sustainability is often depicted as three intersecting circles, with a sweet spot in the center representing the sustainable production system nirvana (Figure 7.1).

One definition of sustainable agriculture is "a way of raising food that is healthy for consumers and animals, does not harm the environment, is humane for workers, respects animals, provides a fair wage to the farmer, and supports and enhances rural communities." While it is hard to find fault in that ideal, it ignores the fact that there are almost always goal conflicts between environmental, social, and economic goals as no one single system can simultaneously fulfil all sustainability goals.

Table 7.1 Biotechnologies Used in Animal Production (Adapted from the Food and Agriculture Organization)

Genetics/Breeding	Nutrition	Health
Artificial insemination	Single-cell proteins	Molecular diagnostics
Progesterone monitoring	Probiotics and prebiotics	DNA vaccines
Estrus synchronization	Recombinant somatotropins	Marker vaccines
In vitro fertilization and embryo transfer	Solid-state fermentation of lignocellulosics	Virus-vectored vaccines
Molecular markers; marker-assisted and genomic selection	Feed additives: amino acids, enzymes & probiotics	Sterile insect technique (SIT)
Cryopreservation	Ionophores	Bioinformatics
Semen and embryo sexing	Molecular gut microbiology	
Cloning	Silage additives (enzymes and microbial inoculants)	
Genetic engineering/transgenesis	Recombinant metabolic modifiers	
Genome editing		

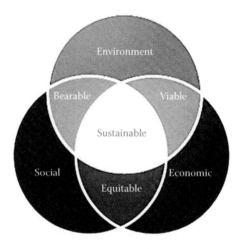

Figure 7.1   The three intersecting components of sustainability.

The best system from an environmental perspective may not conform to the best system from an animal welfare perspective, although disagreements abound as to which metrics define "best" as it relates to all three components. This exposes sustainability goal evaluation to subjective and value-based interpretation. And often special interest groups apply political pressure based on their ranking of one of the sustainability goals as particularly high. There are examples where groups have placed 100% weighting on a single component of the sustainability equation, without regard or consideration of the concomitant tradeoffs on the other aspects of sustainability.

ANIMAL WELFARE ASPECTS OF GENETIC ENGINEERING

Genetic engineering (GE) is a process in which scientists use recombinant DNA (rDNA) technology to introduce desirable traits into an organism. Because the genetic code for all organisms is made up of the same four nucleotide building blocks, this means that a gene encodes the same protein whether it is made in an animal, a plant, or a microbe. Recombinant DNA (rDNA)

Table 7.2 Examples of Transgenic Animals for Agricultural Applications. The Only Product to Obtain Food Regulatory Approval is the AquAdvantage® Fast-Growing Salmon (Bold)

Species	Transgene	Origin	Effect/Goal
Cattle	Lysozyme	Human	Milk composition
	PrP (Prion Protein)	Knockout	Animal health
	α−,κ-Casein	Bovine	Milk composition
	Omega-3	Nematode	Milk composition
	Lysostaphin	Bacterial	Mastitis resistance
Goat	Monosaturated fatty acids	Rat-bovine	Mastitis resistance
	Human beta-defensin 3	Human	Milk composition
Pig	Phytase	*Escherichia coli*-mouse	Feed uptake
	Growth hormone	Human-porcine	Growth rate
	cSKI	Chicken	Muscle development
	Lysozyme	Human	Piglet survival
	Unsaturated fatty acids	Spinach	Meat composition
	Omega-3	Nematode	Meat composition
	α-lactalbumin	Bovine	Piglet survival
	Mx1	Murine	Influenza resistance
Salmon	**Growth hormone**	**Piscine**	**Growth rate**
	Lysozyme	Piscine	Animal health
	wflAFP-6	Piscine	Cold tolerance
Sheep	IGF-1	Ovine	Wool growth
	CsK	Bacterial	Wool growth
	Visna resistance	Viral	Disease resistance
	PrP	Knockout	Animal health

refers to DNA fragments from two or more different sources that have been joined together in a laboratory. The resultant rDNA "construct" is usually designed to express a protein, or proteins, that are encoded by the gene(s) included in the construct. GE involves producing and introducing the rDNA construct into an organism so that new or changed traits can be given to that organism. A GE animal is an animal that carries a known sequence of rDNA in its cells, and passes that DNA on to its offspring. Genetically engineered animals are sometimes referred to as living modified organisms, transgenics, genetically modified organisms (GMOs) or bioengineered animals. Genetically engineered animals were first produced in the late 1970s. Forty years later, transgenic animals have been produced in many different species, including those traditionally consumed as food, although not a single example is has yet been successfully commercialized (Table 7.2).

To date, only a single application has been approved for food purposes, the fast-growing AquAdvantage® Atlantic salmon. This fish was approved for commercialization under specific production conditions by the US Food and Drug Administration in 2015 after a prolonged regulatory evaluation. As of July 2018, its future was still uncertain due to the introduction of a legislative bill that prohibited its sale in the United States pending the publication of final labeling guidelines for informing consumers of GE content. The AquAdvantage® salmon has been approved and is available for commercial sale in Canada.

Many of the goals listed in Table 7.2 are common traits included in the breeding objectives of livestock genetic improvement programs including disease resistance traits. Breeders could conceptually use GE alongside conventional breeding methods to facilitate genetic improvement. To date, the expense of the regulatory process has precluded the commercialization of GE animals for food purposes. There have been some GE animals approved for biomedical pharmaceutical production including goats, rabbits, and chickens, GE fluorescent fish for aquarium purposes, and also some trials using GE insects for pest control applications.

ANIMAL WELFARE ASPECTS OF GENE EDITING

Genome or gene editing refers to the use of site-directed nucleases to precisely introduce a double-stranded break (DSB) at a predetermined location in the genome. The cell can repair that DSB in one of two ways—nonhomologous end joining (NHEJ) or homologous-directed repair (HDR) using a nucleic-acid template that includes the sequences homologous to either side of the DSB. The outcomes of these repair processes result in random mutations or precise gene edits, respectively (Figure 7.2).

As the name "gene editing" suggests, HDR can be employed to precisely add, delete, or replace letters in the genetic code at the location of the break by providing the appropriate template nucleic acid. In the case of NHEJ, although the location of the cut site is very precise, the exact change that occurs when the DNA is repaired is random so a number of different outcomes representing minor sequence insertions (ins) or deletions (del), termed indels, are possible.

Genome-editing technologies enable breeders to efficiently turn off a gene through NHEJ or precisely introduce specific allelic variants. This introduction could be as simple as a single-base-pair change or could conceptually involve entire genes or transgenes, as dictated by the HDR template nucleic-acid sequence, that breeders would like to introduce into their target population using editing.

Gene editing has many potential animal welfare applications. For example, it can be used to correct diseases and disorders that have a genetic basis by altering the error that resulted in the disease phenotype. It could also be used to change a less desirable allele of a gene to a more desirable

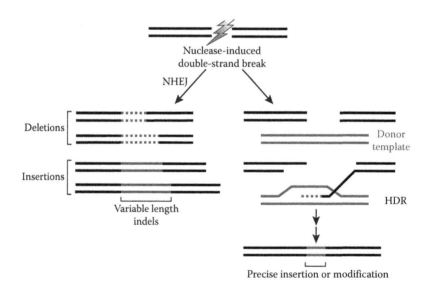

Figure 7.2 Nuclease-induced double-strand breaks can be repaired by nonhomologous end joining (NHEJ) or homology-directed repair (HDR) pathways. Imprecise NHEJ-mediated repair can produce insertion and/or deletion mutations of variable length at the site of the double-strand break. HDR-mediated repair can introduce precise point mutations or insertions from a single-stranded or double-stranded DNA donor template. Image from Sander JD, Joung JK. CRISPR-Cas systems for editing, regulating and targeting genomes. *Nat Biotech* 2014; 32:347–355.

Table 7.3 Examples of Successful Gene-Edited Agricultural Applications in Food Animal Species

Species	Target	Objective	Effect/Goal
Cattle	Polled/hornless	Welfare	No horns
	Myostatin	Productivity	Increased muscle growth
	Beta-lactoglobulin KO	Food composition	Elimination of milk allergen
	Lysostaphin transgene	Disease resistance	Mastitis resistance
	Lysozyme transgene	Disease resistance	Mastitis resistance
	NRAMP1 cisgene	Disease resistance	Resistance to tuberculosis
	Signal peptide of CD1	Disease resistance	Bovine respiratory disease
Chicken	Ovalbumin	Food composition	Elimination of ovalbumin in egg
Goat	Myostatin	Productivity	Increased muscle growth
	Prion protein	Disease resistance	Elimination of prion protein
	Beta-lactoglobulin	Food composition	Elimination of milk allergen
Pig	CD163	Disease resistance	PRRSV resistance
	RELA	Disease resistance	African swine fever resistance
Sheep	Myostatin	Productivity	Increased muscle growth

allele without the need to introgress (repeatedly backcross) or bring in that allele through outcrossing with an animal that carries the desirable allele.

Gene editing has been used to mediate the generation of more than 300 edited pigs, cattle, sheep, and goats. Table 7.3 lists some of those that were directly targeted to agricultural applications including product yield, animal health, and welfare.

One could potentially envision editing several alleles for different traits—such as disease resistance, polled, and to correct a known genetic defect—all while using conventional selection methods to keep making genetic progress toward a given selection objective. It should be remembered that complex traits are typically impacted by many different genes. It is unlikely that all of the genes impacting such traits are known, nor is it typically evident which might be the desirable molecular edits for these genes (i.e., what is the sequence of the desirable allele). It is likely that editing will be focused on large effect loci and known targets to result in discrete changes (e.g., polled), correct genetic defects or decrease disease susceptibility, and conventional selection will continue to make progress in selecting for all of the many small effect loci that impact the complex traits that contribute to the breeding objective.

In the future, it is also possible that genome editing will enable the development of approaches to produce single-gender offspring for industries like laying hens where only the female produces the saleable product. Likewise some groups are working on developing genome-editing approaches to eliminate testes development and the need to castrate males. These applications may address some important welfare concerns such as the fate of male layer chicks and castration of male pigs.

WILL GENE EDITING BE REGULATED?

Animal breeding per se is not regulated by the federal government, although it is illegal to sell an unsafe food product regardless of the breeding method that was used to produce it. Gene editing does not necessarily introduce any foreign genetic rDNA or "transgenic sequences" into the genome, and many of the intended changes would not be distinguishable from naturally occurring alleles and variation. As such, many applications will not fit the classical definition of GE.

For example, many edits are likely to edit alleles of a given gene using a template nucleic acid dictated by the sequence of a naturally occurring allele from the same species. For example, the hornless Holstein (Figure 7.3) carries the polled allele sequence from Angus. As such, there is no

Figure 7.3   Genetically dehorned dairy calf gene edited to carry two copies of the dominant polled allele at the *POLLED* locus. Picture by Hannah Smith Walker, Cornell Alliance for Science.

novel DNA sequence present in the genome of the edited animal that could not otherwise have been produced using traditional breeding techniques. It is not evident what unique risks might be associated with an animal that is carrying the polled allele given the exact same sequence and resulting phenotype that would be observed in the breed from which the allele sequence was derived.

It is possible that nucleases might introduce double-stranded breaks at locations other than the target locus, and thereby introduce alterations elsewhere in the genome. Such off-target events are analogous to spontaneous mutations which occur routinely, and can be minimized by careful design of the gene-editing reagents.

Governments and regulators globally are currently deliberating about how or if gene-edited animals should be regulated. In January 2017, the U.S. Food and Drug Administration's draft guidance for industry (GFI) #187 entitled "Regulation of Intentionally Altered Genomic DNA in Animals" proposes to regulate all animals with "intentionally altered" DNA as drugs (FDA, 2017). The guidance states that "intentionally altered genomic DNA may result from random or targeted DNA sequence changes including nucleotide insertions, substitutions, or deletions"; however, it clarifies selective breeding or other assisted reproductive technologies including random mutagenesis followed by phenotypic selection are not included as triggers. The new draft guidance then goes on to state that

> A specific DNA alteration is an article that meets the definition of a new animal drug at each site in the genome where the alteration (insertion, substitution or deletion) occurs. The specific alteration sequence and the site at which the alteration is located can affect both the health of the animals in the lineage and the level and control of expression of the altered sequence, which influences its effectiveness in that lineage. Therefore, in general, each specific genomic alteration is considered to be a separate new animal drug subject to new animal drug approval requirements

This proposal to regulate genetic variants as drugs if they are intentionally induced seems to trigger regulation of gene editing based on human intention rather than any unique risks associated with the novel characteristics of the end product. It is known that each individual genome harbors many thousands of unique single nucleotide polymorphisms (SNPs), indels, and copy number variants. For example, beef and milk contain enormous numbers of genetic variants that have accumulated within the bovine gene pool because they are being introduced continually through natural mutational processes. In one recent analysis of whole-genome sequence data from 234 taurine cattle representing three breeds, >28 million variants were observed, comprising insertions, deletions, and single-nucleotide variants. A small fraction of these mutations have been selected owing to their beneficial effects on phenotypes of agronomic importance. None of them is known to produce ill effects on the consumers of milk and beef products, and few impact the well-being of the animals themselves.

Although gene editing can be used to introduce virtually any DNA sequence into genomes, many applications will likely result in animals carrying desirable alleles with sequences that originated in other breeds or individuals from within that species, for example, an edit to correct diseases and disorders that have a genetic basis. There is a need to ensure that the extent of regulatory oversight is proportional to the unique risks, if any, associated with novel phenotypes.

ANIMAL WELFARE ASPECTS OF BREEDING

Genetics may not be obviously connected with agricultural sustainability, and yet the importance of animal genetics in contributing to the interplay between the environmental, social, and economic goals of sustainability should not be underrated. Genetic gains are both permanent and cumulative meaning that gains made in 1 year will be transmitted to subsequent generations without further endeavor or expenditure. Genetic improvement has been an important component of the tremendous advances in agricultural productivity that have occurred over the past 50 years.

Traditional breeding programs focused on production traits such as milk yield, growth rate, and meat yield. Key social goals such as food safety, food quality, environmental protection, and animal welfare were not overtly included in breeding objectives. Recently, more selection emphasis has been placed on functional traits that are not directly associated with production outputs including traits that could lead to improved animal welfare. Several authors have discussed approaches to incorporate "sustainability traits" into breeding objectives. As might be predicted from the rather broad definitions of sustainable animal breeding, approaches vary depending upon which components of sustainability are under discussion.

Some important examples where functional traits have been added to breeding objectives include the incorporation of fertility and disease resistance traits into dairy cattle selection indexes, and the inclusion of leg traits into poultry breeding. Table 7.4 shows the ten dairy traits that are currently included in the US dairy selection index with the year that each trait was introduced.

It can be seen that production traits (milk and fat) were the first traits incorporated into selection programs. As time went on, more functional traits were included to widen the scope and redirect the emphasis of breeding programs. At the current time, production traits represent 35% emphasis of the dairy selection index, with the remaining 65% placed on functional traits. Figure 7.4 shows

Table 7.4 Year That Genetic Rankings Began and Emphasis Placed on Dairy Traits in 2010 US National Dairy Selection Index

Trait	Year Trait Added to Index	Current Selection Index Emphasis (%)
1. Milk	1935	0
2. Milk fat	1935	19
3. Milk protein	1977	16
4. Calving ability	1978/2006	5
5. Udder shape and support	1983	7
6. Feet and leg conformation	1983	4
7. Body size/weight	1983	−6
8. Productive life/longevity	1994	22
9. Mastitis susceptibility (somatic cell score)	1994	−10
10. Daughter pregnancy rate/fertility	2003	11

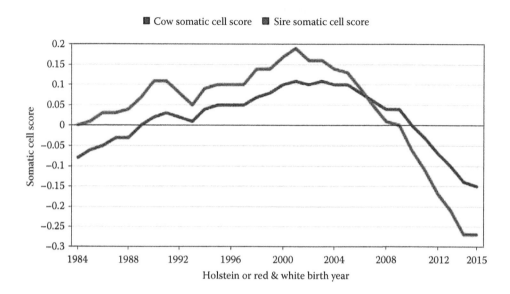

Figure 7.4 Trend in somatic cell score for Holstein or Red & White. Calculated December 2016.

how the genetic trend in somatic cell score, an indicator of mastitis (inflammation of the mammary gland), decreased after inclusion of this trait in the dairy selection index in 1994.

Likewise, because the genetic correlation between body weight and incidence of leg disorders in broiler chickens is positive, appropriate multi-trait selection indexes have been developed to permit genetic improvement in leg health concurrently with continued, though more modest, improvement in growth rate. Cobb-Vantress Inc., a large chicken breeding company based in Arkansas, records 56 individual observations on each pedigree selection candidate in their broiler chicken breeding program. More than 50% of these 56 individual traits are some measure of health and fitness of an individual. This underscores the importance of combined selection for many traits, including robustness, specific and general disease resistance, absence of feet and leg problems, and metabolic defects in the breeding objectives. In recent years, genetic selection has had a major impact on decreasing the incidence of skeletal disorders in broiler chickens.

Some breeders have also incorporated behavioral traits into their selection criteria. It may be considered unethical to select for behaviors that better suit an animal to an agricultural production environment, with some advocating that the production environment should be modified to suit the animal. However, it should be recognized that livestock populations have been selected for behavioral traits since their domestication. Consider the behavior of herding dogs. While altering the environment such as moving to pasture-based systems might be appropriate in some cases, such a change needs to be considered in the context of undesired negative impacts (e.g., increased exposure to pathogens, predators, and weather) on other components of sustainability. It is conceivable that selection to better suit a population of animals to their production environment could improve both animal welfare and productivity, thereby working toward multiple sustainability goals.

An example from the poultry industry illustrates this point well. A 1996 selection experiment was performed on a line of White Leghorns to improve adaptability and well-being of layers in large multiple-bird cages. Feather and vent pecking, and sometimes cannibalism can occur in multi-pen cages, a problem that can be managed with beak trimming young birds. Using a selection approach termed "group selection," offspring from select roosters were housed as a group in multiple-bird cages, and the group was either selected or rejected based on the productivity of the group. An unselected control, with approximately the same number of breeders as the selected line, was maintained for comparison and housed in single-bird cages. After six generations, annual mortality of the selected line in multiple-bird cages decreased from 68% to 8.8%. Mortality of the selected line in multiple-bird cages was similar to that of the unselected control in single-bird cages. Annual survival improved from 169 to 348 days, eggs per hen per day rose from 52 to 68%, total eggs per hen from 91 to 237 eggs, and total egg mass from 5.1 to 13.4 kg, while average egg weight remained unchanged. The author concluded that these data suggested group selection could eliminate the need to beak-trim to avoid cannibalism by breeding hens to better suit multiple-bird cage production systems. These outcomes from group selection would seem to align with several sustainability goals, including decreased cannibalism and a resultant improvement in animal welfare, better production efficiency, reduced need for beak trimming, and group housing rather than single-bird cages.

In reality, it is likely that moving to alternative production systems (e.g., free range) in the interests of animal welfare will necessitate an associated "reselecting" for animals that are better suited to the new production environment. With too much territory birds become territorial and it is not uncommon to have much greater mortality in floor pens than cages due to the increased area.

CONCLUSION

Biotechnology is a broad term that encompasses many technologies that are routinely used in animal agriculture. Emerging biotechnologies offer great potential, especially in the area of animal breeding. Breeding can contribute to improved animal welfare in two ways. First, welfare traits can be included in breeding objectives to select for improved animal welfare and decreased disease incidence. Second, modern molecular tools offer the opportunity to introduce novel genetic variation into animal genomes to address both welfare concerns like dehorning and castration, in addition to correcting diseases and disorders that have a genetic basis. Developing livestock that are more resilient and less susceptible to disease is an important component of the development of more productive and sustainable animal agricultural systems globally.

ACKNOWLEDGMENTS

The author acknowledges research support from the USDA for grants 2015–67015-23316 and 2015–33522-24106 that involve genome editing in livestock.

PART II

Concepts Relevant to Both Beef and Dairy Cattle

CHAPTER 8

The Meaning of Animal Welfare and Its Application to Cattle

Bernard E. Rollin
Colorado State University

Before embarking upon a discussion of cattle welfare, there are certain presuppositional notions that must be clarified. In particular, we may note the widespread belief that animal welfare is strictly a matter of science, i.e., is a concept that is fully empirical. Throughout the 20th century the notion that science is independent of value judgments in general, and ethical judgments in particular, has been so dominant that I have characterized it as a fundamental principle of "scientific ideology," or "The Common Sense of Science," for it is to science what ordinary common sense is to ordinary experience. Graphic evidence of this ubiquitous belief can be found in the introductions to scientific textbooks, particularly in the field of biology.

But the most vivid example of this belief comes from a statement made by the Director of NIH during a 1990 visit to Michigan State University. There are at least two requirements that any person being considered for the Directorship of NIH must satisfy. First, he or she must be a highly accomplished and respected scientist. Second, given the controversial nature of the position, he or she must be extremely astute politically to navigate the minefields surrounding the funding of research, particularly in the biomedical area. In the course of a lecture delivered to premedical students, students asked the Director to discuss the major ethical issues occasioned by biotechnology and genetic engineering. Astoundingly, in his response, the Director opined that "although scientific advances are always controversial, science should never be hindered by ethical considerations." (When I quote that statement to my freshman students and ask them who in the 20th century is likely to have made it, they invariably reply "Hitler.")

Indirect evidence of the denial of ethics in science was presented to me when I was in the process of drafting what eventually became the 1985 laws governing the use of animals in biomedical research. Attempting to understand and engage the ethical position regnant in the research community regarding the use of animals, I searched extensively for articles, papers, and editorials discussing the ethics of animal use in scientific journals. I found nothing, save one offhanded comment in a British scientific journal affirming that animal use in scientific research does not represent an ethical question, but is rather a scientific necessity—as if it could not be both!

In the face of this ideology, it is not difficult to understand why the concept of animal welfare is viewed as devoid of valuational import by scientists, as well as by those in the business of food animal production, particularly those who deploy CAFOs—Confined Animal Feeding Operations (known in the vernacular as "factory farms.") For example, during the 2.5 years that I served as a member of the Pew Commission, the first systematic study ever done on the effects of industrialized confinement agriculture on animal welfare, the environment, agricultural sustainability, small farms and farmers and the societal basis of agriculture, human health, and animal health, I regularly had occasion to discuss animal welfare with a variety of stakeholders. When one discusses farm animal welfare with industry groups or with the American Veterinary Medical Association, one found the same response—animal welfare is solely a matter of "sound science."

For example, one representative of the Pork Producers, testifying before the Commission, affirmed that while people in her industry were quite "nervous" about the Commission, their anxiety would be allayed were we to base all of our conclusions and recommendations on "sound science." Hoping to rectify the error in that comment, as well as educate the numerous industry representatives present, I responded to her as follows: "Madame, if we on the Commission were asking the question of *how* to raise swine in confinement, science could certainly answer that question for us. But that is *not* the question the Commission, or society, is asking. What we are asking is, *ought* we raise swine in confinement? And to this question, science is not relevant." Judging by her "huh," I assume I did not make my point.

In other words, "animal welfare" is not a concept derived strictly from gathering data. If we wish to know an animal's weight, we place it upon a scale. On the other hand, if we wish to affirm that an animal or person is *obese*, we make explicit or implicit reference to a set of valuational standards based on health, esthetics or whatever, based upon what we think an animal or human *ought to weigh*. By the same token, we cannot build a simple machine to measure whether an animal is possessed of positive well-being, unless we first discuss the question of what counts as well-being for that sort of creature. Or to put it in a way that makes crystal clear the ethical basis of well-being, when we ask about an animal's welfare, especially but not exclusively in the case of animals managed by people, we are making reference to what we believe that we *owe* the animal and to *what extent*. Even if we are talking about an animal in nature, not one under the egis of human beings, to ask about its welfare is to ask whether we feel the animal is getting what we feel it ought to be getting, what it *deserves*, given its needs and nature. (We will return to this latter point.) Precisely the same point can of course be made about human welfare. Thus, animal welfare is in part an *ethical notion*.

Let us illustrate this in a crystal-clear way. Questions of animal welfare are at least partly "ought" questions, questions of ethical obligation. The concept of animal welfare is an ethical concept to which, once understood, science brings relevant data. When we ask about an animal's welfare, or about a person's welfare, we are asking about *what* we owe the animal, and to *what extent*. A document called the CAST report, first published by U.S. Agricultural scientists in the early 1980s, discussed animal welfare, and it affirmed that the necessary and sufficient conditions for attributing positive welfare to an animal were represented by the animals' productivity. A productive animal enjoyed positive welfare; a nonproductive animal enjoyed poor welfare.

This notion was fraught with many difficulties. First of all, productivity is an economic notion predicated of a whole operation; welfare is predicated of individual animals. An operation, such as caged laying hens, may be quite profitable if the cages are severely over-crowded, yet the individual hens do not enjoy good welfare. Second, as we shall see, equating productivity and welfare is, to some significant extent, legitimate under husbandry conditions, where the producer does well if and only if the animals do well, and square pegs, as it were, are fitted into square holes with as little friction as possible. Under industrial conditions, however, animals do not naturally fit in the niche or environment in which they are kept, and are subjected to "technological sanders" that allow for producers to force square pegs into round holes—antibiotics, feed additives, hormones, air handling systems—so the animals do not die and produce more and more kilograms of meat or milk. Without these technologies, the animals could not be productive. We will return to the contrast between husbandry and industrial approaches to animal agriculture.

The key point to recall here is that even if the CAST Report definition of animal welfare did not suffer from the difficulties we outlined, it is still an ethical concept. It essentially says "what we owe animals and to what extent is simply what it takes to get them to create profit." This in turn would imply that the animals are well-off if they have only food, water, and shelter, something the industry has sometimes asserted. Even in the early 1980s, however, there were animal advocates and others who would take a very different ethical stance on what we owe farm animals. Indeed, the famous five freedoms articulated in Britain by the Farm Animal Welfare Council during the 1970s (even before the CAST Report) represents quite a different ethical view of what we owe animals, when it affirms that:

> The welfare of an animal includes its physical and mental state and we consider that good animal welfare implies both fitness and a sense of well-being. Any animal kept by man, must at least, be protected from unnecessary suffering.
>
> We believe that an animal's welfare, whether on farm, in transit, at market or at a place of slaughter should be considered in terms of **'five freedoms'** (see www.fawc.org.uk).
>
> 1. **Freedom from Hunger and Thirst**—by ready access to fresh water and a diet to maintain full health and vigor.
> 2. **Freedom from Discomfort**—by providing an appropriate environment including shelter and a comfortable resting area.
> 3. **Freedom from Pain, Injury or Disease**—by prevention or rapid diagnosis and treatment.
> 4. **Freedom to Express Normal Behavior**—by providing sufficient space, proper facilities and company of the animal's own kind.
> 5. **Freedom from Fear and Distress**—by ensuring conditions and treatment which avoid mental suffering.

Clearly, the two definitions contain very different notions of our moral obligation to animals (and there is an indefinite number of other definitions). Which is correct, of course, cannot be decided by gathering facts or doing experiments—indeed which ethical framework one adopts will in fact determine the shape of science studying animal welfare.

To clarify: suppose you hold the view that an animal is well-off when it is productive, as per the CAST Report. The role of your welfare science in this case will be to study what feed, bedding, temperature, etc., are most efficient at producing the most meat, milk, or eggs for the least money—much what animal and veterinary science do today. On the other hand, if you take the FAWC view of welfare, your efficiency will be constrained by the need to acknowledge the animal's natural behavior and mental state, and to assure that there is minimal pain, fear, distress and discomfort—not factors in the CAST view of welfare unless they have a negative impact on economic productivity. *Thus, in a real sense, sound science does not determine your concept of welfare; rather, your concept of welfare determines what counts as sound science!*

The failure to recognize the inescapable ethical component in the concept of animal welfare leads inexorably to those holding different ethical views talking past each other. Thus, producers ignore questions of animal pain, fear, distress, confinement, truncated mobility, bad air quality, social isolation, and impoverished environment unless any of these factors impact negatively on the "bottom line." Animal advocates, on the other hand, give such factors primacy, and are totally unimpressed with how efficient or productive the system may be.

I recently received a question from the editor of my ethics column for *The Canadian Veterinary Journal*. Here is the query:

> Recent studies have shown that taking urban visitors on tours of carefully chosen and well-run livestock operations does not always produce a positive response regarding the care of farm animals. For example, when members of the public visit free stall dairy barns where cows are clean and comfortable with only rare cases of lameness and other "production diseases," the response of the urban visitor is not always positive. Standard farm practices such as the removal of the calf at birth from its mother and the inability of cows to graze at pasture are considered both unnatural and disturbing to many urban visitors. **Is educating the public regarding modern livestock production practices the correct approach to convincing the public that current industry practices ensure the welfare of farm animals?**

One must never forget that standard agricultural practices absolutely depend on public acceptance. A succinct comment in the June 23rd edition of *Pig Progress* by the CEO of a pork industry company makes this crystal clear: "Consumers have the power to change every aspect of the animal livestock industry. Their concerns, their acceptance of production methods, should be critical in how our customers and we ourselves organise our businesses."

A major question obviously arises here. If the notion of animal welfare is inseparable from ethical components, and people's ethical stance on obligations to farm animals differ markedly across a highly diverse spectrum, whose ethic is to predominate and define, in law or regulation, what counts as "animal welfare"? It is to this issue we now turn. The answer is tantalizingly obvious, and is in fact implicit in the previous paragraph: it is the consumer!

There is a not uncommon tendency to affirm that ethics is simply a matter of individual opinion. But a moment's reflection will make us realize that that is simply not the case. In a world where the only ethics was that of individuals, we would live in in a situation of permanent anarchy which would inevitably devolve into what Thomas Hobbes called "the war of each against all." Therefore, in addition to one's personal ethics, which in our society guides such personal behavior as what one reads, one's dietary habits, one's religious belief or lack thereof, there must also be rules that are binding on everyone. Such rules are usually, though not always, encoded in law and regulation. Paradigmatic among these rules are moral/legal principles forbidding robbery, murder, assault, etc. One can call the body of such rules *the social consensus ethic*, which is universally binding on all citizens.

In our society, respect for the individual and the needs and interests of human individuals are the basis for our democratic society, and the test according to which other consensus ethical rules are judged. We will return to this point.

Historically, for the majority of human civilization, very little of the social consensus ethic dealt with the treatment of animals. That is largely because, throughout our history, the predominant use of animals was agricultural—food, fiber, locomotion, and power. When domestication of animals began some 13,000 years ago, successful use of animals dependent upon principles of *good husbandry*. "Husbandry" is derived from the Old Norse words "hus" and "bond"; the animals were bonded to one's household. The essence of husbandry was *care*. Humans put animals into the most ideal environment possible for the animals to survive and thrive, the environment for which they had evolved and been selected. In addition, humans provided them with sustenance, water, shelter, protection from predation, such medical attention as was available, help in birthing, food during famine, water during drought, safe surroundings and comfortable appointments.

Eventually, what was born of necessity and common sense became articulated in terms of a moral obligation inextricably bound up with self-interest. In the Noah story, we learn that even as God preserves humans, humans preserve animals. The ethic of husbandry is in fact taught throughout the Bible; the animals must rest on the Sabbath even as we do; one is not to seethe a calf in its mother's milk (so we do not grow insensitive to animals' needs and natures); we can violate the Sabbath to save an animal. Proverbs tells us that "the wise man cares for his animals." The Old Testament is replete with injunctions against inflicting unnecessary pain and suffering on animals, as exemplified in the strange story of Balaam who beats his ass, and is reprimanded by the animal's speaking through the grace of God.

The true power of the husbandry ethic is best expressed in the 23rd Psalm. There, in searching for an apt metaphor for God's ideal relationship to humans, the Psalmist invokes the good shepherd:

> The Lord is My shepherd; I shall not want. He leadeth me to
> green pastures, He maketh me to lie down beside still waters,
> He restoreth my soul.

We want no more from God than what the good shepherd provides to his animals. Indeed, consider a lamb in ancient Judaea. Without a shepherd, the animal would not easily find forage or water, would not survive the multitude of predators the Bible tells us prowled the land—lions, jackals, hyenas, birds of prey and wild dogs. Under the egis of the shepherd, the lamb lives well and safely. In return, the animals provide their products and sometimes their lives, but while they live, they live well. And even slaughter, the taking of the animal's life, must be as painless as possible, performed with a sharp knife by a trained person to avoid unnecessary pain. Ritual slaughter was, in antiquity, a far kinder death than bludgeoning; most importantly, it was the most humane modality available at the time.

The metaphor of the Good Shepherd is emblazoned in the western mind. Jesus is depicted both as shepherd and lamb from the origin of Christianity until the present in paintings, literature, song, statuary, and poetry as well as in sermons. To this day, ministers are called shepherds of their congregation, and pastor derives from "pastoral." When Plato discusses the ideal political ruler in the *Republic*, he deploys the shepherd—sheep metaphor: The ruler is to his people as the shepherd is to his flock. Qua shepherd, the shepherd exists to protect, preserve, and improve the sheep; any payment tendered to him is in his capacity as wage-earner. So too the ruler, again illustrating the power of the concept of husbandry on our psyches.

The singular beauty of husbandry is that it was at once an ethical and prudential doctrine. It was prudential in that failure to observe husbandry inexorably led to ruination of the person keeping animals. Not feeding, not watering, not protecting from predators, not respecting the animals' physical, biological, physiological, and psychological needs and natures, what Aristotle called their *telos*—the "cowness of the cow" the "sheepness of the sheep"—meant your animals did not survive and thrive, and thus neither did you. Failure to know and respect the animal's needs and natures had the same effect. Indeed, even Aristotle, whose worldview was fully hierarchical with humans at the top, implicitly recognized the contractual nature of husbandry when he off-handedly affirmed that though the natural role of animals is to serve man, domestic animals are "preserved" through so doing. The ultimate sanction of failing at husbandry—erosion of self-interest—obviated the need for any detailed ethical exposition of moral rules for husbandry: Anyone unmoved by self-interest is unlikely to be moved by moral or legal injunctions! Thus, one finds little written about animal ethics and little codification of that ethic in law before the 20th century, with the bulk of what is articulated aimed at identifying overt, deliberate, sadistic *cruelty*, hurting an animal for no purpose or for perverse pleasure, or not providing food or water.

Since husbandry was inextricably bound up with self-interest, a husbandry approach to animals was sanctioned by the strongest possible motivator. Thus, until the late 19th century, the only articulated ethical prohibitions regarding animals were the forbidding of cruelty. First articulated in law in Great Britain in the late 18th century, the anti-cruelty ethic and laws soon became omnipresent in all civilized society. In addition, as St. Thomas Aquinas wrote in the Middle Ages, people who were disposed to the infliction of unnecessary suffering on animals will be likely to "graduate" to hurting people. Indeed, social scientific research in the mid-20th century recognized that cruelty to animals was sentinel behavior betokening future psychopathy (along with bedwetting and fire-starting).

As I have described elsewhere, not only was husbandry agriculture beneficial to humans, and highly congenial to animal welfare, it was also sustainable, preserving the environment in the manner of a balanced aquarium. But as ethical and rational as good husbandry was, it only endured as long as it was necessary for productivity. With the coming of the Industrial Revolution, the positive state of affairs represented by husbandry quickly began to disappear. The development of animal agriculture based in husbandry was presuppositional to the development of civilization and culture, which is itself presuppositional to the development of industry and technology. And in this surely resides one of the most profound ironies in human history. For it is the very fact of husbandry undergirding civilization that created the possibility of the undoing of husbandry-based agriculture! As Thomas Hobbes once remarked, "leisure is the mother of philosophy" in the broadest sense of the term, including science and technology. And surely a secure food supply provided by a portion of the population is a presupposition of leisure.

Though the ancient contract with domestic animals was inherently sustainable, it was not in fact sustained with the coming of industrialization. Husbandry was born of necessity, and as soon as necessity vanished, the contract was broken. The industrial revolution allowed farmers to produce far more plant and animal products than one needed to survive. The industrialization of transportation allowed farmers to access new markets well outside of their local area. The traditional agricultural values of stewardship of land, agriculture as a way of life, and animal husbandry were rather rapidly supplanted by capitalistic values of efficiency and productivity, ushering in an unprecedented vision of technological agriculture.

In essence, the key to success in husbandry agriculture was putting square pegs into square holes, round pegs into round holes, while creating as little friction as possible doing so. Technology, beginning with the Industrial Revolution, created what I have called "technological sanders," allowed us to force square pegs into round holes, round pegs into square holes, so that even though animal welfare is no longer respected, animals remain productive. These "sanders" include antibiotics, bacterins, air-handling systems and vaccines without which the animals would, in addition to being miserable, sicken and die and fail to produce. (There are, of course, legitimate uses of these modalities.)

Not only was the fair contract between animals and humans that had lasted over 10,000 years cavalierly broken, as we mentioned earlier, sustainability, agricultural way of life, small rural communities, small independent farmers, environmental health, animal health, and human health were also eroded. From the late 19th century until the early 1960s, the public did not seem to notice these problems, content to reap the benefits of cheap and plentiful food.

It was well into the 20th century before society became concerned about the treatment of animals in confinement agriculture, and consequently about animal welfare. Arguably, this concern began in Great Britain with the publication of Ruth Harrison's groundbreaking book, *Animal Machines*, in 1964. This book illustrated in a striking way how far animal agriculture had strayed from good husbandry. It was obvious even then, to what extent good husbandry was iconically embedded in the social mind. People were shocked by the crowding of animals into small spaces and unnatural environments. The image of cows grazing on green grass in pastoral conditions was emblazoned in the minds of the general public. Similarly with chickens raised in the open, under extensive conditions. When Harrison exposed the reality—veal calves kept essentially immobile in wooden crates; chickens crowded into battery cages; sows kept in cages sometimes smaller than they were—the British public was horrified, forcing the government to charter the Brambell Commission. Though the Brambell commission had no authority to effect change, it became a moral lighthouse all across the world. Most famously, it was responsible for the concept of animal welfare embedded in the Five Freedoms we discussed earlier. This way of thinking reached its culmination in Europe with the Swedish law of 1989, essentially abolishing confinement agriculture as we know it in North America, and guaranteeing in the Swedish Constitution that cattle must eternally have "the right to graze."

Because such a significant percentage of the United States population lives in urban and suburban situations, most people have no contact with farm animals and thus little awareness of the fact that the values of husbandry had been replaced by the values of industry. But by the mid-1980s, the US public became ever increasingly aware of animal suffering, particularly in research, safety and efficacy testing, and agriculture. Other societal factors also served to potentiate the demand for a new ethic dealing with animal suffering.

As we mentioned earlier, society historically had little need for animal ethics, since good husbandry was presuppositional to self-interest, as poor treatment of animals led to diminished productivity. The ethic of anti-cruelty sufficed to deal with those irrational people, i.e., sadists and psychopaths, who hurt animals for twisted reasons of self-gratification. However, that ethic was totally inadequate for constraining animal suffering resulting from such putatively laudable uses of animals as research or industrialized agriculture. In other words, there was no way to expand the logic of anti-cruelty to cover the ever-increasing situations where animals were suffering as a byproduct of socially legitimate use. Cases brought before the courts attempting to subsume research and agriculture under the umbrella of cruelty were consistently rejected in the justice system. Inevitably, as concern for animal treatment grew, the battle cry became "go to the legislature." For that reason, in 2004, no fewer than 2,100 legislative efforts to protect animals were brought before federal, state, and municipal legislatures.

As I recounted in a report prepared for the USDA in the early 1990s, the growth of societal concern for animal treatment resulted from the convergence of numerous factors. In today's world, less than 1% of the population make their living from animal use, in contradistinction to the situation roughly 100 years ago, where half the public was engaged in producing animal products, primarily food, for our total population. Thus the conditions are created wherein the vast majority of the public can see animals for what they are, rather than tools for making a living.

In addition, the paradigm for animals in the social mind has moved away from what it was 100+ years ago when, if one asked the ordinary person in the street to state the first word that comes into their mind when I say "animal," the reply would undoubtedly have been "horse," "cow," "food," "work." Today, the answer would be "pet—dog, cat, member of the family." Repeated studies have shown that a minimum of 80% and a maximum of almost 100% profess to view their animals as members of the family. Friends of mine who are divorce lawyers have indicated that in many acrimonious divorce settlements, custody of the dog can literally be a more contentious issue than custody of the children! When asked if they have children, many young couples during the last half century have responded "no, we have dogs."

The mass media, which by its very nature is attuned to changes in social thought and trends, has discovered, as one California reporter told me, that "animals sell papers." One cannot channel-surf across normal television service without being bombarded with animal stories, real and fictional. (A *New York Times* reporter told me that more time on cable TV in New York City is devoted to animals than to any other subject.) Recall, for example, the extensive media coverage a couple of decades ago of some whales trapped in an ice-floe, and freed by a Russian ice-breaker. This was hardly an overflowing of Russian compassion—an oxymoronic notion applied to a people who gave us pogroms, the Gulag, and Stalinism. Rather, someone in the Kremlin was bright enough to realize that liberating the whales was an extremely cheap way to score points with U.S. public opinion.

The rise of concern with the treatment of animals is also pretty much coextensive with the rise of concern for the environment, and with societal concern for the disenfranchised—women, African-Americans, the handicapped, migratory workers and so on. Indeed, many of the leaders in the animal movement cut their teeth in campaigns for civil rights, migratory workers, and other people traditionally socially ignored.

Not least is the fact that many books aimed at a popular audience have been written by pioneers in animal ethics—Peter Singer, Tom Regan, and myself, and the interest in animals on the part of philosophers and some progressive scientists such as Jane Goodall, is unabated. We have gone from essentially no books on animal mind to dozens, including the recent *What Fish Think*.

So how does society generate a new ethic for animals? A wonderful clue to answering this question can be found in Plato. Plato points out that when one is attempting to change ethical opinions, or by implication, to articulate a new ethic, it is far more effective to *remind* than to *teach*. To supplement and elucidate Plato's notion of reminding versus teaching, I created my own metaphorical explanation of the strategy I deployed in terms of martial arts. There are two distinct and antithetical approaches to hand-to-hand combat. One is a *sumo* approach, wherein one exerts one's force against the force of one's opponent, in the manner of offensive versus defensive linemen in football. This is a viable approach if you and your opponent are of equal size and strength; ideally you are larger. It is a recipe, however, for certain defeat if you are fighting someone of superior size and strength. In such a case, one is far better advised to use an opponent's strength against that opponent, so that you redirect that strength to unbalance the opponent, or to throw them. The logic similarly obtains in ethical debate. Particularly if one is arguing against a more powerful opponent, one fares far better by showing that opponent that your ethical position is implicit in their own ethical assumptions, albeit in a hitherto unnoticed way, rather than attempting to force your position upon them. (I call this approach *judo*.) Hence, the remarkable effectiveness of the civil rights movement, reminding Americans, even segregationists, that they were committed to the belief that all humans should be treated equally, as well as to the additional notion that black people were human. When Lyndon Johnson "wrote that large," as Plato said, in the law, most people acquiesced to it. This is judo. Prohibition, which we all know was remarkably ineffective, was sumo.

This in turn brought me to a new realization regarding animal ethics. If, as appeared to be the case, Western society was moving steadily toward greater moral concern and moral status for animals, it would not do so by creating a totally new ethic for animals *ex nihilo*. Rather, it would look to our extant ethic for the treatment of human beings, and export it, *mutatis mutandis,* appropriately modified, to the treatment of animals.

What aspect of our ethic for people is being so extended? One that is, in fact, quite applicable to animal use is the fundamental problem of weighing the interests of the individual against those of the general welfare. Different societies have provided different answers to this problem. Totalitarian societies opt to devote little concern to the individual, favoring instead the state, or whatever their version of the general welfare may be. At the other extreme, anarchical groups such as communes give primacy to the individual and very little concern to the group—hence, they tend to enjoy only transient existence. In our society, however, a balance is struck. Although most of our decisions are made to the benefit of the general welfare, fences are built around individuals to protect their fundamental interests from being sacrificed to the majority. Thus, we protect individuals from being silenced even if the majority disapproves of what they say; we protect individuals from having their property seized without recompense even if such seizure benefits the general welfare; we protect individuals from torture even if they have planted a bomb in an elementary school and refuse to divulge its location. We protect those interests of the individual that we consider essential to being human, to *human nature*, from being submerged, even by the common good. Those moral/legal fences that so protect the individual human are called *rights* and are based on plausible assumptions regarding what is essential to being human.

It is this notion to which society in general is looking in order to generate the new moral notions necessary to talk about the treatment of animals in today's world, where cruelty is not the major problem but where such laudable, general human welfare goals as efficiency, productivity, knowledge, medical progress, and product safety are responsible for the vast majority of animal suffering. People in society are seeking to "build fences" around animals to protect the animals and their interests and natures from being totally submerged for the sake of the general welfare, and are trying to accomplish this goal by going to the legislature. In husbandry, this occurred automatically; in industrialized agriculture, where it is no longer automatic, people wish to see it legislated.

It is necessary to stress here certain things that this ethic, in its mainstream version, is *not* and does not attempt to be. As a mainstream movement, it does not try to give human rights to animals. Since animals do not have the same natures and interests flowing from these natures as humans do, human rights do not fit animals. Animals do not have basic natures that demand speech, religion, or property; thus according them these rights would be absurd. On the other hand, animals have natures of their own (what I have, following Aristotle, called their *telos*) and interests that flow from these natures, and the thwarting of these interests matters to animals as much as the thwarting of speech matters to humans. The agenda is not, for mainstream society, making animals "equal" to people. It is rather preserving the common-sense insight that "fish gotta swim and birds gotta fly," and suffer if they don't.

Nor is this ethic, in the minds of mainstream society, an abolitionist one, dictating that animals cannot be used by humans. Rather, it is an attempt to constrain *how* they can be used, so as to limit their pain and suffering. In this regard, as a 1993 *Beef Today* article points out, the thrust for protection of animal natures is not at all radical; it is very conservative, *asking for the same sort of husbandry that characterized the overwhelming majority of animal use during all of human history, save the last fifty or so years.* It is not opposed to animal use; it is opposed to animal use that goes against the animals' natures and tries to force square pegs into round holes, leading to friction and suffering. If animals are to be used for food and labor, they should, as they traditionally did, live lives that respect their natures. If animals are to be used to probe nature and cure disease for human benefit, they should not suffer in the process. Thus this new ethic is *conservative,* not radical, harking back to the animal use that necessitated and thus entailed respect for the animals' natures. It is based on the insight that what we do to animals *matters* to them, just as what we do to humans matters to them, and that consequently we should respect that mattering in our treatment and use of animals as we do in our treatment and use of humans. *And since respect for animal nature is no longer automatic as it was in traditional husbandry agriculture, society is demanding that it be encoded in law.*

Granted, there are activists who do not wish to see animals used in any way by humans, and in the eyes of many animal users, the activists *are* the "animal rights people." Yet to focus on them is to eclipse the main point of the animal rights thrust in society in general—it is an effort to constrain *how* we use animals, not an attempt to stop all animal use. Indeed, it is only in the context of animal use that constraints on use make any sense at all! Thus, the new mainstream ethic is not an ethic of abolition; it is an effort to reaffirm that the interests of the animals count for themselves, not only in terms of how they benefit us. Like all rights ethics, it accepts that some benefits to be gained by unbridled exploitation will be lost and that there is a cost to protecting the animals' natures. In agriculture, for example, the cost may be higher food prices. But as the Federation of European Veterinarians asserted more than 20 years ago, that is a small price for a society to pay to ensure proper treatment of objects of moral concern.

In my view, the most important result emerging from our discussion is the realization that *an animal's welfare lies in its ability to satisfy the needs and interests emerging from its telos.* By starting with *telos,* we can address such insults to animals as high confinement where they cannot move, or removing a baby calf from its mother. Obviously, since animals evolved to function under extensive circumstances, replication of those circumstances as much as possible remains the ideal.

It is unquestionable that appeal to an animal's nature captures the essence of what society worries about in animal use. When Smithfield, the world's largest pork producer, polled its customers at my suggestion, they found that 78% of the public despised sow stalls because in them animals could not express their natures. As I have said elsewhere, common sense tells us that an animal born with bones and muscles and legs needs the opportunity to move. The same societal idea led to the elimination of sow stalls across the industry, with over 100 food stores and restaurants demanding meat from pigs raised in more natural systems. In the past year, virtually the same story has occurred in the egg industry, with consumers demanding the abolition of battery cages. In the 12 states that HSUS sponsored referenda aimed at eliminating battery cages, veal crates, and sow stalls, all the referenda passed by a significant margin. The veal industry has moved to open systems for veal calves in the face of massive consumer rejection of crates for veal. Zoos are moving away from being prisons for animals, and public concern has led to Sea World eliminating killer whale shows, and California forbidding captivity for these animals. Social concern has led to circuses eliminating elephant shows, despite the public's love of seeing such animals. The list goes on and on.

It is not hyperbole to affirm that just as human rights are derived from human nature, and are protected strongly by the Bill of Rights, society is demanding protection for animal nature, and the legal encoding of those protections in the equivalent of rights. That is a consequence of drawing animal ethics from our consensus social ethic for humans. In support of our claim, let us recall the thousands of bills promulgated across the US in 2004, and the fact that the Gallup poll conducted in 2001 indicated that 70% of US public wishes to see legislated guarantees of farm animal well-being. By 2012, the number was an astonishing 94%!

This book focuses upon the welfare of cattle, both cattle used for meat and those used for dairy. In the course of our narrative, we will discuss the *telos* of cattle and how it is maximally accommodated under extensive conditions. It is of course very unlikely that the cattle industry will move more toward extensive rearing of cattle. In fact, given burgeoning populations, increasing income in many Third World countries such as China and India, and above all an almost insatiable demand for meat and milk, it is very likely that intensification will continue and very possibly increase. Thus, we will also discuss, given cattle natures and the inexorable movement toward further intensification, what is the best way in which intensified cattle production systems can maximize animal welfare by assuring respect for *telos*.

Assuring the well-being of animals utilized for human purposes is a moral imperative. We have presented the rational argument emerging in society for assuring morally sound animal treatment. But there is another argument coming more perhaps from the heart than from the head. As I stated in my first book on animal ethics published in 1981.

> just as we morally expect fair and benevolent treatment at the hands of those capable of imposing their wills upon us, so ought we extend similar treatment to those vulnerable to us.

For those among us who are not moved by moral arguments, there exists an undeniable argument from self-interest. In their account of the nature of human life, the ancient Stoic philosophers articulated a powerful metaphor that admirably fits the requirement that we all act in accord with societal ethics. Paraphrasing them, I point out that social ethics can be schematized as an oxcart on its way to a nearby town. You are chained to the oxcart. You have two choices—you can dig in your heels and resist, in which case you will arrive at the town broken and bleeding. Or you could walk when the oxcart walks, rest when it rests, in which case you will be unscathed.

Dealing with ethical issues raised by your activities is very much analogous to the Stoic point. It is far easier to self-regulate and do the right thing than to wait for society to impose it upon you. You understand your own activities far better than society does. You are far more concerned with preserving your freedom of action than the rest of society is. And most important, you understand what can practically be done without imposing great harm on the regulated area. In most cases, society wants to see a problem solved and will impose a putative solution with a heavy hand. Self-regulation, if permitted, will accomplish far more in a far less onerous manner. The emergence of an industry devoted to creating "artificial meat," in which even meat companies are investing, is a harbinger of where society may turn if the animal industry fails to acquiesce to the demands of society.

CHAPTER 9

Sense and Sensibility: Understanding How Cattle Perceive Our Collective World

Lily N. Edwards-Callaway
Colorado State University

CONTENTS

INTRODUCTION

Senses create an internal representation of the outside world for the person or animal receiving the information (Chandrashekar, Hoon, Ryba, & Zuker, 2006). Although cattle and humans are often exposed to similar sights, sounds, and smells in their environments, the way in which they sense, perceive, receive, and understand the stimuli around them is often different. Although recognizing and understanding these differences between perceptual worlds of livestock and humans is a key component to successful human–animal interactions, many people that work within the cattle industry are not exposed to this type of information prior to handling animals. It is critical to understand the world from the cow's point of view to have successful experiences for both the human handler and the animal. The anthropomorphic approach of "if we can see it, why can't they?" does not provide an effective or accurate framework for approaching work with livestock (Entsu, Dohi, & Yamada, 1992).

Understanding *Umwelten* of Cattle

Jakob von Uexkull, a twentieth-century German biologist, introduced the concept of an "umwelt," simply defined as, the sensory world specific to an organism (von Uexkull, 1909). Von Uexkull was interested in the perceptual worlds of animals, well-known for his studies of ticks, and theorized that all organisms have a unique umwelt. In one of von Uexkull's descriptions of umwelten he instructs his readers to go on a "stroll through the worlds of animals" literally but also figuratively by putting a bubble around the creatures one sees and imagining their world within; the bubble represents "worlds strange to us but known to other creatures, manifold and varied as the animals themselves" (von Uexkull, 1957). An animal's sense organ physiology, the signals it receives, the animal's response to the situation, and the resulting perception of the environment all play a role in defining an animal's umwelt (Partan & Marler, 2002).

Most humans are likely participants in an animal's perceptual world on a daily basis whether it be with the squirrel darting across the front yard as the homeowner walks to her car or with the family pet that is excited when the children come home from school or with the mosquito that is swatted away to avoid a bite. Despite this relatively consistent interaction with animals, many people do not fully appreciate the importance of understanding an animal's perceptual world and its significance to being able to effectively work with animals. There are other people that work with animals more closely as part of their profession whether it be as a veterinary professional, a livestock producer, any type of animal researcher, or one of the many other careers that involve being in close and constant contact with animals. There is anecdotal evidence that suggests that many of the people employed within the livestock industry have not had previous experience working and interacting with livestock, likely coming to the job without much background knowledge of how animals sense and perceive their environment. While it is not necessarily essential to understand the intricate physiological workings of the senses to work humanely and safely with cattle, it is obligatory to understand on a basic level how cattle perceive and understand their environment and ultimately how the animal handler can influence, both positively and negatively, the animal's response.

Predator vs. Prey

One key component to understanding cattle perception is that cattle have evolved as a prey species and thus the way they interact with human handlers and other animals in their environment is as a prey animal. This distinction will be important in the discussion of differences in sensory systems. As a prey species, cattle are reactionary as a mechanism for escape from potentially dangerous situations and they use the information gathered from their sensory systems to make these determinations, sometimes rather quickly (Fraser, 1983). Both wild and domesticated cattle spend varying amounts of time scanning their environments for predators (Partan & Marler, 2002), i.e., engaging in vigilance behavior, a behavior that is motivated by fear (Grandin, 2014). Their sensory systems have evolved to enable cattle to collect information from their surrounding environment as a means of survival. Although normally humans do not instill as much fear in domesticated cattle as a predator in the wild does, such as a wolf or a mountain lion, a human handler can be perceived as a predator/intruder in the environment of a cow. It is this (often muted) prey response that enables trained individuals to skillfully move cattle through handling facilities. The concept of the flight zone and point of balance are foundational components of animal handling and are based upon the animal's response to people in their environment (Grandin, 2014).

SENSORY RECEPTORS AND THEIR CONNECTION TO THE BOVINE BRAIN

Each of the five major senses of cattle, olfaction, gustation, vision, hearing and touch, have specialized cells that respond to the different types of sensory information received and transduce that information into electrical signals that communicate with the brain. There are two different types of sensory receptors, chemoreceptors and mechanoreceptors. Although these receptor cells respond to different types of stimuli, chemical vs mechanical, they have morphological and functional similarities. Chemoreceptors, found in the olfactory (smell) and gustatory (taste) systems, respond to chemicals that are either airborne in the case of olfaction or dissolved in a liquid as in the case of gustation. Mechanoreceptors, found in the vision, hearing and somatosensory (touch) systems, respond to mechanical stimuli, including sound waves, photons, and pressure. The sensory systems have evolved to capture the chemical and mechanical stimuli through a specialized system that ultimately transforms the information so it is understandable to the brain and the recipient can use it. For example, vision begins when a photon is absorbed by a photopigment found in a specialized photoreceptor cell within the eye. The photopigments initiate an electrical change that sends a message to the visual cortex of the brain. The other sensory systems all function in a similar manner. There are two different types of sensory receptors: one type of receptor cell is an afferent nerve fiber containing a specialized region that captures sense information. The other type of receptor synapses with the afferent nerve fiber to send the electrical message to the central nervous system. The subsequent sections describe the specialized sensory cells within system.

Simply, each of the sense systems is connected via nerve fibers to a particular part in the brain's cerebral cortex that processes the information. In the parietal lobe, located in the cortex from the middle to the back of the skull, the somatosensory cortex is responsible for processing information about touch. In the occipital lobe, located in the cortex at the back of the skull, the visual cortex processes visual information. In front of the occipital lobe, the temporal lobe, the auditory cortex is responsible for processing hearing. The gustatory cortex and the olfactory cortex are found under the frontal, temporal and parietal lobes.

Fear and Sensory Inputs

When thinking about sensory processing of cattle, it is also important to understand how sensory stimuli interact with fear processing in the brain. The amygdala is a brain structure that is anatomically complex made up of distinct regions called nuclei and is responsible in complex vertebrates, such as cattle, for processing fear. From an evolutionary perspective, the growth and differentiation of the amygdala and the mammalian cortex have occurred in parallel (Stephan & Andy, 1977). The neural network that is involved in fear processing is complex and involves many different circuits that process different types of fear; it has been suggested that there are actually distinct circuits for fear of pain, fear of predators, and fear of conspecifics (Gross & Canteras, 2012).

The amygdala has been referred to as the "sensory gateway to the emotions" (Aggleton & Mishkin, 1986) and is critical for producing appropriate behavioral responses to sensory stimuli. Visual, olfactory, gustatory, somatosensory, and auditory information is transmitted via mapped projections to the amygdala (McDonald, 1998). McDonald (1998) provides a detailed overview of the cortical pathways projecting to the amygdala from the sensory cortices, highlighting the amygdala as an essential link between the cerebral cortex, where sensory information is processed and other brain regions that are integral in emotional (e.g., fear) and motivational responses of animals.

WHAT'S THAT SMELL?—THE OLFACTORY SYSTEM

Olfaction, the sense of smell, is a chemical sense utilized by cattle to gain information about their environment. Smell is a primitive sense, emerging as a result of the evolutionary development of land animals (Damask, 1981). Strict marine animals do not differentiate between gustation and olfaction as all chemicals they are confronted with are in an aqueous solution. With the rise of airborne creatures the need to be able to sense both chemicals in the air (smell) and chemicals in solution (taste) became necessary and thus two different chemoreceptor systems evolved. Cattle use their sense of smell to find food, identify individuals within their social network, find potential mates, communicate, and be aware of danger. While humans use olfaction for similar purposes, it is often accepted that other animals have a more highly developed olfactory system, being able to differentiate and identify smells at a much farther range and concentration. Interestingly, sensory system scientists have noted that studying olfactory perception in non-human species can sometimes be advantageous, despite the fact that animals can't verbally share the things they smell, because there are no outside lifestyle factor impacts such as smoking, eating, and sinus conditions (Amoore, 1991).

Not All Noses Are Created Equal

Not only are noses and snouts morphologically different between species, different species also have distinctive abilities to detect odors. Species can be categorized by their level of olfactory function as microsmatic or macrosmatic, classification attributed to Turner (1890). Macrosmatic species, such as carnivores and ungulates are "keen-scented" (Negus, 1958) as compared to other species such as humans and other primates who have evolved to rely more heavily on other senses (e.g., vision) and thus do not have as great a level of olfactory function. Primates, as representative microsmatic animals, have proportionately smaller olfactory bulbs (area in the brain that receives olfactory information) (Baron, Frahm, Bhatnagar, & Stephan, 1983) and less olfactory epithelia surface area (area containing the odorant receptors) in the nasal cavity as compared with macrosmatic mammals (Le Gros Clark, 1960), such as cattle.

Form and Function—What's in a Nose?

A cow's nose, technically called a muzzle, is its gateway into the olfactory world. Looking closely, one will notice that the cow's nose is covered in grooves and ridges; interestingly, these muzzle dermatoglyphics can serve as a unique animal identifier, a cow's nose print being like a human's finger print with no two animals having identical patterns (Noviyanto & Arymurthy, 2013; Petersen, 1922). Inside the cow's muzzle lies all the olfactory specific structures that capture the chemical information being delivered to the cow, i.e., the main olfactory system. The gross anatomy of the nasal passage differs greatly between species (Harkema, 1991).

Comparatively, humans have rather simple noses with the primary function being respiratory, while other mammals have more complex systems due to the primary function being olfaction

(Harkema, 1991; Reznik, 1990). Due to this distinction in nasal complexity, humans can breathe both nasally and oronasally, while other mammals, such as cattle, are obligate nose breathers (Harkema, Carey, & Wagner, 2006; Proctor & Chang, 1983). As air travels into the cow's nostrils, it passes through the nasal vestibule and into the main nasal chamber. The distance from the nostrils to the nasopharynx is proportional to the length of the animal's snout. Additionally, the surface area of the nasal epithelium varies between species influenced by size of the animal and the specialization (respiratory vs olfactory) which varies by species' olfactory ability. In macrosmatic species like cattle, deep in the nasal cavity is a space called the olfactory recess filled with convoluted scroll-like bone structures called turbinates lined by the well-vascularized mucosal tissue (Eiting, Smith, Perot, & Dumont, 2014; Moore, 1981). The complexity of the olfactory recess is thought to be related to olfactory ability, and is a structure essentially absent in humans (Smith, Eiting, & Bhatnagar, 2015) and highly developed in canines (Craven et al., 2007). In cattle and other ungulate species, this area is a double scroll-like structure resembling a "T" in cross-section with double ends coiled (Reznik, 1990). The function of this olfactory recess is to sequester air so that the odor-laden air can circulate throughout the extended surface area increasing the chances that odors will be identified (Yang, Scherer, & Mozell, 2007).

The percentage of nasal airway that is lined with olfactory epithelium (OE) is species specific (Harkema, Carey, & Wagner, 2006), microsmatic species having less surface area covered by OE. To provide perspective, the OE in rats covers 50% of the nasal cavity and the OE in humans covers 3% (Gross, Swenberg, Fields, & Popp, 1982; Weiss, 1988); there is a paucity of information detailing these comparisons in cattle but they are likely closer to rats than humans. There are millions of olfactory sensory cells (often called olfactory sensory neurons) within the OE, the total number variable by species and directly proportional to the surface area of the OE. The olfactory sensory cells extend above the epithelial surface and have cilia that increase the surface area for the reception of odorant molecules; the actual olfactory receptors (ORs) are in the cilial membranes (Harkema, Carey, & Wagner, 2006). Each OR cell extends through the basal lamina of the OE and joins with other fibers forming olfactory nerves that extend across the cribiform plate out of the nasal cavity forming the outer nerve layer of the olfactory bulb of the brain.

Multiple different types of OR cells will respond to the same odorant but one odor will elicit the greatest response from one type of receptor cell (Møller, 2003). The brain interprets the impulses from many different receptors to provide a distinct sensation of one specific odor (Firestein, Picco, & Menini, 1993). The threshold of perception of a certain odor varies with the odor itself and with the species, and even with the individual (Møller, 2003). The number of odors that an individual can recognize is determined by training/learning and experience in addition to inherited ability. Additionally, the surrounding environment, such as the speed and direction of the wind, temperature and humidity, may change the strength of a scent.

In addition to the OE in the nasal cavity, the vomeronasal organ (VNO; secondary olfactory system) also contains ORs. The VNO is situated bilaterally at the base of the nasal septum and its function is greatly determined by species. This sensory organ is responsible for detecting pheromones which are, for certain species like cattle, very significant in intraspecies communication and reproduction. Cattle (and other ungulates and felids) will exhibit the Flehmen response which is the curling of the upper lip exposing the front teeth to assist facilitating the transfer of pheromones into the VNO. The significant differences in size of the VNO receptor gene family between species are considered to be the largest variation in size in all mammalian gene families (Grus, Shi, Zhang, & Zhang, 2005). Additionally, there is a relationship between the size of VNO receptor gene family and the complexity and size of the VNO morphology of a species suggesting that the latter may be indicative of the sophistication of species pheromone communication (Grus, Shi, Zhang, & Zhang, 2005). Rats and mice have a relatively large VNO functional gene family, cows having fewer but humans having an extremely small gene family in comparison Grus, Shi, Zhang, & Zhang, 2005; Shi, Bielawski, Yang, & Zhang, 2005; Zhang & Webb, 2003).

Genetics and Olfaction

Traditionally, olfactory research has not been as common as research in some of the other senses like vision. In the early 1990s, two scientific accomplishments changed the approach to olfactory research: (1) the discovery of genes encoding odorant receptors (Buck & Axel, 1991) and (2) the completion of sequencing of human and other animal genomes (Glusman, Yanai, Rubin, & Lancet, 2001; Keller & Vosshall, 2008; Zozulya, Echeverri, & Nguyen, 2001). Genomic and genetic tools have enabled researchers to begin to identify how smell evolved to fit the specific needs of a variety of species (Keller & Vosshall, 2008). Understanding species differences is interesting but also helpful to provide perspective on the physiological and genetic reasons the animals under human care perceive and behave the way they do when exposed to certain stimuli.

Odor perception is mediated by OR genes and the number of OR genes, both functional genes and pseudogenes, varies greatly between species (Keller & Vosshall, 2008). (Note: Pseudogenes, gene copies that have coding deficiencies but resemble functional genes, were once thought to be "junk" but now are considered to be important for gene regulation although their mechanism is still not well understood) (Niimura & Nei, 2007; Tutar, 2012). The cow has approximately 1,000 functional OR genes, the range between species quite broad being approximately 300 functional genes in humans and other microsmatic species and approximately 1,500 functional genes in mice and other macrosmatic species (Niimura & Nei, 2003; 2007). The number of OR genes has changed significantly during mammalian evolution, OR genes both gained and lost, perhaps due to mammals' need to adapt to their environment for survival, i.e., needing to be able to identify many odors vs relying more heavily on other senses, respectively (Fleischer, Breer, & Strotmann, 2009; Niimura & Nei, 2007). Although in most cases this seems highly likely, there are some instances in which the survival factor and the number of genes are not congruent, i.e., the dog which has approximately 800 functional genes, less than the mouse, for example, but a highly sensitive sense of smell.

Olfaction and Human–Cattle Interaction

Recognizing odorous substances is essential to cattle survival, to recognize predators, food, and mating partners. As mentioned, pheromones are an important component in cattle communication influencing social and sexual responses. Cattle have many odiferous glands suggesting the importance of olfaction in social structure. Interestingly, cattle can be trained to recognize conspecifics from olfactory cues (the urine) alone (Baldwin, 1977). Cattle behavior associated with pheromone communication, i.e., the Flehmen response, is often observed as a reproductive management tool at production facilities. In specific relation to cattle handling, smell may impact how cattle move through their environment but the chemical signals they receive are more difficult for humans to control, alter and even recognize, as compared with visual and auditory cues, and therefore often times less of a focus.

THAT TASTES GOOD, DOESN'T IT?—GUSTATION

Taste, or gustation, is another chemical sense. Spector and Glendinning (2009) assert that there are three categories of taste processing: (1) stimulus identification, (2) ingestive motivation, and (3) digestive preparation. Stimulus identification refers to the animal's ability to differentiate tastes and ultimately link that information to other outcomes, i.e., at a basal level, those outcomes that influence survival. Ingestive motivation encompasses things like palatability and reward that can activate physiological pathways that can positively and negatively impact ingestion. Digestive preparation includes activation of physiological components that aid in the digestive processes and help maintain homeostasis. This categorization of taste can be applied when thinking about the foraging behavior of cattle: looking for food, eating the meal, and digesting it; gustation plays an important role in all components of cattle feeding behavior.

Taste can be influenced by the other senses; think about when someone has a head cold and he complains about not being able to fully taste his food. It is likely that the future approach of studying gustation will take into consideration the role of other senses in determining taste (Spector & Glendinning, 2009). When studying humans, taste sensations are categorized by the subjects into specific classifications, traditionally sweet, sour, bitter, salty or umami (and the taste of fat, the sixth taste modality?; Bernard, 1964; Besnard, Passilly-Degrace, & Khan, 2016) and when studying animals, researchers are ultimately looking at acceptance, selection, rejection, aversion and/or indifference (Bernard, 1964). Studying taste alone can be somewhat misleading as olfaction and even somatosensation (temperature of food in the mouth and texture) can impact how something "tastes."

A Cow's Tongue

The tongue is the cow's main vehicle for tasting. A cow's tongue is also used for prehension, mastication, and food manipulation. The tongue is covered with lingual papillae, projections on the surface of the tongue of different shapes and sizes by function and species. Papillae are categorized as mechanical (filiform, conical, and lenticular papillae) or gustatory (circumvallate and fungiform) based on their function (Chamorro, De Paz, Sandoval, & Fernandez, 1986; de Paz Cabello, Chamorro, Sandoval, & Fernandez, 1988). If you have ever been licked by a cow you have likely seen and felt the filiform papillae, the rougher projections on the surface of the cow's tongue that help with prehension. Along the surface area of the gustatory papillae are taste buds, onion-shaped structures which are made up of clusters of taste cells (usually 50–100 taste cells per cluster; Bachmanov & Beauchamp, 2007; Lindemann, 2001), the gustatory receptor cells or chemoreceptors. [Interesting fact: the chemoreceptors of fish, similar to taste buds in humans, are on the outside of their body, helping to chemically scan their environment for food (Kare, 1971).] In order to "taste" something, the chemical molecule must be liquid or be dissolved in saliva so that it can be transported to the taste cell microvilli positioned at taste pores on the papillae. Once in contact with the chemical, the taste cell transduces a signal to the gustatory neural pathway; there are different theories regarding how the taste is encoded in this neural pathway (Chandrashekar, Hoon, Ryba, & Zuker, 2006).

On the cow's tongue, the highest concentrations of taste buds are found on the posterior portion of the tongue on the circumvallate papillae (Davies, Kare, & Cagan, 1979; Kare, 1971). The circumvallate papillae on a cow's tongue contain approximately 90% of all the taste buds (Davies, Kare, & Cagan, 1979). The fungiform papillae which contain taste buds as well but significantly fewer per papilla are found in high concentration on the tip of the tongue with few scattered on the mid-section (Davies, Kare, & Cagan, 1979). The size, number, distribution on the papillae, and number of taste receptors vary greatly between species (Kare, 1971). To provide some comparison of the range of number of taste buds identified in several studies in the 1940s to the 1960s, calves have 25,000 taste buds (Weber, Davies, & Kare, 1966), compared with 9,000 in humans (Cole, 1941) and 24 in the chicken (Lindenmaier & Kare, 1959). It is necessary to note that the total numbers of taste buds do not correlate to the gustatory abilities of the species, i.e., some species have a relatively low number of taste buds but can differentiate certain chemicals at levels imperceptible to species with a larger number of taste buds (Kare, 1971).

It has been suggested that domestication has impacted the function of taste and an animal's ability to taste (Belyaev, 1969). Some studies in animals have explored preference for and intake of sugary or toxic solutions. Maller and Kare (1965) conducted a study in which they provided wild and domestic rats with nonnutritive and potentially toxic sugar solutions (different experimental treatments) in addition to a nutritive balanced diet. Although both the wild and domestic rats preferred the sugar solutions, the domestic rats almost doubled their fluid intake by drinking the sugar solution and the wild rats did not. The results of their study suggested that the wild rats were more reactive to the nutritional aspects of the food, while the domestic rats were self-indulgent. A similar phenomenon was seen in a study with domestic and jungle fowl, in which it was speculated that domestication has produced an animal with diminished sensitivity to energy regulation (Kare & Maller, 1967). Whereas these findings are not necessarily directly related to an animal's taste ability, they do suggest a change in ingestive motivation, one of Spector and Glendinning's (2009) categories of taste processing.

Taste and Human–Cattle Interaction

Although a cow's ability to taste does not play a direct role in animal handling, taste is considered one of the main determinants of food choice (Drewnowski & Rock, 1995; Leterme, Brun, Dittmar, & Robin, 2008) and therefore is significant to other aspects of cattle management such as health and nutrition. Taste perception of a species is linked to its diet type and environment. Bitter taste, in part due to its significant role in identifying toxins in humans and other mammals, has been studied more extensively than some of the other taste categories. In herbivores like cattle, the bitter taste is complex; it is hypothesized that bitter-taste perception evolved in these animals to prevent the consumption of plant toxins yet plants are also characteristically bitter and comprise the majority of a herbivore's diet (Garcia-Bailo, Toguri, Eny, & El-Sohemy, 2009). So how do cattle perceive bitter tastes? Behavioral studies have indicated that carnivores and omnivores are more sensitive than herbivores to certain bitter compounds (Glendinning, 1994). Glendinning (1994) predicts that animals (like cattle) who are constantly exposed to bitter and potentially toxic compounds in their diet have evolved to have a high threshold for bitter taste (or a reduction in bitter sensitivity) and a higher tolerance for toxic substances. Cattle cannot reject bitter foods or they would risk compromising their diet, whereas a carnivorous animal can more readily reject bitter food as bitter substances are not a large component of their diet so there is no malnutrition risk. Cattle sensitivity to bitter compounds is directly related to their selectivity of the plants for consumption, their ruminal fermentation, and their feeding behavior when they are identifying plants (i.e., they eat samples of plants) (Freeland & Janzen, 1974; Li & Zhang, 2014). Neural communication between what a cow tastes and smells and the subsequent reactions in the viscera enable ruminants, like cattle, to sense the consequences of food ingestion (Provenza, 1995). Thus, the role of gustation in cattle feeding and digestive behavior and subsequent nutritional status is significant.

EYE SEE YOU—THE VISION SYSTEM

As cattle are a prey species, their sense of vision has evolved to facilitate their need to continually scan the environment around them for predators. As a naturally grazing animal, they need to have the capacity to be able to survey the fields around them while grazing, capturing a panoramic view of their environment. Although vision is also critical to survival for predatory species, they do not use it in the same way that cattle and other ungulates do and this can be seen in some of the physical differences in the vision systems between species. Visual cues are often used with other sensory cues such as hearing and olfaction in many different aspects of cattle behavior, such as foraging, communication, and reproductive behaviors.

Cow's Eye View

A cow's eyes are on the side of its head as with many other grazing prey species. In nature, cattle need to have the ability to see far off into the horizon to identify predators. The wide set eyes provide cattle with approximately 330° of vision (Phillips, 2002). While cattle have extensive panoramic monocular vision, they do not have significant binocular vision. A cow's binocular range is only 25°–30° (Prince, 1956). In contrast, predatory species have eyes in the middle and front of their head providing them with good binocular vision and limited panoramic vision. Due to the relatively narrow binocular field of vision characteristic of many grazing species, cattle do not have good depth perception. This is one reason that cattle balk at shadows. The dark shadow on the ground appears to be a large hole to cattle as they are unable to accurately recognize its depth, or lack thereof. Notice when handling cattle that they often will lower and raise their heads at the edge of the shadow trying to assess the safety for moving forward. It should also be noted that cattle cannot see directly behind them as is also seen in many other animals. When working with cattle, it is important to be aware of the blind spot and make sure to not surprise an animal while approaching from directly behind them.

A cow's pupil is horizontal and oval in shape. People who do not take the time to watch cattle closely may not have observed this, particularly in dark-eyed cattle as most are. On the contrary, many nocturnal animals, like cats and prosimians, have vertical, slit pupils; it is thought that slit pupils are better at contracting to protect the inner eye from too much light, a protection needed in nocturnal animals as they have highly sensitive retinas (Charman, 1991).

The pupil's function is to protect the eye from letting in too much light but additionally helps with visual acuity. Rehkämper, Perrey, Werner, Opfermann-Rüngeler, and Görlach (2000) determined that their research bulls were better able to discriminate images when the images were presented vertically rather than horizontally, supposedly due to the shape of cows' pupils (Rehkämper, Perrey, Werner, Opfermann-Rüngeler, & Görlach, 2000). Considering that cattle scan the horizon for predators that would likely have a vertical aspect as they approach from the distance, suggesting that this difference in visual acuity based on orientation may be an adaptive characteristic of the eye of grazing animals. Additionally, ungulates have a "visual streak" on the retina, which is an elongated area of high ganglion cell density (Hebel, 1976; Hughes, 1977). It has been theorized that the visual streak found on the retina of ungulates is a visual adaptation that enables the prey species to better identify predators on the horizon (Hughes, 1977). The placement on the head and topography of a cow's eye are characteristic of grazing species and essential to survival.

Colors—To See or Not to See

It is often asked if animals, including cattle, can see colors as humans can. To see color, the eye has to be able to respond to different spectral sensitivities (i.e., different wavelengths of light). Color vision adds contrast between objects seen and therefore increases the visibility of objects within its surroundings. In the mammalian eye, there are different types of photoreceptors that have different photopigments and an associated neural network that can detect changes within this photosensitive system. Rods and cones are the two types of photoreceptors found in the retina of vertebrates. Rods assist with vision in dim light and do not perceive color and cones assist with vision in bright light and do perceive color. Early research suggested that cattle have a high proportion of rods in the retina (ranging between 3:1 rods:cones to 6:1) (Rochon-Duvigneaud, 1943) which supports cattle's behavior pattern, diurnal with crepuscular grazing activity.

Humans have trichromatic color vision, meaning that they have three different types of cone receptors (S-cone, short wavelength; M-cone, medium wavelength; L-cone, long wavelength) and each type reacts specifically within a certain spectral range with a peak spectral sensitivity. Measurements of photopigments in bovine and other ungulate retina have indicated two unique photopigments with peak sensitives in the short wavelengths (S cone; 439–456 nm) and the medium to long wavelengths (M/L cone; 537–557 nm), indicating that cattle are dichromats (Jacobs, 1993; Jacobs, Deegan, & Neitz, 1998). The two bovine cones specifically were shown to have peak sensitivities at 455 nm (S-cone) and 554 nm (M/L-cone) (Jacobs, Deegan, & Neitz, 1998) (for reference, the peak sensitivity for human S-, M-, and L-cones are 419 nm, 531 nm, and 538 nm, respectively; Dartnall et al., 1983). Interestingly, many of the papers discussing dichromatic vision in cattle and other ungulates do not draw conclusions about what colors cattle can actually perceive. The photopigment absorption curves developed from measuring spectral sensitivity of cones can potentially be impacted by light absorption of the lens and/or light reflection by the tapetum (a tapetum lucidum is a feature in the eyes of some vertebrates that enhances visual sensitivity at low-light levels) (Jacobs, Deegan, & Neitz, 1998). It is clear in the literature that to make conclusions about what colors cattle can actually perceive, behavioral studies must be designed to measure a cow's ability to discriminate between different colors. Several studies have been conducted to determine which colors cattle can discriminate and although the studies are successful in identifying that cattle have some ability to discriminate between colors there have been some issues in research design that have limited the application of results (Gilbert & Arave, 1986; Phillips & Lomas, 2001; Riol, Sanchez, Eguren, & Gaudioso, 1989). It is difficult to assess a cow's ability to discriminate colors as researchers have a limited perspective of what cattle actually are seeing but this is an area of potential research. Interestingly, in dichromatic vision, when a certain wavelength (color) stimulates both cone types equally there are varying theories as to what the cow actually perceives. Some theories suggest the cow will not be able to distinguish what color the object is at the neutral point and thus the object will appear achromatic (i.e., white or gray) others suggest the cow sees a continuum of colors and the object is a desaturated version of the two colors (Carroll, Murphy, Neitz, Ver Hoeve, & Neitz, 2001; Roth, Balkenius, & Kelber, 2007).

Although there are some limitations associated with dichromacy, it is thought that dichromatic animals are better adapted to identifying things that are camouflaged (Morgan, Adam, & Mollon, 1992); they use cues other than color to differentiate. Additionally, dichromats, like cattle, are more effective than trichromats at grazing in the shade (Caine, Osorio, & Mundy, 2009). It has also been suggested that dichromats outperform trichromats in low-light situations (Verhulst & Maes, 1998). This is a characteristic of arrhythmic species, like cattle, that are active during the day but can successfully navigate difficult terrain in the evening as well.

Vision and the Human–Cattle Interaction

Understanding the visual perception of cattle is a critical component to working with and raising cattle. They "see" the world from a different perspective than humans. Cattle are prey animals and humans are not and therefore cattle and their human caretakers do not approach and/or perceive situations in the same manner. The differences in cattle vision can aid in cattle handling but the differences can also hinder cattle handling if not managed properly. As noted, cattle have a wide angle of vision and therefore see and respond to animal handlers entering their field of vision, likely differently that a predatory species would. A cow's lack of depth perception can impact how they are handled. As discussed extensively by Dr. Temple Grandin, shadows and sharp contrasts in light make it difficult to move cattle due to their limited ability to determine depth. When confronted with shadows on the ground cattle will often refuse to move forward or will walk around the dark shadow contrast if they are able. It is essential to understand and consider visual perception of cattle when designing yards and various types of cattle handling facilities (Grandin, 1978).

Due to cattle's large number of rod photoreceptors they are better equipped than humans are to see in low-light scenarios but this limits their acuity. Objects are often blurry and not as well defined. Fast moving objects that come into their field of vision are sometimes not easily identifiable and therefore can often scare cattle. It is important to move calmly and deliberately when moving into a cow's field of vision.

DO YOU HEAR WHAT I HEAR?—THE AUDITORY SYSTEM

For cattle, hearing and locating a sound are critically important to survival. In a natural habitat, a sound is often made by another animal and therefore an animal's ability to hear and localize that sound enables the animal to approach or avoid it. Young calves can recognize their mothers from auditory cues alone. Cattle use vocalizations to communicate with each other, to find mates, and express distress. Hearing is one of the senses that humans caretakers have an easier time understanding or recognizing the behavioral response to because in most cases the human and the animal she is working with both hear the sound. When working with cattle (and other livestock species), the position of their ears can provide a helpful clue to understand what noise they are listening to and identifying.

Inner Workings of an Ear

Ears, or pinnae, come in difference shapes and sizes. Humans have short, fixed pinnae while cattle and other mammals have pinnae that extend farther from the head and usually have some degree of mobility with the ability to alter orientation. Unlike olfaction and gustation, hearing is a mechanical sense. A sound source creates waves or pressure changes in the air and it is these alterations in pressure that the animal senses. The ear has several distinct parts: the outer ear, the middle ear, and the inner ear. The pinnae (outer ear) function to help funnel the sound into the ear. The sound waves travel through the outer ear into the auditory canal which is often elongated in mammals. The ear drum is at the start of the middle ear and it responds to the sound wave with vibrations. The auditory canal amplifies sound and the ear drum is very sensitive to changes in pressure both of which act as natural hearing aids. The sound wave transformed into the vibration of the ear drum transfers to three bones in the middle ear that propagate the information further to the oval window. The oval window propagates the sound information into the inner ear which is a maze of tubes and fluid. The cochlea is the structure in the inner ear in which the sound wave information is transferred into electrical impulses that send information to the brain via the auditory nerve. In the fluid-filled cochlea there are membranes that contain microscopic hair-like fibers (hair cells) that are connected to the auditory nerve. The fibers move in response to the movement of the cochlear fluid in response to the sound and depending on the degree of movement different fibers will react and transduce information to the auditory nerve.

Sound Localization

Sound localization is a significant component of hearing. There is a large variation in sound localization acuity between species. The location of a sound is described by its azimuth (direction left to right relative to a listener), elevation (direction up and down relative to the listener), and distance from the listener. The directionality of an animal's pinnae provide information that assist in vertical sound localization, reduce front–back confusion and assist in locating in the horizontal plane (Butler, 1975; Musicant & Butler, 1985).

Mammals use several binaural acoustic cues, i.e., cues based on the comparison of the signals received by the left and right ears, to help identify the location of the sound. To identify sounds moving in the vertical dimension, the comparison of sound energy across different sound wave frequencies arriving at each ear, i.e., spectral analysis, is utilized (Grothe, Pecka, & McAlpine, 2010). Typically, mammals use differences in time of sound arrival (interaural temporal difference, ITD) and differences in received sound pressure level (interaural level differences, ILD) (Heffner & Heffner, 1992a). The ITD is the difference between the times sounds reach the two ears; the sound is "heard" in one ear and after it travels around the surface of the head it is "heard" by the other ear. The ILD is the difference in pressure reaching the two ears. This occurs because the head of the person/animal hearing casts an acoustic shadow which changes the intensity of the sound that reaches each of the ears. This is most applicable with high-frequency sounds as the "shadow effect" is more pronounced. Head width determines at which sound frequencies the ILD becomes significant (Grothe & Pecka, 2014). Humans are able to locate sound within 1° of its location as compared with cattle that localize sound within 30° of its actual location (Phillips, 2002). There was some speculation that auditory abilities have been negatively impacted by evolutionary adaption but an analysis of sound localization of species divided into groups based on domestication status indicated that domesticated animals are not less accurate localizers than wild animals (R. S. Heffner & Heffner, 1992b). Less-accurate sound localizers tend to be prey animals as opposed to predatory species. Animals that are poor sound localizers also tend to have their most acute vision in a horizontal streak providing them with a wide range of vision. It is speculated that cattle don't need to be good sound localizers because by the time they hear a predator approaching on the horizon, they have already seen the approach (Phillips, 2002).

Do Cattle Hear What Their Human Caretakers Hear?

The ability to hear high-frequency sounds is characteristic of the mammalian auditory system (Heffner & Heffner, 2008); high-frequency meaning above 10 kHz as that is the upper limit of other animal families. Researchers had observed that generally smaller mammals were more adept at hearing high frequencies. To use the binaural spectral difference discussed above to locate sounds, it is necessary to have a head large enough to attenuate the sound and change its intensity level at each ear. Heffner and Heffner (2008) have shown a significant negative relationship ($r = -0.79$, $p < 0.0001$) between head size and high-frequency hearing for over 60 species. Although cattle have a bigger head size than humans, they can hear higher frequency sounds than humans can but do not hear particularly high frequencies as compared to other species such as cats and mice (Heffner & Heffner, 2008).

Heffner and Heffner (1983) determined that cattle have a well-defined "best frequency" for hearing at 8 Hz, with a range from 23 to 35 kHz. Generally, it has been demonstrated that ungulates are not sensitive to high-frequency sounds as compared with other mammals. The audiogram developed for cattle indicates that cattle are sensitive to low-frequency noises, more sensitive than humans and horses (Heffner & Heffner, 1983). Cattle vocalizations average between 50 and 1,250 Hz (Kiley, 1972). Studies have been conducted on cattle exploring vocalizations in response to various management procedures such as painful procedures, weaning, and isolation (Boissy & Le Neindre, 1997; Schwartzkopf-Genswein, Stookey, & Welford, 1997; Watts & Stookey, 1999). There is opportunity for further research relating the frequency of cattle vocalizations in response to certain stimuli in relation to the cattle auditory system.

Hearing and the Human–Cattle Interaction

Understanding the auditory sensitivity of cattle is helpful because it provides handlers and managers with some insight as to what types of sound may be aversive to the cattle they are working with. Typically, animals do not like loud sounds—this has been documented both in research studies and publications on how to handle animals humanely. Grandin (1980) discussed the impacts that noise can have on cattle handling, particularly noise associated with equipment (e.g., motors from hydraulic chute) and handlers, and recommends that for optimum handling, noise should be minimized. It is important for animal handlers to recognize that their voices alone can provoke fear or stress responses in cattle when they are moving them. Some studies have indicated altered heart rate and behavior in cattle indicative of a fear response when exposed to recordings of animal handlers' voices (Waynert, Stookey, Schwartzkopf-Genswein, Watts, & Waltz, 1999). The mobility of animals' pinnae varies from no mobility in humans to 180° pinnae movement in horses, with cattle on the same end of the spectrum. It is difficult for humans to comprehend the benefit and/or result of pinnae movement to localize sound because human pinnae are not motile (Heffner & Heffner, 1992a). Cattle, however, do move their ears and this is noticeable when working with cattle. Reading cattle's ears is one way to understand what they may be listening to or distracted by. It is important to note though that although an animal may have motile ears, it does not necessarily mean they are good at localizing sound, as is the case with some livestock species, including cattle (Heffner & Heffner, 1992a). When animals hear a sound the subsequent response is to orient to the sound with the best part of the visual field. In humans, this would be the fovea which does not have a large field of vision (several degrees so the sound localization must be fairly accurate). In cattle, the best field of vision is in the horizontal streak which encompasses a large angle and therefore the sound localization can be more general, just orienting them to the general direction. When working with cattle observe their ears and their subsequent head orientation to observe this response to a noise. Overall, noise is a critical component of animal handling both because humans use noise to garner the attention and response of cattle but also that the noises that come along with handling impact cattle behavior and physiology, not always in a positive manner.

THE BOVINE TOUCH—SOMATOSENSATION

When thinking about the sense of touch what often comes to mind is how we as humans use our hands to understand our environment by determining what things are, based on how their texture and weight feel, determining the temperature of something, or responding to pain. These types of tactile signals combined with other external cues help humans understand and interact with their surroundings. Humans obviously use their hands extensively to touch things and as cattle are not able to access the tactile world the same way as humans do they rely on other methods of gathering tactile information, primarily doing so with their noses and mouths. Cattle use touch and cutaneous sensitivity in a number of ways to both gain information about and respond to their environment. In cattle, touch plays a role in exploratory behavior, in sensing and responding to aversive stimuli, and in communication between herd-mates.

How Do Cattle Feel Touch?

Skin receptors detect pressure, movement, temperature, and some damaging pathological conditions such as inflammation—and this is how cattle feel "touch." The cattle sense of touch is varied in that the sense provides cattle with feedback to a range of different types of mechanical stimuli. The somatosensory system of cattle, and mammals in general, is able to differentiate between noxious stimuli that cause pain and innocuous or pleasant stimuli. In the skin, there are mechanoreceptors, thermoreceptors, and nociceptors. There are five different types of mechanoreceptors: two found in the superficial skin and three found deep in the skin. The receptors are actual nerve endings that have specialized areas on the cell membranes that respond to mechanical distortion. There are several different kinds of thermoreceptors, some that respond to innocuous cool and warm stimuli and others that respond to heat and cold that are nociceptors, responding to painful/noxious stimuli. The somatosensory system is a complex network delivering varying sensations to the brain. Mechanoreceptors differ in surrounding tissues (which alters the response identified), adaptation to stimuli (slow or fast), and the receptive field of the receptor (Møller, 2003). In the somatosensory cortex of the brain, there is a "map" of the body as neurons are anatomically organized by the part of the body to which they are connected. The representation of the body surface is not uniform. For example, in humans the area dedicated to the fingers, hands, and face are much larger than the rest of the body. Considering what is known about cattle, and how they "touch" their environment, the mouth, nose, and tongue would have a larger representation in the somatosensory cortex than other parts of the body.

Touch between Herdmates

Tactile interactions, such as grooming, play a critical role in forming and stabilizing the social networks of many species, including cattle (Gutmann, Špinka, & Winckler, 2015). Allogrooming is defined as the licking of one cow on the body surface of another primarily in the forequarter and head region excluding anal and udder licking (Wood, 1977). Although allogrooming does serve a physiological function in skin and hair hygiene in cattle (Sato, Sako, & Maeda, 1991; Simonsen, 1979; Val-Laillet, Guesdon, Von Keyserlingk, De Passillé, & Rushen, 2009), allogrooming has a critical role in forming and stabilizing the social network within cattle herds (Boissy et al., 2007; Sato, Sako, & Maeda, 1991; Sato, Tarumizu, & Hatae, 1993). Tactile interactions can also have positive impacts on cattle well-being. Multiple studies have indicated that allogrooming can have a calming effect on those being groomed as indicated by decreased heart rates (Laister, Stockinger, Regner, Zenger, Knierim & Winckler, 2011) in addition to positively impacting milk yield and weight gain (Arave & Albright, 1981; Sato, Sako, & Maeda, 1991). Social grooming is complex and can be used in a variety of ways to express hierarchy position, build intragroup relationships, reinforce positive social bonds, and ameliorate negative group dynamics.

Sensory Input of Food—Texture

The physical properties of food in the mouth can impact the palatability of or aversion to the food itself. Many of the senses already discussed (olfaction, gustation, vision) impact food choices but somatosensory input does as well, often discussed as texture. The idea of "dynamic contrast," the changes in sensations during chewing, swallowing, and ruminating, can impact the memory of an eating experience and thus impact subsequent food choice (Forbes, 2007). Cattle, and other animals, learn to associate sensory characteristics of food such as texture with unpleasant or pleasant metabolic experiences (Forbes, 2007). Food aversion and preference is complex, involving both sensory input and post-ingestive feedback (Scott & Provenza, 1998).

Can't Touch This—Aversive Stimuli

The way in which cattle respond to aversive stimuli, such as pain associated with management procedures such as castration and dehorning, has been studied extensively. Nociception, the sense of pain, occurs in response to chemical, mechanical, or thermal stimulation of nociceptors (sensory nerve cells) that produce an electric impulse in response to a noxious stimulus that travels to the spinal cord and ultimately the brain. This triggers an experience of pain in the animal and is associated with behavioral and physiological changes (see Coetzee, 2013 for a review of pain assessment associated with castration in cattle). Recognition of pain in cattle is essential to proper management. A contemporary welfare issue in the cattle industry is developing methods to properly assess and alleviate pain resulting from standard management procedures.

The Role of Touch in Human–Cattle Interactions

When humans handle animals, they often touch them with their hands or handling tools. Thus, the application of touch in human–cattle interactions during handling can be direct. A negative handling experience in which a cow is hit, kicked, or struck with a handling tool can have a lasting impression on that animal. As discussed, the negative sensory input could cause an association of fear with the location, the handler, and the procedure. That type of behavior is inexcusable when working with animals as it impacts animal well-being, causes fear and distress, can potentially impact handling safety depending on the animal's response and can have detrimental effects on efficiency and production. Hemsworth and Coleman (2010) provide a comprehensive review on the impacts of human–animal interaction on animal welfare and productivity.

Tactile interactions can also have positive impacts on cattle well-being. Multiple studies have indicated that allogrooming can have a calming effect on those being groomed as indicated by decreased heart rates (Laister, Stockinger, Regner, Zenger, Knierim & Winckler, 2011; Sato,

Sako, & Maeda, 1991; Sato, Tarumizu, & Hatae, 1993) in addition to positively impacting milk yield and weight gain (Arave & Albright, 1981; Sato, Sako, & Maeda, 1991). Oftentimes research focuses on the impact that negative human–animal interactions have on cattle behavior and well-being but there have been some studies that focus on the impacts of positive tactile interactions between cattle and humans (Boissy & Bouissou, 1988; Boivin, Le Neindre, & Chupin, 1992; Schmied, Boivin, Scala, & Waiblinger, 2010; Schmied, Boivin, & Waiblinger, 2008). Schmied, Boivin, and Waiblinger (2008; 2010) have conducted several studies exploring the impacts of stroking different body areas performed by humans on the subsequent behavior of dairy cattle. The studies determined that stroking in certain areas, particularly the neck, reduced avoidance behavior and stress reactions and increased approach behavior in dairy cattle, providing evidence that positive tactile interactions with humans can impact cattle well-being.

SUMMARY

The paradox of novelty in the context of cattle behavior is a cow's desire to explore something new in its environment slowly, that if otherwise confronted with quickly would evoke fear. Cattle are curiously afraid of novel items in their environment and their sensory systems allow them to see, hear, smell, touch, and even taste objects they are confronted with. Often times when working with cattle we expose them to new things whether it be facilities, handlers, working animals, conspecifics, or locations. Understanding how cattle gather sensory information about their surroundings often explains why they behave in certain ways, e.g., refusing to move when confronted with a sharp contrast in flooring, getting agitated by yelling, being spooked by people that move quickly into their field of vision. Some aspects of human and bovine sensory systems are similar but others are very different and even though understanding these sensory systems provides some appreciation for the umwelt of a cow, human handlers will never be able to fully grasp how a cow experiences the handler-cow collective world. Sensory systems are complex but by understanding the basics, the safety and well-being of both the cattle and human handlers can be improved.

REFERENCES

Aggleton, J., & Mishkin, M. (1986). The amygdala: Sensory gateway to the emotions. In R. Plutchik & H. Kellerman (Eds.), *Emotion: Theory, research, experience* (Vol. 3, pp. 281–299). New York: Academic Press.

Amoore, J. E. (1991). Specific anosmias. In T. V. Getchell, Doty, R. L., Bartoshuk, L. M., Snow, J. B. Jr (Eds.), *Smell and taste in health and disease*. New York: Raven Press.

Arave, C., & Albright, J. (1981). Cattle behavior. *Journal of Dairy Science, 64*(6), 1318–1329.

Bachmanov, A. A., & Beauchamp, G. K. (2007). Taste receptor genes. *Annual Review of Nutrition, 27*(1), 389–414. doi:10.1146/annurev.nutr.26.061505.111329.

Baldwin, B. (1977). Ability of goats and calves to distinguish between conspecific urine samples using olfaction. *Applied Animal Ethology, 3*(2), 145–150.

Baron, G., Frahm, H., Bhatnagar, K., & Stephan, H. (1983). Comparison of brain structure volumes in insectivora and primates. III. Main olfactory bulb (MOB). *Journal Fur Hirnforschung, 24*(5), 551–568.

Belyaev, D. K. (1969). Domestication of animals. *Science Journal, 5*(1), 47–52.

Bernard, R. A. (1964). An electrophysiological study of taste reception in peripheral nerves of the calf. *American Journal of Physiology—Legacy Content, 206*(4), 827–835.

Besnard, P., Passilly-Degrace, P., & Khan, N. A. (2016). Taste of fat: A sixth taste modality? *Physiological Reviews, 96*(1), 151–176. doi:10.1152/physrev.00002.2015.

Boissy, A., & Bouissou, M.-F. (1988). Effects of early handling on heifers' subsequent reactivity to humans and to unfamiliar situations. *Applied Animal Behaviour Science, 20*(3), 259–273.

Boissy, A., & Le Neindre, P. (1997). Behavioral, cardiac and cortisol responses to brief peer separation and reunion in cattle. *Physiology Behavior, 61*(5), 693–699.

Boissy, A., Manteuffel, G., Jensen, M. B., Moe, R. O., Spruijt, B., Keeling, L. J.,... Langbein, J. (2007). Assessment of positive emotions in animals to improve their welfare. *Physiology Behavior, 92*(3), 375–397.

Boivin, X., Le Neindre, P., & Chupin, J. (1992). Establishment of cattle-human relationships. *Applied Animal Behaviour Science, 32*(4), 325–335.

Buck, L., & Axel, R. (1991). A novel multigene family may encode odorant receptors: A molecular basis for odor recognition. *Cell, 65*(1), 175–187.

Butler, R. A. (1975). The influence of the external and middle ear on auditory discriminations. In: Keidel, W.D. & Neff, W.D. *Handbook of sensory physiology,* Vol V/2: Auditory system (pp. 247–260). New York: Springer-Verlag.

Caine, N. G., Osorio, D., & Mundy, N. I. (2009). A foraging advantage for dichromatic marmosets (Callithrix geoffroyi) at low light intensity. *Biology Letters.* doi:10.1098/rsbl20090591.

Carroll, J., Murphy, C. J., Neitz, M., Ver Hoeve, J. N., & Neitz, J. (2001). Photopigment basis for dichromatic color vision in the horse. *Journal of Vision, 1*(2). doi:10.1167/1.2.2.

Chamorro, C., De Paz, P., Sandoval, J., & Fernandez, J. (1986). Comparative scanning electron-microscopic study of the lingual papillae in two species of domestic mammals (Equus caballus and Bos Taurus). I Gustatory Papillae. *Cells Tissues Organs, 125*(2), 83–87.

Chandrashekar, J., Hoon, M. A., Ryba, N. J., & Zuker, C. S. (2006). The receptors and cells for mammalian taste. *Nature, 444*(7117), 288–294. doi:10.1038/nature05401.

Charman, W. (1991). The vertebrate dioptric apparatus. *Vision and Visual Dysfunction, 2*, 82–117.

Coetzee, J. F. (2013) Assessment and management of pain associated with castration in cattle. *Veterinary Clinics: Food Animal Practice, 29*(1), 75–101. doi:10.1016/j.cvfa.2012.11.002.

Cole, E. C. (1941). *Text-book of comparative histology.* Philadelphia: The Blakiston Company.

Craven, B. A., Neuberger, T., Paterson, E. G., Webb, A. G., Josephson, E. M., Morrison, E. E., & Settles, G. S. (2007). Reconstruction and morphometric analysis of the nasal airway of the dog (Canis familiaris) and implications regarding olfactory airflow. *The Anatomical Record (Hoboken), 290*(11), 1325–1340. doi:10.1002/ar.20592.

Damask, A. C. (1981). CHAPTER 3-Olfaction. In *Medical physics: External senses* Vol 2 (pp. 41–68). New York: Academic Press.

Dartnall, H. J. A., Bowmaker, J. K., & Mollon, D. (1983). Human visual pigments: microspectrophotometric results from the eyes of seven person. *Proceedings of Royal Society, London B, 220*(1218): 115–130.

Davies, R. O., Kare, M. R., & Cagan, R. H. (1979). Distribution of taste buds on fungiform and circumvallate papillae of bovine tongue. *The Anatomical Record, 195*(3), 443–446.

de Paz Cabello, P., Chamorro, C., Sandoval, J., & Fernandez, M. (1988). Comparative scanning electron-microscopic study of the lingual papillae in two species of domestic mammals (Equus caballus and Bos taurus). II Mechanical Papillae. *Cells Tissues Organs, 132*(2), 120–123.

Drewnowski, A., & Rock, C. L. (1995). The influence of genetic taste markers on food acceptance. *The American Journal of Clinical Nutrition, 62*(3), 506–511.

Eiting, T. P., Smith, T. D., Perot, J. B., & Dumont, E. R. (2014). The role of the olfactory recess in olfactory airflow. *The Journal of Experimental Biology, 217*(10), 1799–1803. doi:10.1242/jeb.097402.

Entsu, S., Dohi, H., & Yamada, A. (1992). Visual acuity of cattle determined by the method of discrimination learning. *Applied Animal Behaviour Science, 34*(1), 1–10. doi:10.1016/S0168-1591(05)80052-8.

Firestein, S., Picco, C., & Menini, A. (1993). The relation between stimulus and response in olfactory receptor cells of the tiger salamander. *The Journal of Physiology, 468*, 1–10.

Fleischer, J., Breer, H., & Strotmann, J. (2009). Mammalian olfactory receptors. *Frontiers in Cellular Neuroscience, 3*, 9. doi:10.3389/neuro.03.009.2009.

Forbes, J. M. (2007). *Voluntary food intake and diet selection in farm animals,* 2nd edn: Cabi. Wallingford, UK.

Fraser, A. F. (1983). The behavior of maintenance and the intensive husbandry of cattle, sheep and pigs. *Agriculuture, Ecosystems, and Environment, 9*, 1–23.

Freeland, W. J., & Janzen, D. H. (1974). Strategies in herbivory by mammals: The role of plant secondary compounds. *The American Naturalist, 108*(961), 269–289.

Garcia-Bailo, B., Toguri, C., Eny, K. M., & El-Sohemy, A. (2009). Genetic variation in taste and its influence on food selection. *Omics, 13*(1), 69–80. doi:10.1089/omi.2008.0031.

Gilbert, B., & Arave, C. (1986). Ability of cattle to distinguish among different wavelengths of light. *Journal of Dairy Science, 69*(3), 825–832.

Glendinning, J. I. (1994). Is the bitter rejection response always adaptive? *Physiology and Behavior, 56*(6), 1217–1227.

Glusman, G., Yanai, I., Rubin, I., & Lancet, D. (2001). The complete human olfactory subgenome. *Genome Research, 11*(5), 685–702. doi:10.1101/gr.171001.

Grandin, T. (1978). Design of lairage, yard and race systems for handling cattle in abattoirs, auctions, ranches, restraining chutes and dipping vats. *Paper Presented at the 1st World Congress on Ethology applied to Zootechnics, Madrid.*

Grandin, T. (1980). Observations of cattle behavior applied to the design of cattle-handling facilities. *Applied Animal Ethology, 6*(1), 19–31. doi:10.1016/0304-3762(80)90091-7.

Grandin, T. (2014). Behavioural principles of handling cattle and other grazing animals. In T. Grandin (Ed.), *Livestock handling and transport* (4th ed., pp. 94–115). CAB International. Wallingford, UK.

Gross, C., & Canteras, N. (2012). The many paths to fear. *Nature Reviews Neuroscience, 13*(9), 651–658. doi:10.1038/nrn3301.

Gross, E., Swenberg, J., Fields, S., & Popp, J. (1982). Comparative morphometry of the nasal cavity in rats and mice. *Journal of Anatomy, 135*(1), 83.

Grothe, B., & Pecka, M. (2014). The natural history of sound localization in mammals – a story of neuronal inhibition. *Frontiers in Neural Circuits, 8*, 116. doi:10.3389/fncir.2014.00116.

Grothe, B., Pecka, M., & McAlpine, D. (2010). Mechanisms of sound localization in mammals. *Physiological Reviews, 90*(3), 983–1012. doi:10.1152/physrev.00026.2009.

Grus, W. E., Shi, P., Zhang, Y.-P., & Zhang, J. (2005). Dramatic variation of the vomeronasal pheromone receptor gene repertoire among five orders of placental and marsupial mammals. *Proceedings of the National Academy of Sciences of the United States of America, 102*(16), 5767–5772.

Gutmann, A. K., Špinka, M., & Winckler, C. (2015). Long-term familiarity creates preferred social partners in dairy cows. *Applied Animal Behaviour Science, 169*, 1–8.

Harkema, J. R. (1991). Comparative aspects of nasal airway anatomy: Relevance to inhalation toxicology. *Toxicologic Pathology, 19*(4-1), 321–336.

Harkema, J. R., Carey, S. A., & Wagner, J. G. (2006). The nose revisited: A brief review of the comparative structure, function, and toxicologic pathology of the nasal epithelium. *Toxicologic Pathology, 34*(3), 252–269. doi:10.1080/01926230600713475.

Hebel, R. (1976). Distribution of retinal ganglion cells in five mammalian species (pig, sheep, ox, horse, dog). *Anatomy and Embryology, 150*(1), 45–51.

Heffner, H. E., & Heffner, R. S. (2008). High-frequency hearing. In P. Dallos, D. Oertel and R. Hoy (Eds.) *Handbook of the senses: Audition* (pp. 55–60). New York: Elsevier.

Heffner, R. S., & Heffner, H. E. (1983). Hearing in large mammals: Horses (Equus caballus) and cattle (Bos taurus). *Behavioral Neuroscience, 97*(2), 299.

Heffner, R. S., & Heffner, H. E. (1992a). Evolution of sound localization in mammals. In: D.B. Webster, A.N. Popper, R.R. Fay (Eds.) *The evolutionary biology of hearing* (pp. 691–715). New York: Springer.

Heffner, R. S., & Heffner, H. E. (1992b). Hearing in large mammals: Sound-localization acuity in cattle (Bos taurus) and goats (Capra hircus). *Journal of Comparative Psychology, 106*(2), 107.

Hemsworth, P., & Coleman, G. (2010). *Human–livestock interactions: The stockperson and the productivity and welfare of intensively farmed animals.* Oxford: CAB International.

Hughes, A. (1977). The topography of vision in mammals of contrasting life style: Comparative optics and retinal organisation. In F. Crescitelli (Ed.) *The visual system in vertebrates. Handbook of Sensory Physiology* Vol 7/5(pp. 613–756). Berlin, Heidelberg, Springer.

Jacobs, G. H. (1993). The distribution and nature of colour vision among the mammals. *Biological Reviews of the Cambridge Philosophical Society, 68*(3), 413–471.

Jacobs, G. H., Deegan, J. F., & Neitz, J. (1998). Photopigment basis for dichromatic color vision in cows, goats, and sheep. *Visual Neuroscience, 15*(3), 581–584.

Kare, M. (1971). Comparative study of taste. In T. E. Acree, J. Atema, J. E. Bardach, L. M. Bartoshuk, L. M. Beidler, R. M. Benjamin, R. M. Bradley, Z. Bujas, H. Burton, L. P. Cole, A. I. Farbman, L. Guth, H. Kalmus, M. Kare, K. Kurihara, D. H. McBurney, R. G. Murray, M. Nachman, C. Pfaffmann, M. Sato, R. S. Shallenberger, Y. Zotterman, & L. M. Beidler (Eds.), *Taste* (pp. 278–292). Berlin and Heidelberg: Springer.

Kare, M. R., & Maller, O. (1967). Taste and food intake in domesticated and jungle fowl. *Journal of Nutrition, 92*, 191–196.

Keller, A., & Vosshall, L. B. (2008). Better smelling through genetics: Mammalian odor perception. *Current Opinion in Neurobiology, 18*(4), 364–369. doi:10.1016/j.conb.2008.09.020.

Kiley, M. (1972). The vocalizations of ungulates, their causation and function. *Ethology, 31*(2), 171–222.

Laister, S., Stockinger, B., Regner, A.-M., Zenger, K., Knierim, U., & Winckler, C. (2011). Social licking in dairy cattle—Effects on heart rate in performers and receivers. *Applied Animal Behaviour Science, 130*(3), 81–90.

Le Gros Clark, W. (1960). *The antecedents of man. An introduction to the evolution of primates.* Chicago: Quadrangle Books.

Leterme, A., Brun, L., Dittmar, A., & Robin, O. (2008). Autonomic nervous system responses to sweet taste: Evidence for habituation rather than pleasure. *Physiology and Behavior, 93*(4), 994–999. doi: 10.1016/j.physbeh.2008.01.005.

Li, D., & Zhang, J. (2014). Diet shapes the evolution of the vertebrate bitter taste receptor gene repertoire. *Molecular Biology and Evolution, 31*(2), 303–309. doi:10.1093/molbev/mst219.

Lindemann, B. (2001). Receptors and transduction in taste. *Nature, 413*(6852), 219.

Lindenmaier, P., & Kare, M. R. (1959). The taste end-organs of the chicken. *Poultry Science, 38*(3), 545–550.

Maller, O., & Kare, M. R. (1965). Selection and intake of carbohydrates by wild and domesticated rats. *Proceedings of the Society for Experimental Biology and Medicine, 119*(1), 199–203.

McDonald, A. J. (1998). Cortical pathways to the mammalian amygdala. *Progress in Neurobiology, 55*(3), 257–332. doi: 10.1016/S0301-0082(98)00003-3.

Møller, A. R. (2003). *Sensory systems: Anatomy and physiology.* Amsterdam, Academic Press.

Moore, W. (1981). *The mammalian skull.* Cambridge: University Press.

Morgan, M. J., Adam, A., & Mollon, J. D. (1992). Dichromats detect colour-camouflaged objects that are not detected by trichromats. Paper presented at the Proceedings of the Royal Society of London B.

Musicant, A. D., & Butler, R. A. (1985). Influence of monaural spectral cues on binaural localization. *The Journal of the Acoustical Society of America, 77*(1), 202–208.

Negus, V. (1958). *Comparative anatomy and physiology of the nose and paranasal sinuses E&S Livingstone.* Edinburgh and London: E. & S. Livingstone Ltd, 214.

Niimura, Y., & Nei, M. (2003). Evolution of olfactory receptor genes in the human genome. *Proceedings of the National Academy of Sciences, 100*(21), 12235–12240. doi:10.1073/pnas.1635157100.

Niimura, Y., & Nei, M. (2007). Extensive gains and losses of olfactory receptor genes in mammalian evolution. *PLoS One, 2*. doi:10.1371/journal.pone.0000708.

Noviyanto, A., & Arymurthy, A. M. (2013). Beef cattle identification based on muzzle pattern using a matching refinement technique in the SIFT method. *Computers and Electronics in Agriculture, 99*, 77–84. doi:10.1016/j.compag.2013.09.002.

Partan, S., & Marler, P. (2002). The Umwelt and its relevance to animal communication: Introduction to special issue. *Journal of Comparative Psychology, 116*(2), 116–119.

Petersen, W. E. (1922). The identification of the bovine by means of nose-prints. *Journal of Dairy Science, 5*(3), 249–258. doi:10.3168/jds.S0022-0302(22)94150-5.

Phillips, C. (2002). Environmental perception and cognition. In C. Phillips (Ed.) *Cattle behaviour and welfare* (2nd ed., pp. 49–61). Oxford: Wiley-Blackwell.

Phillips, C., & Lomas, C. (2001). The perception of color by cattle and its influence on behavior. *Journal of Dairy Science, 84*(4), 807–813.

Prince, J. H. (1956). *Comparative anatomy of the eye.* Springfield: Thomas.

Proctor, D. F., & Chang, J. C. (1983). Comparative anatomy and physiology of the nasal cavity. *Nasal Tumors in Animals and Man, 1*, 1–33.

Provenza, F. D. (1995). Postingestive feedback as an elementary determinant of food preference and intake in ruminants. *Rangeland Ecology & Management/Journal of Range Management Archives, 48*(1), 2–17.

Rehkämper, G., Perrey, A., Werner, C. W., Opfermann-Rüngeler, C., & Görlach, A. (2000). Visual perception and stimulus orientation in cattle. *Vision Research, 40*(18), 2489–2497.

Reznik, G. K. (1990). Comparative anatomy, physiology, and function of the upper respiratory tract. *Environmental Health Perspectives, 85*, 171–176. doi:10.2307/3430681.

Riol, J., Sanchez, J., Eguren, V., & Gaudioso, V. (1989). Colour perception in fighting cattle. *Applied Animal Behaviour Science, 23*(3), 199–206.

Rochon-Duvigneaud, A. (1943). *Les yeux et la vision des vertébrés*. Paris: Masson.

Roth, L. S., Balkenius, A., & Kelber, A. (2007). Colour perception in a dichromat. *Journal of Experimental Biology, 210*(16), 2795–2800.

Sato, S., Sako, S., & Maeda, A. (1991). Social licking patterns in cattle (Bos taurus): Influence of environmental and social factors. *Applied Animal Behaviour Science, 32*(1), 3–12.

Sato, S., Tarumizu, K., & Hatae, K. (1993). The influence of social factors on allogrooming in cows. *Applied Animal Behaviour Science, 38*(3–4), 235–244.

Schmied, C., Boivin, X., Scala, S., & Waiblinger, S. (2010). Effect of previous stroking on reactions to a veterinary procedure: Behaviour and heart rate of dairy cows. *Interaction Studies, 11*(3), 467–481.

Schmied, C., Boivin, X., & Waiblinger, S. (2008). Stroking different body regions of dairy cows: Effects on avoidance and approach behavior toward humans. *Journal of Dairy Science, 91*(2), 596–605.

Schwartzkopf-Genswein, K., Stookey, J., & Welford, R. (1997). Behavior of cattle during hot-iron and freeze branding and the effects on subsequent handling ease. *Journal of Animal Science, 75*(8), 2064–2072.

Scott, L. L., & Provenza, F. D. (1998). Variety of foods and flavors affects selection of foraging location by sheep. *Applied Animal Behaviour Science, 61*(2), 113–122. doi:10.1016/S0168-1591(98)00093-8.

Shi, P., Bielawski, J. P., Yang, H., & Zhang, Y.-P. (2005). Adaptive diversification of vomeronasal receptor 1 genes in rodents. *Journal of Molecular Evolution, 60*(5), 566–576.

Simonsen, H. (1979). Grooming behaviour of domestic cattle. *Nordisk Veterinaermedicin, 31*(1), 1–5.

Smith, T., Eiting, T., & Bhatnagar, K. (2015). Anatomy of the nasal passages in mammals. *Handbook of Olfaction and Gustation, 3*, 37–62.

Spector, A. C., & Glendinning, J. I. (2009). Linking peripheral taste processes to behavior. *Current Opinion in Neurobiology, 19*(4), 370–377. doi:10.1016/j.conb.2009.07.014.

Stephan, H., & Andy, O. (1977). Quantitative comparison of the amygdala in insectivores and primates. *Cells Tissues Organs, 98*(2), 130–153.

Turner, W. (1890). The convolutions of the brain: A study in comparative anatomy. *Journal of Anatomy and Physiology, 25*(1), 105–153.

Tutar, Y. (2012). Pseudogenes. *Comparative and Functional Genomics, 2012*, 424526(1–4).

Val-Laillet, D., Guesdon, V., Von Keyserlingk, M. A., De Passillé, A. M., & Rushen, J. (2009). Allogrooming in cattle: Relationships between social preferences, feeding displacements and social dominance. *Applied Animal Behaviour Science, 116*(2), 141–149.

Verhulst, S., & Maes, F. (1998). Scotopic vision in colour-blinds. *Vision Research, 38*(21), 3387–3390.

von Uexkull, J. (1909). *Umwelt und Innenwelt der Tiere*. Berlin: J. Springer.

von Uexkull, J. (1957). A stroll through the worlds of animals and men: A picture book of invisible worlds. In C. H. Schiller (Ed.), *Instinctive behavior: The development of a modern concept* (pp. 5–80). New York: International Universities Press.

Watts, J. M., & Stookey, J. M. (1999). Effects of restraint and branding on rates and acoustic parameters of vocalization in beef cattle. *Applied Animal Behaviour Science, 62*(2), 125–135.

Waynert, D. F., Stookey, J. M., Schwartzkopf-Genswein, K. S., Watts, J. M., & Waltz, C. S. (1999). The response of beef cattle to noise during handling. *Applied Animal Behaviour Science, 62*(1), 27–42. doi:10.1016/S0168-1591(98)00211-1.

Weber, W., Davies, R., & Kare, M. (1966). Distribution of taste buds and changes with age in the ruminant. Unpublished data.

Weiss, L. (1988). *Cell and tissue biology, a textbook of histology* (6th ed. pp. 1–17). Munich: Urban & Schwarzenberg.

Wood, M. T. (1977). Social grooming patterns in two herds of monozygotic twin dairy cows. *Animal Behaviour, 25*, 635–642.

Yang, G. C., Scherer, P. W., & Mozell, M. M. (2007). Modeling inspiratory and expiratory steady-state velocity fields in the Sprague-Dawley rat nasal cavity. *Chemical Senses, 32*(3), 215–223.

Zhang, J., & Webb, D. M. (2003). Evolutionary deterioration of the vomeronasal pheromone transduction pathway in catarrhine primates. *Proceedings of the National Academy of Sciences, 100*(14), 8337–8341.

Zozulya, S., Echeverri, F., & Nguyen, T. (2001). The human olfactory receptor repertoire. *Genome Biology, 2*(6), research0018.0011. doi:10.1186/gb-2001-2-6-research0018.

CHAPTER **10**

Breeding and Welfare: Genetic Manipulation of Beef and Dairy Cattle

Courtney Daigle, Andy D. Herring, and Fuller W. Bazer
Texas A&M University

CONTENTS

WHAT IS ANIMAL WELFARE?

While different sectors of society debate what constitutes good animal welfare, what really matters in this argument is the experience of the individual animal. Humans have the capacity to evaluate the physiology, behavior, and mental activity of animals, but we cannot truly understand what it is like to be that animal because we are equipped with different morphology and physiology, thus limiting our capacity to understand their perspective. This presents animal managers with the challenge of needing to provide animals with good welfare, identifying and recording objective metrics that ensure good welfare is being promoted, and making breeding and genetic selection choices that enhance animal welfare, all while providing assurances to consumers that their expectations regarding what is good welfare are being met.

Further complicating this scenario is that multiple—and sometimes conflicting—scientific interpretations of animal welfare and strategies to measure animal welfare are found throughout the scientific literature (see Fraser, 2003). As the fields of ethology and animal welfare science have evolved over the past 60 years, so have the definitions, approaches, and relative importance of the different components of animal welfare (Table 10.1). Furthermore, the approaches and techniques available to measure animal welfare (e.g., productivity, motivation testing (Kirkden and Pajor, 2006), thermal imaging (Nääs et al., 2014), brain activity (Perentos et al., 2017)), and the interpretations of the scientific data collected [e.g., endocrine markers (Mormède et al., 2007)] have evolved alongside the development of the scientific field of animal welfare science. Therefore, scientists have a larger and more diverse arsenal of tools and data available to use as part of the multifactorial evaluation of the animal that is required to understand welfare.

Table 10.1   Chronological History of Approaches Employed to Measure and Define Animal Welfare

Year	Approach to Welfare	Defining Characteristics of Approach
1965	Five freedoms Brambell (1970)	1. Freedom from thirst and hunger 2. Freedom from discomfort 3. Freedom from pain, injury, and disease 4. Freedom to express normal behavior 5. Freedom from fear and distress
1986	Ability to cope Broom (1986)	The animal's state as regards its attempts to cope with its environment
1991	Cognition & emotion Duncan and Petherick (1991)	Dependent solely on the cognitive needs of the animal concerned
1994	Five domains Mellor and Reid (1994)	1. Nutrition 2. Environment 3. Health 4. Behavior 5. Mental State
2003	Two questions Dawkins (2003)	1. Is the animal physically healthy? 2. Does it have what it wants?
2008	Three circles Fraser (2008)	1. Basic health and functioning 2. Affective states 3. Natural living
2010	Welfare quality Quality (2009)	1. Good feeding 2. Good housing 3. Good health 4. Appropriate behavior
2016	Quality of life Mellor (2016)	1. A good life 2. A life worth living 3. Point of balance 4. A life worth avoiding 5. A life not worth living

The parameters set forth by the Brambell Committee (e.g., The Five Freedoms) highlighted outcome-based targets for agricultural animals, but they did not set clear definitions regarding how best to meet those targets from the animal's perspective. As our scientific knowledge regarding biological, physiological, and neurological functioning has increased, so has our ability to quantify the affective state—thus providing scientists the opportunity to quantify, and subsequently emphasize, animal emotion (or the behavioral proxies of emotion) in the metrics evaluated during welfare assessment. As such, the positive emotional state of the animal has increasingly become an integral component of animal welfare assessment to the point at which animals are expected to experience pleasure—not simply have an absence of pain and suffering—in order to have a "life worth living." Because animal welfare addresses the intersection between science and ethics, having a solid scientific foundation with which to make ethical choices is imperative to sustainable management of these animals.

Today, the World Animal Health Organization (OIE) defines animal welfare as

…how an animal is coping with the conditions in which it lives. An animal is in a good state of welfare if (as indicated by scientific evidence) it is healthy, comfortable, well nourished, safe, able to express innate behavior, and if it is not suffering from unpleasant states such as pain, fear and distress. Good animal welfare requires disease prevention and appropriate veterinary treatment, shelter, management and nutrition, humane handling and humane slaughter or killing. Animal welfare refers to the state of the animal; the treatment that an animal receives is covered by other terms such as animal care, animal husbandry, and humane treatment. (OIE, 2016)

This definition put forth by the OIE provides a definition in which objective metrics (e.g., the capacity for an animal to cope) are propelling us toward a quantifiable metric of animal welfare.

Utilizing an objective metric of animal welfare will be imperative as we continually reassess our expectations and moral framework regarding food animals. As human societies change, the ethical frameworks surrounding the interpretation of the scientific evidence as to what constitutes appropriate animal care and welfare will change. Meanwhile, the constant component, irrespective of human interpretation of the science, is the importance of the individual animal's experience. Many of the factors that are considered important to good animal welfare (e.g., emotions, quality of life, injury) are either (1) only possible to experience at the individual level or (2) can change drastically throughout the course of a day, week, or lifetime. Therefore, it is important to identify and propagate individuals that are well adapted to and have the skills to thrive in their current management environment.

Implications of Genetic Selection and Biotechnology on Animal Welfare

In the livestock industry, welfare may be defined as the physical and mental health of the animal as expressed through biological functioning and behavior. This requires the producer and members of society to think about welfare from the perspective of the animal since it is the animal's perception and subsequent interactions with the environment that impact its health and well-being. The impact of conventional breeding, genomic selection, and other advances in biotechnology on animal welfare are explored in this chapter with respect to how they influence health and well-being, and production efficiencies in animal agriculture. *Biotechnology is* defined as technology based on biology and the application of scientific and engineering *principles to the processing or production of materials by biologic agents to provide goods and services.*

Genetic selection in animal agriculture is a double-edged sword. Targeted selection of traits in agricultural animals can either challenge or enhance animal welfare by emphasizing traits associated with (1) productivity, (2) health and disease control, (3) social tolerance, (4) physical size and morphology, (5) behavior and stress responsivity, and (6) heat tolerance. Agricultural animals are housed in increasingly larger groups, are physically larger, are more efficient, are being provided with more complex and dynamic environments with which to interact, and are being managed on a planet that is experiencing climate change. Therefore, identifying phenotypes, and their respective genotypes, that promote harmony between the animal and its production environment will ultimately enhance animal welfare while supporting the sustainability of our food system.

The emphasis on selecting solely for increased productivity has unintentionally yielded undesirable consequences that impair animal welfare. Selection for productivity in broiler chickens has inadvertently, and negatively, impacted survival (Havenstein et al., 1994), reproduction (Liu et al., 1995), and immune performance (Miller et al., 1992). Boars selected for high-lean tissue growth rates (Sather, 1987) and those selected for greater fat depth (Webb et al., 1983) had weaker legs. Laying hens selected for high productivity are less likely to thrive in large groups because they are more likely to perform feather pecking and cannibalism (Rodenburg and Koene, 2007). Pigs selected for fast growth are more likely to perform tail biting (Breuer et al., 2005) and are more aggressive during mixing events (Løvendahl et al., 2005). Selection for dairy cattle with higher milk yields has been associated with reduced fertility, higher rates of mastitis, decreased longevity (Oltenacu and Broom, 2010), and reduced tolerance for heat (Ravagnolo and Misztal, 2000). Therefore, while breeding choices are made in an effort to enhance the sustainability and efficiency of the food system, in some instances we have unintentionally created welfare challenges for the animals in our care.

Fortunately, genetic selection is a tool to optimize welfare and enhance productivity. Aggression, social tolerance, the prevalence, and the location of skin lesions are heritable in swine (Turner et al., 2009); thus, this phenotype can be used to select for individuals that are more socially tolerant and will fight less during and after mixing. Dairy cattle have been selected for high tolerance to human handling (Boissy et al., 2005) which can subsequently promote a positive human–animal relationship, thus reducing stress and enhancing productivity (Hemsworth et al., 2000). Resistance to Bovine Respiratory Disease, the most prevalent and costly disease in feedlot cattle (USDA-APHIS, 2013), is a trait that may be enhanced through genetic selection (Snowder et al., 2006; Schneider et al., 2010; Cockrum et al., 2016) and would be beneficial to cattle welfare, yet more research is needed to elucidate this relationship. Laying hens housed in aviary systems are being selected for bone strength to reduce keel bone damage in commercial settings (Stratmann et al., 2016) which is painful (Nasr et al., 2012b), and is associated with increased mortality (McCoy et al., 1996) and reduced egg production (Nasr et al., 2012a). By evaluating the heritability of species-specific behavioral, physical, and physiological traits that may enhance animal welfare, animal managers may be able to harness the genetic potential of animals that are optimized mentally, physically, and emotionally for food production systems.

Feed additives. Direct-fed microbials (DFMs) (i.e., probiotics) and β-agonists have been developed to promote rumen health, productivity, and efficiency of both beef and dairy cattle. In young calves, probiotics are administered to stabilize the gut microbiota, reduce the risk of pathogen colonization, and reduce the risk and severity of diarrhea. Thus, administration of probiotics during stressful events may mitigate some of the consequences of the stressor. In adult beef and dairy cattle, many probiotics are given to stabilize rumen pH and improve fiber digestion by rumen microorganisms, and probiotics increase milk-fat yield in dairy cows (Chiquette et al., 2008). Investigations into the impact of probiotics on average daily gain, final weight, feed intake, and feed efficiency have yielded contradicting results (Uyeno et al., 2015). β-Agonists fed to cattle during the final phase of the growing period promote muscle growth and feed efficiency. Research opportunities to better understand the impact of directly fed microbials and β-agonists on cattle health and productivity are abundant, as their impact on cattle productivity can be influenced by multiple factors including diet, age, stress, management practices, and adaptive response of the rumen.

The gut microbiome component of the gut–brain axis is an untapped resource for harnessing its ability to influence behavior, stress responsivity, and productivity in animals (Wiley et al., 2017). The field of microbial neuroendocrinology is in its infancy, and recent discoveries into the brain–gut axis suggest behaviors are influenced directly by communication from the gut microbiome (Rhee et al., 2009; Ezenwa et al., 2012). The gut microbiome is beginning to be considered an endocrine organ capable of influencing behavior (O'Callaghan et al., 2016) and physiology, including milk-fat yield (Jami et al., 2014), ADG, and feed efficiency (Myer et al., 2017). By increasing our understanding of the impact the gut microbiome has on animal physiology and emotion, we expect to identify biotechnologies that enhance the mental and physical aspects of animal welfare.

The gut microbiome influences anxiety, depression, and stress-induced corticosterone release in mice and humans (Desbonnet et al., 2008; Rao et al., 2009; Bravo et al., 2012; Cryan and Dinan, 2012; Foster and McVey Neufeld, 2013). The villi of the small intestine have a high concentration of enteric nerves which connect directly to the vagal nerve and the brain. This direct connection between the gut and the brain suggests that brain activity is influenced by gut contents, and potentially biotechnologies developed to enhance gut health and promote productivity. DFMs are being utilized in poultry production to promote productivity, decrease morbidity, and enhance profitability (Flint and Garner, 2009). Probiotic supplementation in mice stimulates the gut immune system and protect against infection during a stress challenge (Martin Manuel et al., 2017). Therefore, the gut microbiome has the potential to be used as a management tool designed to reduce the stress response in animals to a known stressor (e.g., weaning, transportation, handling).

However, the brain–gut axis is a two-way street. Not only can intestinal microbiota influence behavior but also the intestinal microflora can be influenced by the experiences of the host (Bailey et al., 2011). Therefore, understanding the temporal relationships between microbiome stability and behavior is an important component of this development in biotechnology. Because the relationship between the intestinal microflora and host appears to be symbiotic and egalitarian (Rhee et al.,

2009), understanding how quickly, and what types of bacteria are effective in facilitating a positive welfare state are critical pieces of knowledge needed for the development of this biotechnology.

Improving cattle temperament during common stressors (e.g., transportation and handling) may provide economic and animal welfare benefits. Temperament in cattle is a behavioral characteristic that impacts profitability and worker safety. Cattle with excitable temperaments have reduced growth rates (Voisinet et al., 1997), decreased carcass quality (King et al., 2006), reduced immune function (Burdick et al., 2011), and higher cortisol levels during handling (Curley et al., 2006), yet more aggressive maternal behavior can be associated with increased calf survival (Sandelin et al., 2005). From an animal management perspective, cattle with excitable temperaments are more difficult to handle, present greater risk to animal managers, and may influence the behavior of herdmates. Cattle temperament has been observed to be consistent over time and context and influences animal responsivity to stress and human handling (Curley et al., 2006). Cattle temperament is moderately to highly heritable in many breeds (Stricklin et al., 1980; Gauly et al., 2001; Riley et al. 2014), indicating selection potential to alter it. Most studies have also shown worse temperament scores for females than males (Shrode and Hammack, 1971; Stricklin et al., 1980; Voisinet et al., 1997 Gauly et al., 2001; Riley et al., 2014). It is also known that genetics influences feed efficiency (Archer et al., 1997; Arthur et al., 2001; Robinson and Oddy, 2004; Nkrumah et al., 2007) indicating the potential to alter nutrient utilization in cattle, and there are interrelationships between cattle temperament, feed efficiency, and feeding behaviors. Whether the gut microbiome can influence temperament in cattle or the plasticity of the relationships with other traits remains unexplored. Therefore, we need to understand production- and behavioral-based aspects of the brain–gut axis in cattle to identify biotechnologies designed to promote animal welfare.

Biotechnologies

Many of the biotechnologies developed for cattle have the capacity to enhance productivity and profitability in both the beef and dairy industries. The type of commodity being produced (e.g., milk or meat) will dictate what types of technologies are applied, when during the production cycle in which they are implemented, and how management practices are altered to accommodate the implementation of these biotechnologies. The following sections identify management practices that may be implemented to enhance the well-being of cattle.

Dehorning (including disbudding). Cattle are dehorned in order to reduce injuries to themselves and conspecifics, minimize hide damage, improve human safety, reduce damage to facilities, and facilitate safe transport and handling. Genetically selecting for polled cattle is a viable option, but this is a long-term process. When polled genes or seed stock are not readily available, dehorning (disbudding) is achieved by removal of the horn buds with a knife, thermal cautery of the horn buds, or the application of chemical paste to cauterize the horn. It is recommended that producers use analgesia or anesthesia for dehorning of beef cattle, particularly for older cattle with more advanced horn development. However, it is also recommended that dehorning be performed in cattle at the earliest age possible. Biotechnology offers a better path to increasing the frequency of the polled gene in cattle. In cattle, the poll allele (P) is dominant to the horn allele (p). If a calf inherits a single polled allele from either parent then it will be born without horns. Animals with two polled alleles are homozygous polled. Animals with single polled allele and single horn allele are heterozygous and carry the horn gene. Therefore, 25% of offspring from matings between heterozygous animals will be horned.

In the United States, about 80% of all dairy calves (4.8 million per year) and 25% (8.75 million animals) of beef cattle are dehorned yearly (Spurlock et al., 2014). Dairy cattle are less likely to be born polled because the dominant *POLLED* gene exists at a much lower frequency in dairy cattle (e.g., Holsteins), while it is well fixed in beef cattle such as the Angus breed. This is partly due to the fact that polled Holstein sires have lower estimated breeding values for milk production when the polled allele is introgressed and that is estimated to result in a loss of $252 per lactation cycle.

In the Holstein breed, only 6% of dairy sires produce commercially available seedstock carrying the *POLLED* gene and it could take more than 20 years of breeding to achieve a frequency of 50% polled animals (Dorshorst, 2014). Therefore, genetic biotechnology is being implemented to accelerate efforts to achieve hornless cattle, particularly in dairy cattle (see Carlson et al., 2016). The need to utilize biotechnology to enhance welfare is also receiving pressure from corporations like Wal-Mart, Starbucks, Nestle, and Kroger because they have prioritized a reduction of dehorning practices into their animal welfare policies and supply chain producer requirements (Swanson, 2015).

Genome editing using transcription activator-like effector nucleases (TALENs) to introgress the polled allele into the genome of bovine embryonic fibroblasts has been used to produce a genotype identical to what is achievable using natural mating, but without negative effects on lactation in dairy cows (Carlson et al., 2016). The researchers used TALEN-stimulated homology-dependent repair to produce four cell lines either homozygous or heterozygous for the polled allele10. Each of the four lines was cloned by somatic cell nuclear transfer and full- term pregnancies were established for three of the four lines. Five calves were polled and the homozygous introgression of the polled allele into the calves was confirmed and there were no "off-target" effects associated with introgression of the polled allele that might affect lactation.

Carlson et al. (2016) propose that genetic improvement of livestock using TALENs, or other genome-editing methods, is an alternative to transgenic methods for genetic improvement of livestock using variation that is already present in species without the admixture that can result from classic breeding methods. They caution that their results do not demonstrate that the introgression of polled alleles into elite animals would be without risk to the economics of milk production by dairy cows in the United States.

Castration. Castration of cattle is performed to reduce inter-animal aggression, improve human safety, avoid the risk of unwanted pregnancies, and enhance product quality and production efficiency. Castration may be accomplished by surgical removal of the testes, use of a rubber band to inhibit blood flow to testes or by crushing the spermatic cord that includes blood vessels to the testes. However, immunological castration is also possible. One approach was to immunize calves against Gonadotropin Releasing Hormone (GnRH) that stimulates the pituitary gland to release Luteinizing Hormone (LH) and Follicle Stimulating Hormone (FSH). Robertson et al. (1982) immunized 10 Holstein calves against GnRH and found that five calves responded poorly in terms of producing antibodies to GnRH. However, the other five calves responded with high-antibody titers, low concentrations of testosterone in serum, involuted testes, reduced libido and semen production, and docile behavior. The immunocastration effect lasted approximately 6 months before the calves exhibited behaviors of intact males. Temporary immunocastration did not affect weight gain that was improved over that for castrated steers. Carcass traits were similar for immunocastrated and castrated calves, but the lean content in all ten calves subjected to immunocastration was greater than for steer carcasses.

Marti et al. (2017) reported on an immunocastration study that initially included 493 bulls of which 476 remained in the study for 133 days. Meat quality (carcass fat-cover, marbling, tenderness score, pH), marbling, and other meat quality characteristics were assessed and related to testicular function and concentrations of testosterone in serum in immunized bulls. The results of Marti et al. (2017) have been corroborated by others who reported that bulls immunized against GnRH had carcasses more likely to grade choice. Cook et al. (2000) did not report an improvement in meat tenderness when bulls were vaccinated with another GnRH vaccine; rather they found that as the response to GnRH immunization increased, meat tenderness also increased. Similarly, Ribeiro et al. (2004) collected rib dissection data and found that the carcass composition of vaccinated animals and castrated animals had more fat and less muscle than carcasses of intact bulls.

Vaccinating bulls against GnRH is an effective alternative to surgical castration as demonstrated by the suppressed serum testosterone levels for over 100 days. Performance of the immunologically castrated animals was intermediate between bulls and surgically castrated animals. The reduction in testosterone production by testes in bulls also reduces their aggressive behavior and reduces risks of injury to other animals and human handlers. Immunological castration using GnRH vaccine is a welfare friendly alternative to achieve the same meat quality as for surgically castrated cattle.

Biotechnology and resistance to disease. Genetic engineering has potential to minimize and control animal diseases—a critical component of animal welfare. Swine research efforts are underway to free the swine industry of porcine reproductive and respiratory syndrome virus (PRRSV). Vaccines have not reduced the prevalence of this viral disease in pigs which results in producers having to depopulate their farm of all pigs after an outbreak of PRRSV (see Niu et al., 2017; Prather et al., 2017). CRISPR-Cas9 (Clustered Regularly Interspaced Short Palindromic Repeats) and Cas9 (Cas9 is a biotechnology that allows modification of DNA to correct genetic mutations associated with diseases including Down syndrome, spina bifida, anencephaly, and Turner and Klinefelter syndromes in humans). CRISPR Cas9 is a biotechnology used to delete proteins on cell membranes to make the cell and whole animal resistant to infection. Thus, genetic engineering is a valuable tool to create animals resistant to disease. These technologies decrease production costs, enhance sustainable agriculture and food security, and improve animal welfare.

Biotechnologies for enhancing animal health and animal production. The goal of sequencing and mapping genomes of livestock is to establish linkages between inheritance of a desirable trait (e.g., milk yield), and segregation of specific genetic markers coupled to that trait. The genetic "tools" to accomplish this are increasingly sophisticated and include marker-assisted selection based on quantitative trait loci (QTL), identification of a single-nucleotide polymporphism (SNP) within QTL, gene editing, and genetic modification. A QTL may serve as a marker associated with a gene(s) of interest, for resistance to disease or a production trait. For example, the difference in size of dogs (e.g., Great Dane versus Chihuahuas) is due, in part, to differences in frequency of a single allele of insulin-like growth factor 1 (IFG1) (Sutter et al., 2007). Thus, minor changes in gene expression can have large impacts on animal growth, health, productivity, and behavior.

There are also genetic markers (QTL) for production traits in dairy cattle (see Weller and Ron, 2011), litter size in swine (King et al., 2003), and twinning rate in beef cows (Allan et al., 2007). Current technologies coupled with biopsy and genetic analyses of preimplantation blastocysts allows for selection of blastocysts with the desired genotype to enhance genetic progress in breeding programs. In addition, sexing of embryos and sorting of X-bearing and Y-bearing sperm will also allow for producers to select the gender of their offspring—which can have large implications for welfare—particularly in dairy cattle, as dairy steers require different management.

Genomics biology has moved beyond sequencing of the genome to defining desirable gene products across different tissues or conditions (Mortazavi et al., 2008). Therefore, it is used to monitor gene expression for cell growth and differentiation, track gene expression changes during development, and assess differences in gene expression among different tissues. That information generated is used to advance understanding of genes associated with development, normal physiological changes, differences between diseased and normal tissues, and classification of disease states (Wang et al., 2009). Copy number variation (CNVs) refer to differences in the number of copies of a gene due to deletions or duplications of genes. Knowledge of CNVs influences the amount of a gene product that may influence resistance to diseases or desired production traits in livestock (Conrad et al., 2010).

The use of genome-based biotechnologies to enhance both animal health and animal production characteristics is desirable because natural biological variation can be harnessed and used to its full potential within a population. This is done by comparing genomics among animals that are resistant to or susceptible to disease or that have high- versus low-production traits (e.g., milk yield) and then using genetic markers to select for the desired phenotype. Then the use of genomic markers in selection of genotypes that favor the desired phenotype (health status or production traits) avoids controversy associated with cloning and transgenic animals. Genome marker-assisted selection can be applied to large populations of animals used in food production enterprises. This approach benefits the animal as well as the social acceptability of animal agriculture.

Alternative Interpretations of Biotechnologies: Exploiting the Microbiome in Cattle

Bioactive molecules in colostrum and the gut microbiome: effects on health of the neonate. Calves acquire passive immunity and resistance to diseases through the first milk or colostrum that is consumed within 24–48 hours after birth. The volume and quality of colostrum ingested is important during the first 36–48 hours after birth to promote passive immunity and to affect maturation of the gut. Adequate colostrum intake will decrease morbidity and mortality, as well as increase gut maturation and development of reproductive organs. This is due to the establishment of a population of bacteria in the gut referred to as the gut mass microbiome and those bacteria are favorable for the health of calves.

The gut microbiome contributes to the nutritional status, immune function, and psychological well-being of the newborn calf, for example, because the bacteria metabolize dietary nutrients, inhibit pathogen colonization, regulate immune processes, and produce neural signals (see Hinde and Lewis, 2015). Microbes present in breast milk affect development of the neonatal gut and those derived from surface skin of the mother's mammary gland are taken into the gut while the neonate is suckling. Human milk, for example, has a great diversity of sugars that are preferred nutrients for a species of bacteria known to promote a healthy gut in infants. These sugars are metabolized to fatty acids and other molecules used as energy by bacteria in the neonatal gut to aid the infant's immune system and create an environment that is hostile to some pathogens.

Maternal effects mediated via colostrum also influence development and phenotype of animals (see Bartol et al. 2017). For example, bioactive molecules in colostrum have significant long-term effects on the infants even though they are only transferred from mother to neonate during the first 24–36 hours after birth. In pigs, sheep, cows, and mice, for example, disruption of colostrum-dependent development during neonatal life can have lasting effects on uterine growth and reproductive performance as adults. Maturation of the male reproductive tract of pigs is also stimulated by the intake of colostrum as is gut maturation in both sexes of mammals.

Weaning refers to the transfer of the calf to a solid and more fibrous diet consumed by either self-feeding or grazing. Weaning is more stressful for beef than dairy calves, but should be done in beef and dairy calves being shifted to a solid or fibrous diet only when the ruminant digestive system has developed sufficiently to enable the calf to maintain growth and health from a solid feed or forage based diet. Weaning may be accomplished by abrupt separation of cow and calf, fence-line separation of cow and calf or placement of a device on the nose of the calf to discourage suckling. Abrupt weaning is especially stressful and often associated with additional stressors including transportation, vaccination, dietary changes, and social mixing—all of which increase the risk of morbidity and mortality. Ranchers and dairy managers should consider these factors when implementing best practices as to the time and method of weaning for their type of cattle and production system. Sound management decisions will decrease morbidity and mortality rates for calves, promote calm and safe cattle movement, physical appearance, and reduce adverse effects on health and productivity.

PUBLIC PERCEPTIONS AND ETHICS OF ANIMAL BIOTECHNOLOGY

Animal biotechnology is not a new concept, nor is it a new tool in our quest to feed the future and create a sustainable food system. Precision breeding is commonplace and the capacity to generate transgenic animals has existed for over 30 years. Irrespective of the trans-generational implementation of biotechnology in animal agriculture, the general public is severely uninformed regarding the role of biotechnology in their food (Hallman et al., 2016). Public acceptance of biotechnology varies across cultures and continents, and many criticisms of biotechnology include the perception that biotechnology is either intervening in "nature's order" or that the product of the biotechnology presents a risk to the environment and the humans that consume the product. The human response to products of biotechnology may be influenced by cultural issues or affluence and the level of attention to biotechnology and food.

In his 1988 paper "Genetic Engineering Biotechnology: Animal Welfare and Environmental Concerns," Michael Fox describes the development of "super animals" that are genetically engineered to create cattle weighing over 10,000 pounds, pigs that are 12-feet long and 5-feet high through the use of genetic engineering. Contrary to his 1988 predictions, the use of biotechnology has not created "super animals" and scientists are not "playing God"; conversely, many biotechnological advancements have been developed and refined to optimize the animal's biological system. However, this imagery propagates the misconceptions regarding the role of biotechnology in agriculture, and has been recently depicted in the movie "Okja," a story about a genetically modified (GM) hippo-sized pig developed out of corporate greed to provide more food to more people across the globe. While the story is fictional, the premise behind the film propagates the myth that humans are utilizing biotechnology for corporate greed and without regard for the animal. Increased production per unit of land based on larger animals is possible, but more production units of smaller sized animals may also increase production per unit of land. The misrepresentation of how biotechnology is integrated into our food system and the public lack of understanding of how agricultural biotechnology benefits the individual consumer impacts consumer acceptance of these practices.

Public acceptance of biotechnology in agriculture varies across continents, countries, and cultures. Northern and central European countries do not believe the benefits of biotechnology outweigh the risks of implementing them, while most of the public will accept transgenic food in southern European countries (Costa-Font et al., 2008). Attitudes toward biotechnology are less negative in the United States and Canada compared to European attitudes (Gaskell et al., 1999), and countries (e.g., China) in which the middle class is growing and there is an increasing need to produce more high-quality protein and use of devoted governmental resources to further develop agricultural biotechnologies (Tizard et al., 2016). Attitudes and acceptance of biotechnologies appear to be influenced by the country's need for food safety and security. This suggests that even if there are opponents to the implementation of agricultural biotechnology, their basic needs to establish food security outweigh any concerns about the risks and ethical implications of utilizing these new technologies.

Biotechnology and society. There are at least five overarching societal concerns regarding the role of biotechnology in animal agriculture. The concerns are stated as follows: (1) can anything theoretically go wrong with any of the technologies; (2) are food and other products of animal biotechnology, whether genetically engineered or not, or from clones, substantially different from those derived using traditional management of food animals; (3) what are the effects of biotechnologies on the environment; (4) how do animal biotechnologies affect animal welfare; and (5) do governmental regulatory agencies have the expertise and capacity to fully assess the risks associated with implementation of new biotechnologies? Animal biotechnology has enhanced production animal agriculture and aquaculture including the development of transgenic salmon soon to be commercialized. The question is whether these scientific advancements will be approved by governmental regulatory agencies and whether they will be accepted by society.

EnviroPig. Judicious implementation of biotechnology has the potential to enhance animal welfare, protect the environment and promote the overall sustainability of animal agriculture. However, commercial adoption of these biotechnologies is slow. The Precautionary Principle dictates that a new methodology or product not be implemented on a wide scale without intense scrutiny to ensure that any risks associated with adopting this technology are objectively evaluated and balanced against all other alternatives. This ideology slows the progress of biotechnology implementation in agriculture, and is particularly slow to be adopted in animal agriculture as the *telos* of the animal must be included in the assessment of risk.

For example, a strain of pigs developed in 2000 makes better use of phosphorus with the goal of reducing the concentrations of phosphorus in waters downstream of commercial piggeries. High concentrations of phosphorus in water deplete oxygen levels, kill aquatic life, and create algae blooms in lakes and rivers. The researchers identified a gene that increases recovery of phosphorus in the digestive tract of pigs and, therefore, decreases phosphorus in feces of pigs. These GM pigs were named EnviroPigs (see Taylor, 2000) and the responsible gene encoded for a phytase enzyme found in bacteria. The phytase enzyme in the digestive tract of pigs is very active in the acidic conditions of the gut and break down the phosphorus in feed grain prior to reaching the intestine.

Subsequent research efforts confirmed that the transgenic pigs were healthy and that the phytase transgene was transmitted to eight subsequent generations. Studies of EnviroPigs through the eighth generation revealed that the phystase transgene was transferred to all subsequent generations without change in the structure of the gene. The EnviroPig addresses both an environmental concern and a societal challenge of pig farming. The EnviroPig can help pig farmers comply with "zero discharge" rules in the United States that do not allow nitrogen or phosphorus runoff from animal operations (see Minard, 2010). However, the use of this strain of pig in commercial pork production is not yet approved. Pork producers are waiting to learn whether EnviroPig passes safety tests and if the cost-benefit analysis is such that Enviropig will have a long-term positive impact on the swine industry. Pork producers favor technologies and biotechnologies that increase their competitiveness; however, to date—no transgenic mammal has been approved for consumption in the United States.

AquAdvantage salmon. AquaAdvantage salmon is a GM Atlantic salmon developed by AquaBounty Technologies (see Ahrens et al., 2011). A growth hormone- regulating gene from a Pacific Chinook salmon, with a promoter from an ocean pout, was added to the Atlantic salmon's 40,000 genes. This gene enables AquaAdvantage salmon to grow all year rather than just spring and summer. The transgene increases the growth rate of the AquaAdvantage salmon to its mature size, but without affecting its ultimate size or other qualities compared to non-transgenic salmon. The fish grows to market size in 16–18 months rather than 3 years.

Society and, therefore, regulatory agencies and legislative committees have deemed the use of biotechnology to be more widely accepted in salmon than in pigs. On November 25, 2013, AquaBounty, Technologies announced that the Canadian Government decided that AquAdvantage® Salmon is not harmful to the environment or human health when produced in confined facilities. Furthermore, it was recognized that AquaBounty's hatchery producing sterile, all female eggs is no longer solely a research facility, an approved facility to produce eggs on a commercial scale without harm to the environment or human health.

The United States of America's Food and Drug Administration (FDA) approved AquaBounty Technologies' application to sell the AquAdvantage salmon to U.S. consumers on November 19, 2015. However, a rider to a spending bill signed into law on December 18, 2015 by President Obama bans its import until the FDA mandates labels for the GM product (see Dennis, 2015). Thus, for the first time, a GM animal has been approved to enter the food supply in the United States. It took nearly 20 years for the FDA to approve AquAdvantage salmon as equally safe and nutritious to eat as non-transgenic Atlantic salmon. The FDA reviewed data from the company and concluded that the allergenic potency of AquaAdvantage salmon was not significantly different from that of unmodified salmon and a proposed federal spending bill was created that requires consumer notification that the fish is GM.

The differences in social acceptability between the EnviroPig and the AquaAdvantage Salmon is surprising. The EnviroPig provides substantial benefits to the collective good and the probability that the EnviroPig would interbreed with wild hog populations is extremely low. However, humans have a greater capacity to empathize and communicate with pigs, which may be contributing to the delay in product approval.

Empathizing with fish is more challenging as they do not have similar facial expressions or morphology and, therefore, humans have difficulty communicating with these animals. Not until recently have scientists been able to demonstrate that fish are capable of feeling pain. Recent reports that GM salmon have been escaping from fisheries into environments where they are not native, thus causing a GM animal to also become an invasive species is not true as the transgenic salmon are sterile. Significant biotechnological advancements that will enhance the food supply, animal welfare, and the environment are possible, yet ironically, their adoption and subsequent implementation into commercial food production is challenging due to social resistance to modifying mammals, while these societal hesitations are less pronounced with regard to fish. Legislators must objectively evaluate the impact of these biotechnologies on the environment, wildlife, animal welfare, food security, and productivity in the context in which they are implemented—not just within the context of how well they can empathize with the animal's experience.

Biotechnology and animal welfare. Detractors to the implementation of biotechnology express concerns regarding the risks to the environment and wildlife populations should GM animals either interbreed or interact with wildlife. Concerns have been raised regarding the unknown health risks to humans that consume GM animals and animal products, and some argue that genetic modification compromises the genetic integrity and inherent value of the animals, thus intentionally altering them is morally wrong. Statements such as this are true, and, get people upset. It has been documented that beef from animals implanted with growth promoting hormone implants has 40% or more estrogen than beef from non-implanted animals; however, what is not usually provided is that the levels of estrogen per 3-ounce beef serving may range from 0.85 to 1.20 ng. Blair (2015) reported estrogenic activity from several foods and showed that 3-ounce servings of peanuts provided 17,010 ng and tofu provided 19,306,004 ng of estrogenic activity from isoflavins. Most birth control pills provide 20,000–50,000 ng of estrogen daily (FDA, 2017). Stanford et al. (2010) found estradiol equivalents per serving to be 0.19 ng in cow milk, 1.0 ng in coffee, 50 ng in beer, and 1,000 ng in soy milk. It may be that many naturally occurring hormones in non-animal foods need as much or more scrutiny as hormone levels in animal-based foods.

Many of these ethical concerns regard the impact of GM animals on human and environmental health, yet few realize how their own bodies metabolize ingested compounds. The irony of this is that we eat a lot of DNA daily, some people express concern about eating meat from a hormone-treated animal yet give little attention to the dosage of hormones in birth control pills with hundreds of time the active ingredient dosage as might occur in animal products. Therefore, concerns regarding the implementation of biotechnologies should not be applied carte blanche. Context-specific, objective, and biologically appropriate considerations should be made regarding the impact of these technologies on the animal itself, and its subsequent welfare when determining whether implementation of this technology is suitable for commercialization.

The role of biotechnology in food security should be addressed in this chapter. Without the advancements available from biotechnology, the costs of animal health and disease management will increase. Entire industries (e.g., the citrus industry utilizes biotechnology to eliminate greening in oranges thus increasing the efficiency of the citrus industry) may be completely eliminated. Furthermore, abandonment of biotechnological use will limit our ability to optimize production efficiencies and mitigating production costs in developing countries. Therefore, without the advancements and benefits of biotechnology, we may have trouble meeting the food needs of the future.

Again, considerations regarding the implications of biotechnology and animal welfare are not new (Broom, 1993) and the scientific literature is rampant with examples of how biotechnology has influenced our expectations regarding animal husbandry and welfare. In the 1980s, much interest and effort was devoted to cloning of cattle by embryo nuclear transfer. In several instances, extreme variation was associated with birth weights in the clones resulting in a large offspring (fetal overgrowth) syndrome and extreme variation was also possible within clone-mates (Wilson et al., 1995; Young et al., 1998). This phenomenon was also reported for calves produced through *in vitro* fertilization. More recently, variability in methylation patterns has been associated with fetal overgrowth syndrome in cattle (Hiendleder et al., 2004), and wide discrepancies in methylation patterns of genes have been shown among cattle clones (Bourc'his et al., 2001; de Montera et al., 2010). Development of *in vitro* cultured embryos has also resulted in "Large Offspring Syndrome" calves where gestation lengths are longer. Thus, the offspring grow to a size that makes calving difficult, and many of the offspring possess congenital abnormalities. The implementation of this biotechnology before its components were understood presented welfare risks to both the offspring and the dam. Animals that are developed for disease modeling may become intentionally sick which is contrary to "Freedom from disease and injury." Furthermore, if the concept of animal welfare is moving toward providing a "life worth living," then breeding animals to be chronically ill or to knowingly undergo a difficult pregnancy is in direct contradiction to efforts to enhance animal welfare.

We are just beginning to understand factors such as epigenetics and modification of DNA that influence both animal production and animal health. Genetic modification can result in a wide variety of phenotypic consequences. While scientists can intentionally alter genetic material, they have limited control on how genetic material is altered and when the alteration goes into effect. This unpredictability of when and how genetic modifications will be presented phenotypically can have implications for animal managers, the efficacy of the technology, and the impact of the genetic modification on future generations. This lack of knowledge regarding when and if welfare problems will manifest provides animal managers additional challenges.

Animals modified by biotechnology may require different management, may have different capacities to cope with stress and may have to work harder to cope with stressors. Conversely, they may be able to cope better with stressors, may respond less severely to these stressors, and may subsequently have better welfare. The unknown consequences of genetic manipulation or biotechnology implementation on the sensory functioning, bone or muscle structure, hormone production, and neural control presents a scenario in which animal managers must evaluate these animals objectively and be willing to make management changes as needed. Therefore, animal managers must evaluate animals that are either GM or treated with biotechnology differently than those that are not—as their biological systems may be operating differently. When necessary, management practices, human behavior toward animals, and resource provision should be altered to accommodate any observed changes so that these animals can experience a welfare state comparable to or better than their unmodified counterparts.

REFERENCES

Ahrens, R. N., and R. H. Devlin. 2011. Standing genetic variation and compensatory evolution in transgenic organisms: a growth-enhanced salmon simulation. *Transgenic Research* 20(3):583–597.

Allan, M., R. Thallman, R. Cushman, S. Echternkamp, S. White, L. Kuehn, E. Casas, and T. Smith. 2007. Association of a single nucleotide polymorphism in with growth traits and twinning in a cattle population selected for twinning rate. *Journal of Animal Science* 85(2):341–347.

Archer, J., P. Arthur, R. Herd, P. Parnell, and W. Pitchford. 1997. Optimum postweaning test for measurement of growth rate, feed intake, and feed efficiency in British breed cattle. *Journal of Animal Science* 75(8):2024–2032.

Arthur, P., G. Renand, and D. Krauss. 2001. Genetic and phenotypic relationships among different measures of growth and feed efficiency in young Charolais bulls. *Livestock Production Science* 68(2):131–139.

Bailey, M. T., S. E. Dowd, J. D. Galley, A. R. Hufnagle, R. G. Allen, and M. Lyte. 2011. Exposure to a social stressor alters the structure of the intestinal microbiota: implications for stressor-induced immunomodulation. *Brain Behavior and Immunology* 25(3):397–407.

Bartol, F., A. Wiley, A. George, D. Miller, and C. Bagnell. 2017. Physiology and endocrinology symposium: postnatal reproductive development and the lactocrine hypothesis. *Journal of Animal Science* 95(5):2200–2210.

Blair, A. 2015. Hormones in beef: myth vs. fact. South Dakota State University Extension. http://igrow.org/livestock/beef/hormones-in-beef-myth-vs.-fact/.

Boissy, A., A. D. Fisher, J. Bouix, G. N. Hinch, and P. Le Neindre. 2005. Genetics of fear in ruminant livestock. *Livestock Production Science* 93(1):23–32.

Bourc'His, D., D. Le Bourhis, D. Patin, A. Niveleau, P. Comizzoli, J.-P. Renard, and E. Viegas-Pequignot. 2001. Delayed and incomplete reprogramming of chromosome methylation patterns in bovine cloned embryos. *Current Biology* 11(19):1542–1546.

Brambell, F. W. R. 1970. Report of the technical committee to enquire into the welfare of animals kept under intensive livestock husbandry systems: presented to parliament by the secretary of state for Scotland and the minister of agriculture, fisheries and food by command of her majesty December, 1965. HM Stationery Office.

Bravo, J. A., M. Julio-Pieper, P. Forsythe, W. Kunze, T. G. Dinan, J. Bienenstock, and J. F. Cryan. 2012. Communication between gastrointestinal bacteria and the nervous system. *Current Opinion in Pharmacology* 12(6):667–672.

Breuer, K., M. E. M. Sutcliffe, J. T. Mercer, K. A. Rance, N. E. O'Connell, I. A. Sneddon, and S. A. Edwards. 2005. Heritability of clinical tail-biting and its relation to performance traits. *Livestock Production Science* 93(1):87–94.

Broom, D. 1993. Assessing the welfare of modified or treated animals. *Livestock Production Science* 36(1):39–54.

Broom, D. M. 1986. Indicators of poor welfare. *British Veterinary Journal* 142(6):524–526.

Burdick, N., R. Randel, J. Carroll, and T. Welsh. 2011. Interactions between temperament, stress, and immune function in cattle. *International Journal of Zoology* 2011:9.

Carlson, D. F., C. A. Lancto, B. Zang, E.-S. Kim, M. Walton, D. Oldeschulte, C. Seabury, T. S. Sonstegard, and S. C. Fahrenkrug. 2016. Production of hornless dairy cattle from genome-edited cell lines. *Nature Biotechnology* 34(5):479–481.

Chiquette, J., M. J. Allison, and M. A. Rasmussen. 2008. Prevotella bryantii 25A used as a probiotic in early-lactation dairy cows: effect on ruminal fermentation characteristics, milk production, and milk composition. *Journal of Dairy Science* 91(9):3536–3543.

Cockrum, R. R., S. E. Speidel, J. L. Salak-Johnson, C. C. L. Chase, R. K. Peel, R. L. Weaber, G. H. Loneagan, J. J. Wagner, P. Boddhireddy, M. G. Thomas, K. Prayaga, S. DeNise, and R. M. Enns. 2016. Genetic parameters estimated at receiving for circulating cortisol, immunoglobulin G, interleukin 8, and incidence of bovine respiratory disease in feedlot beef steers1. *Journal of Animal Science* 94(7):2770–2778.

Conrad, D. F., D. Pinto, R. Redon, L. Feuk, O. Gokcumen, Y. Zhang, J. Aerts, T. D. Andrews, C. Barnes, and P. Campbell. 2010. Origins and functional impact of copy number variation in the human genome. *Nature* 464(7289):704–712.

Cook, R., J. Popp, J. Kastelic, S. Robbins, and R. Harland. 2000. The effects of active immunization against GnRH on testicular development, feedlot performance, and carcass characteristics of beef bulls. *Journal of Animal Science* 78(11):2778–2783.

Costa-Font, M., J. M. Gil, and W. B. Traill. 2008. Consumer acceptance, valuation of and attitudes towards genetically modified food: review and implications for food policy. *Food Policy* 33(2):99–111.

Cryan, J. F., and T. G. Dinan. 2012. Mind-altering microorganisms: the impact of the gut microbiota on brain and behaviour. *Nature Reviews Neuroscience* 13(10):701–712.

Curley, K., J. Paschal, T. Welsh, and R. Randel. 2006. Technical note: exit velocity as a measure of cattle temperament is repeatable and associated with serum concentration of cortisol in Brahman bulls. *Journal of Animal Science* 84(11):3100–3103.

Dawkins, M. S. 2003. Behaviour as a tool in the assessment of animal welfare. *Zoology* 106(4):383–387.

Ribeiro, E., J. Hernandez, E. Zanella, M. Shimokomaki, S. Prudêncio-Ferreira, E. Youssef, H. Ribeiro, R. Bogden, and J. Reeves. 2004. Growth and carcass characteristics of pasture fed LHRH immunocastrated, castrated and intact Bos indicus bulls. *Meat Science* 68(2):285–290.

De Montera, B., D. El Zeihery, S. Müller, H. Jammes, G. Brem, H.-D. Reichenbach, F. Scheipl, P. Chavatte-Palmer, V. Zakhartchenko, and O. J. Schmitz. 2010. Quantification of leukocyte genomic 5-methylcytosine levels reveals epigenetic plasticity in healthy adult cloned cattle. *Cellular Reprogramming (Formerly"Cloning and Stem Cells")* 12(2):175–181.

Dennis, B. 2015. FDA must develop plan to label genetically engineered salmon, Congress says *The Washington Post.*

Desbonnet, L., L. Garrett, G. Clarke, J. Bienenstock, and T. G. Dinan. 2008. The probiotic Bifidobacteria infantis: an assessment of potential antidepressant properties in the rat. *Journal of Psychiatric Research* 43(2):164–174.

Dorshorst, B. 2014. Half of Holstein heifer calves could be polled by 2034. Progressive Dairyman.

Duncan, I., and J. C. Petherick. 1991. The implications of cognitive processes for animal welfare. *Journal of Animal Science* 69(12):5017–5022.

Ezenwa, V. O., N. M. Gerardo, D. W. Inouye, M. Medina, and J. B. Xavier. 2012. Animal behavior and the microbiome. *Science* 338(6104):198–199.

FDA. 2017. Drugs@FDA: FDA approved drug products searchable database. U.S. Food & Drug Administration, U.S. Department of Health and Human Services. www.accessdata.fda.gov/scripts/cder/daf/.

Flint, J. F., and M. R. Garner. 2009. Feeding beneficial bacteria: a natural solution for increasing efficiency and decreasing pathogens in animal agriculture. *The Journal of Applied Poultry Research* 18(2):367–378.

Foster, J. A., and K.-A. McVey Neufeld. 2013. Gut–brain axis: how the microbiome influences anxiety and depression. *Trends in Neurosciences* 36(5):305–312.

Fraser, D. 2003. Assessing animal welfare at the farm and group level: the interplay of science and values. *Animal Welfare* 12(4):433–443.

Fraser, D. 2008. Understanding animal welfare. *Acta Veterinaria Scandinavica* 50(1):1.

Gaskell, G., M. W. Bauer, J. Durant, and N. C. Allum. 1999. Worlds apart? The reception of genetically modified foods in Europe and the US. *Science* 285(5426):384–387.

Gauly, M., H. Mathiak, K. Hoffmann, M. Kraus, and G. Erhardt. 2001. Estimating genetic variability in temperamental traits in German Angus and Simmental cattle. *Applied Animal Behaviour Science* 74(2):109–119.

Hallman, W., C. Cuite, and X. Morin. 2016. Public perceptions of animal-sourced genetically modified food products. *Journal of Animal Science* 94(5):216–216.

Havenstein, G., P. Ferket, S. Scheideler, and B. Larson. 1994. Growth, livability, and feed conversion of 1957 vs 1991 broilers when fed "typical" 1957 and 1991 broiler diets. *Poultry Science* 73(12):1785–1794.

Hemsworth, P. H., G. J. Coleman, J. L. Barnett, and S. Borg. 2000. Relationships between human-animal interactions and productivity of commercial dairy cows. *Journal of Animal Science* 78(11):2821–2831.

Hiendleder, S., C. Mund, H.-D. Reichenbach, H. Wenigerkind, G. Brem, V. Zakhartchenko, F. Lyko, and E. Wolf. 2004. Tissue-specific elevated genomic cytosine methylation levels are associated with an overgrowth phenotype of bovine fetuses derived by in vitro techniques. *Biology of Reproduction* 71(1):217–223.

Hinde, K., and Z. T. Lewis. 2015. Mother's littlest helpers. *Science* 348(6242):1427–1428.

Jami, E., B. A. White, and I. Mizrahi. 2014. Potential role of the bovine rumen microbiome in modulating milk composition and feed efficiency. *PLOS One* 9(1):e85423.

King, A.H., Jiang, Z., Gibson, J.P., Haley, C.S. and Archibald, A.L. 2003. Mapping quantitative trait loci affecting female reproductive traits on porcine chromosome 8. *Biology of Reproduction* 68 (6):2172-2179.

King, D., C. S. Pfeiffer, R. Randel, T. Welsh, R. Oliphint, B. Baird, K. Curley, R. Vann, D. Hale, and J. Savell. 2006. Influence of animal temperament and stress responsiveness on the carcass quality and beef tenderness of feedlot cattle. *Meat Science* 74(3):546–556.

Kirkden, R. D., and E. A. Pajor. 2006. Using preference, motivation and aversion tests to ask scientific questions about animals' feelings. *Applied Animal Behaviour Science* 100(1):29–47.

Liu, G., E. Dunnington, and P. Siegel. 1995. Correlated responses to long-term divergent selection for eight-week body weight in chickens: growth, sexual maturity, and egg production. *Poultry Science* 74(8):1259–1268.

Løvendahl, P., L. H. Damgaard, B. L. Nielsen, K. Thodberg, G. Su, and L. Rydhmer. 2005. Aggressive behaviour of sows at mixing and maternal behaviour are heritable and genetically correlated traits. *Livestock Production Science* 93(1):73–85.

Marti, S., J. Jackson, N. Slootmans, E. Lopez, A. Hodge, M. Pérez-Juan, M. Devant, and S. Amatayakul-Chantler. 2017. Effects on performance and meat quality of Holstein bulls fed high concentrate diets without implants following immunological castration. *Meat Science* 126:36–42.

Martin Manuel, P., B. Elena, M. G. Carolina, and P. Gabriela. 2017. Oral probiotics supplementation can stimulate the immune system in a stress process. *Journal of Nutrition and Intermediary Metabolism* 8:29–40.

McCoy, M., G. Reilly, and D. Kilpatrick. 1996. Density and breaking strength of bones of mortalities among caged layers. *Research in Veterinary Science* 60(2):185–186.

Mellor, D. 2016. Updating animal welfare thinking: moving beyond the "Five Freedoms" towards "A Life Worth Living". *Animals* 6(3):21.

Mellor, D., and C. Reid. 1994. *Improving the well-being of animals in the research environment*. Australian and New Zealand Council for the Care of Animals in Research and Teaching (ANZCCART):3–18.

Miller, L., P. Siegel, and E. Dunnington. 1992. Inheritance of antibody response to sheep erythrocytes in lines of chickens divergently selected for fifty-six–day body weight and their crosses. *Poultry Science* 71(1):47–52.

Minard, A. 2010. *Pigs modified to excrete less phosphorus win limited approval in Canada*. National Geographic News.

Mormède, P., S. Andanson, B. Aupérin, B. Beerda, D. Guémené, J. Malmkvist, X. Manteca, G. Manteuffel, P. Prunet, C. G. van Reenen, S. Richard, and I. Veissier. 2007. Exploration of the hypothalamic–pituitary–adrenal function as a tool to evaluate animal welfare. *Physiology and Behavior* 92(3):317–339.

Mortazavi, A., B. A. Williams, K. McCue, L. Schaeffer, and B. Wold. 2008. Mapping and quantifying mammalian transcriptomes by RNA-Seq. *Nature Methods* 5(7):621–628.

Myer, P. R., H. C. Freetly, J. E. Wells, T. P. L. Smith, and L. A. Kuehn. 2017. Analysis of the gut bacterial communities in beef cattle and their association with feed intake, growth, and efficiency. *Journal of Animal Science* 95(7):3215–3224.

Nääs, I. A., R. G. Garcia, and F. R. Caldara. 2014. Infrared thermal image for assessing animal health and welfare. *Journal of Animal Behaviour and Biometeorology* 2(3):66–72.

Nasr, M., J. Murrell, L. Wilkins, and C. Nicol. 2012a. The effect of keel fractures on egg-production parameters, mobility and behaviour in individual laying hens. *Animal Welfare* 21(1):127.

Nasr, M. A., C. J. Nicol, and J. C. Murrell. 2012b. Do laying hens with keel bone fractures experience pain? *PLoS One* 7(8):e42420.

Niu, D., H.-J. Wei, L. Lin, H. George, T. Wang, I.-H. Lee, H.-Y. Zhao, Y. Wang, Y. Kan, and E. Shrock. 2017. Inactivation of porcine endogenous retrovirus in pigs using CRISPR-Cas9. *Science* 357(6357):1303–1307.

Nkrumah, J., D. Crews, J. Basarab, M. Price, E. Okine, Z. Wang, C. Li, and S. Moore. 2007. Genetic and phenotypic relationships of feeding behavior and temperament with performance, feed efficiency, ultrasound, and carcass merit of beef cattle. *Journal of Animal Science* 85(10):2382–2390.

O'Callaghan, T. F., R. P. Ross, C. Stanton, and G. Clarke. 2016. The gut microbiome as a virtual endocrine organ with implications for farm and domestic animal endocrinology. *Domestic Animal Endocrinology* 56(Supplement):S44–S55.

Oltenacu, P. A., and D. M. Broom. 2010. The impact of genetic selection for increased milk yield on the welfare of dairy cows. *Animal Welfare* 19(1):39–49.

Perentos, N., A. U. Nicol, A. Q. Martins, J. E. Stewart, P. Taylor, and A. J. Morton. 2017. Techniques for chronic monitoring of brain activity in freely moving sheep using wireless EEG recording. *Journal of Neuroscience Methods* 279:87–100.

Prather, R. S., K. M. Whitworth, S. K. Schommer, and K. D. Wells. 2017. Genetic engineering alveolar macrophages for host resistance to PRRSV. *Veterinary Microbiology* 209:124–129.

Quality, W. 2009. Welfare quality® assessment protocol for cattle. Welfare Quality® Consortium, Lelystad, Netherlands p180.

Rao, A. V., A. C. Bested, T. M. Beaulne, M. A. Katzman, C. Iorio, J. M. Berardi, and A. C. Logan. 2009. A randomized, double-blind, placebo-controlled pilot study of a probiotic in emotional symptoms of chronic fatigue syndrome. *Gut Pathogens* 1(1): 6.

Ravagnolo, O., and I. Misztal. 2000. Genetic component of heat stress in dairy cattle, parameter estimation. *Journal of Dairy Science* 83(9):2126–2130.

Rhee, S. H., C. Pothoulakis, and E. A. Mayer. 2009. Principles and clinical implications of the brain-gut-enteric microbiota axis. *Nature Review Gastroenterology and Hepatology* 6(5):306–314.

Riley, D., C. Gill, A. Herring, P. Riggs, J. Sawyer, D. Lunt, and J. Sanders. 2014. Genetic evaluation of aspects of temperament in Nellore–Angus calves. *Journal of Animal Science* 92(8):3223–3230.

Robertson, I., H. Fraser, G. Innes, and A. Jones. 1982. Effect of immunological castration on sexual and production characteristics in male cattle. *The Veterinary Record* 111(23):529–531.

Robinson, D., and V. Oddy. 2004. Genetic parameters for feed efficiency, fatness, muscle area and feeding behaviour of feedlot finished beef cattle. *Livestock Production Science* 90(2):255–270.

Rodenburg, T. B., and P. Koene. 2007. The impact of group size on damaging behaviours, aggression, fear and stress in farm animals. *Applied Animal Behaviour Science* 103(3):205–214.

Sandelin, B., A. Brown, Z. Johnson, J. Hornsby, R. Baublits, and B. Kutz. 2005. Postpartum maternal behavior score in six breed groups of beef cattle over twenty-five years. *The Professional Animal Scientist* 21(1):13–16.

Sather, A. 1987. A note on the changes in leg weakness in pigs after being transferred from confinement housing to pasture lots. *Animal Science* 44(3):450–453.

Schneider, M. J., R. G. Tait, M. V. Ruble, W. D. Busby, and J. M. Reecy. 2010. Evaluation of fixed sources of variation and estimation of genetic parameters for incidence of bovine respiratory disease in preweaned calves and feedlot cattle. *Journal of Animal Science* 88(4):1220–1228.

Shrode, R., and S. Hammack. 1971. Chute behavior of yearling beef cattle. *Journal of Animal Science* 33(1):193.

Snowder, G. D., L. D. Van Vleck, L. V. Cundiff, and G. L. Bennett. 2006. Bovine respiratory disease in feedlot cattle: environmental, genetic, and economic factors. *Journal of Animal Science* 84(8):1999–2008.

Spurlock, D., M. Stock, and J. Coetzee. 2014. The impact of 3 strategies for incorporating polled genetics into a dairy cattle breeding program on the overall herd genetic merit. *Journal of Dairy Science* 97(8):5265–5274.

Stanford, B. D., S. A. Snyder, R. A. Trenholm, J. C. Holaday, and B. J. Vanderford. 2010. Estrogenic activity of US drinking waters: A relative exposure comparison. *Journal of the American Water Works Association* 102:1–11.

Stratmann, A., E. K. F. Fröhlich, S. G. Gebhardt-Henrich, A. Harlander-Matauschek, H. Würbel, and M. J. Toscano. 2016. Genetic selection to increase bone strength affects prevalence of keel bone damage and egg parameters in commercially housed laying hens. *Poultry Science* 95(5):975–984.

Stricklin, W., C. Heisler, and L. Wilson. 1980. Heritability of temperament in beef cattle. *Journal of Animal Science* 51(1):109.

Sutter, N. B., C. D. Bustamante, K. Chase, M. M. Gray, K. Zhao, L. Zhu, B. Padhukasahasram, E. Karlins, S. Davis, and P. G. Jones. 2007. A single IGF1 allele is a major determinant of small size in dogs. *Science* 316(5821):112–115.

Swanson, A. 2014. Wanted: more bulls with no horns NPR. *The Salt*. www.npr.org/sections/the-salt/2015/08/03/429024245/wanted-more-bulls-with-no-horns.

Taylor, D. A. 2000. A less polluting pig. *Environmental Health Perspective* 108(1):A14.

Tizard, M., E. Hallerman, S. Fahrenkrug, M. Newell-McGloughlin, J. Gibson, F. de Loos, S. Wagner, G. Laible, J. Y. Han, and M. D'Occhio. 2016. Strategies to enable the adoption of animal biotechnology to sustainably improve global food safety and security. *Transgenic Research* 25(5):575–595.

Turner, S. P., R. Roehe, R. B. D'Eath, S. H. Ison, M. Farish, M. C. Jack, N. Lundeheim, L. Rydhmer, and A. B. Lawrence. 2009. Genetic validation of postmixing skin injuries in pigs as an indicator of aggressiveness and the relationship with injuries under more stable social conditions1. *Journal of Animal Science* 87(10):3076–3082.

USDA-APHIS. 2013. Types and costsof respiratory disease treatments in U.S. Feedlots. In: *Veterinary Services Info Sheet*, Safeguarding American Agriculture

Uyeno, Y., S. Shigemori, and T. Shimosato. 2015. Effect of probiotics/prebiotics on cattle health and productivity. *Microbes and Environments* 30(2):126–132.

Voisinet, B., T. Grandin, J. Tatum, S. O'connor, and J. Struthers. 1997. Feedlot cattle with calm temperaments have higher average daily gains than cattle with excitable temperaments. *Journal of Animal Science* 75(4):892–896.

Wang, Z., M. Gerstein, and M. Snyder. 2009. RNA-Seq: a revolutionary tool for transcriptomics. *Nature Review Genetics* 10(1):57–63.

Webb, A., W. Russell, and D. Sales. 1983. Genetics of leg weakness in performance-tested boars. *Animal Science* 36(1):117–130.

Weller, J., and M. Ron. 2011. Invited review: quantitative trait nucleotide determination in the era of genomic selection. *Journal of Dairy Science* 94(3):1082–1090.

Wiley, N. C., T. G. Dinan, R. P. Ross, C. Stanton, G. Clarke, and J. F. Cryan. 2017. The microbiota-gut-brain axis as a key regulator of neural function and the stress response: implications for human and animal health. *Journal of Animal Science* 95(7):3225–3246.

Wilson, J., J. Williams, K. Bondioli, C. Looney, M. Westhusin, and D. McCalla. 1995. Comparison of birth weight and growth characteristics of bovine calves produced by nuclear transfer (cloning), embryo transfer and natural mating. *Animal Reproduction Science* 38(1–2):73–83.

Young, L. E., K. D. Sinclair, and I. Wilmut. 1998. Large offspring syndrome in cattle and sheep. *Reviews in Reproduction* 3(3):155–163.

CHAPTER 11

Providing Assurance That Cattle Have a Reasonably Good Life

Jennifer Walker
Director Milk Quality, Danone North America

Marina A. G. von Keyserlingk
University of British Columbia

CONTENTS

INTRODUCTION

The topic of farm animal welfare is a consequence of a social movement that began in the early 1960s in the United Kingdom following the publication of Ruth Harrison's book "Animal Machines" in 1964. In her book, she voiced concerns about the lives led by farm animals, particularly laying hens, broilers and veal calves, and housing environments where animals were prevented from seeing the sun. The UK public's outcry in response to her book resulted in the UK government initiating an investigation that culminated in the publication of The Brambell Report (1965) titled "Report of the Technical Committee to Enquire into the Welfare of Animals Kept under Intensive Livestock Husbandry Systems." In his report Professor Brambell argued that farm animal agriculture was based on standards of care and housing that were morally unacceptable by the public and that farm animals should have the freedom "to stand up, lie down, turn around, groom themselves and stretch their limbs." Approximately 30 years later The Brambell Report was used as the basis for developing the Five Freedoms that state that all animals must have: (1) freedom from thirst and hunger, (2) freedom from discomfort, (3) freedom from pain, injury, and disease, (4) freedom to express normal behavior and, (5) freedom from fear and distress (FAWC, 1992).

Since then, the topic of farm animal welfare has continued to gain considerable traction in the academic community, and in North America, particularly over the last 30 years (von Keyserlingk and Weary, 2017). This increased interest is due almost entirely to the increased demand for science-based solutions that address concerns raised by stakeholders, including the public. However, despite the wealth of knowledge generated by scientists on ways to improve welfare of farm animals, and that various food producer and retail groups have developed animal-welfare standards for US farms (Mench, 2008), there is evidence that many welfare issues continue to plague the food animal industries. For example, despite lameness in dairy cattle production systems being the focus of numerous scientific studies, including the identification of known risk factors (Chapinal et al., 2013; Chapinal et al., 2014a), it continues to be one of the global dairy industry's greatest welfare challenges (Cook, 2017). Equally worrisome is the fact that despite it being well-documented that routine management practices such as dehorning (Stafford and Mellor, 2011), castration (Thuer et al., 2007, Marti et al., 2010, Becker et al., 2012), and branding (Schwartzkopf-Genswein et al., 1998) are painful and that pain mitigation strategies exist, only a modest number of producers implement best management practices that address the issue of pain. For example, in 2014 only 30% of dairy farmers used pain control while dehorning calves (USDA, 2017). On a positive note, this is almost double that reported in the previous National Animal Health Monitoring survey completed in 2007 (USDA, 2009). The beef industry has taken a somewhat different approach to dehorning with a large portion of the producers using polled genetics as a means to replace the need for this painful procedure (Long and Gregory, 1978; Hoeschele, 1990), with more than 85% of beef calves born without horns in 2007, a fivefold increase since 1992 (USDA, 2008a). It is also well established that management practices such as abrupt weaning (Haley et al., 2005) combined with the stresses of transport (Flint et al., 2014) and comingling (Step et al., 2008) negatively impact the welfare of beef calves. Despite the well-known adverse effects of abrupt weaning, 50% of beef cow–calf operations sold calves immediately at weaning and only 40% vaccinated calves prior to sale (USDA, 2008).

The status of farm animal welfare has been further challenged over the last 20 years by the use of undercover videos by special interest groups. Video footage depicts excessive cruelty such as workers dragging, kicking, and electrically shocking dairy cattle unable to walk (i.e., Hallmark-Westland Meat Packing Company in Chino, California, 2008; Miller, 2014; Paul, 2015) but also routine practices that the public find abhorrent, such as tail docking and dehorning (Schecter and Ross, 2010). Given these controversies, it is not surprising that public trust in the care of farm animals is eroding (Robbins et al., 2016).

In the case of animal welfare, the values of farmers differ from those of the general public, with farmers tending to emphasis different dimensions of animal welfare than the public (Te Velde et al., 2002; Lassen et al., 2006; Vanhonacker et al., 2008; Bergstra et al., 2015). For instance, farmers and their frequently trusted advisor, the veterinarian, have historically emphasized biological functioning and health as primary determinants of animal welfare (Ventura et al., 2013; Weary et al., 2016). In contrast, the public's perspective places emphasis on the ability of an animal to live a reasonably natural life (Lassen et al., 2006; Prickett et al., 2010; Verbeke et al., 2010; Cardoso et al., 2016; see review by Clark et al., 2016). The fact that the farmer and general public's values are different could be one reason why practices persist on farms that fail to resonate with societal values and can result in distrust.

Given the growing discussions about what type of life farm animals lead it is not surprising that questions arise on how animals on a particular farm are cared for—in essence, do they have good welfare? The answer to this question is not simple. Additional questions that arise include: who sets the standards of care? Who applies the standards to a farm (and the animals) and what happens if the farm fails to meet the standard? Equally important is: how does one motivate farmers to implement best practices on farms that result in improved animal welfare?

To answer these questions, we envision a three-pronged approach. First, the development of standards and audits that reflect societal and industry concerns based on an understanding of farm animal-welfare requirements and recommended practices is crucial. These standards must reflect the latest scientific evidence and be developed and executed with transparency to be sustainable—maintaining public trust now and in the future. To maintain public trust third-party audits are needed. Results that are kept confidential will not be sufficient; animal agriculture will need to approach and embrace the topic of transparent third-party audits as these will provide external stakeholders confidence that standards are being met (Olynk et al., 2010; Wolf and Tonsor, 2017). Second, audits must result in improved animal welfare where needed. It is here where rigorous first- and second-party audits will play an important role in preparing farmers to meet animal-welfare standards (Weary and von Keyserlingk, 2017). Third, efforts must focus on producer engagement, such as benchmarking, to improve the adoption of implementing proven welfare solutions on farm.

DEVELOPING ANIMAL WELFARE STANDARDS AND PROVIDING ASSURANCE

Animal Welfare Legislation

Different countries have used different vehicles to develop farm animal-welfare standards and ensure compliance (see von Keyserlingk and Hötzel, 2015) and there are various approaches to ensuring welfare (Fraser, 2006). For example, in some parts of the world such as in the European Union and New Zealand legislation plays a central role (e.g., European Commission Directive 2001/93/EC and the 1999 Animal Welfare Act, Parliamentary Counsel Office of New Zealand (1999)).

In North America, the role of legislation has traditionally played a minor role in improving animal welfare. This is likely due in part to "animal care" being primarily governed at the province or state level, which has resulted, at least in Canada, in a non-harmonized approach to the issue of animal welfare (Fraser et al., in press). Interestingly, six Canadian provinces (British Columbia, Manitoba, New Brunswick, Newfoundland and Labrador, Prince Edward Island, and Saskatchewan) make reference to one or more of the industry-led national codes for farm animals in their regulations (Fraser et al., 2016). To the best of our knowledge, no US state makes reference to industry-led standards in state legislation.

In the US, many food animal industry groups have argued strongly against animal-welfare regulations (Fraser, 2001): a position that has frequently led to polarized debates between food animal industry lobby groups and animal advocates (Cantrell et al., 2013). This intense debate has resulted in the introduction of a number of farm animal-welfare laws via legislation or ballot initiatives in some US states. The first of this type was passed in 2002 (and took effect in 2008) in the state of Florida and resulted in the banning of gestation stalls in pigs. Since then, this political vehicle has been effective in introducing animal-welfare legislation in numerous states— banning a variety of standard industry practices. One of the most discussed ballot initiatives was California's "Proposition 2," passed in November 2008 prohibiting the use of the conventional battery cage for hens and crates for gestating sows and veal calves (and took effect in 2015). The use of the ballot initiative route in the US will likely continue in states where allowed, particularly if the animal industries are not seen as adopting practices that resonate with widely held public values. For instance, the Human Society of the United States has recently introduced a new ballot initiative in California that calls for stricter animal-welfare standards (compared to Proposition 2) in addition to requiring out-of-state producers to comply with the California regulations when selling their product in California (Duggan, 2017). Whether this ballot initiative is voted into law in California remains to be seen but regardless the pressure for changes on how farm animals are cared for will do doubt continue. In contrast to the US use of the legal system, industry-led voluntary compliance programs have played a much greater role in Canada (von Keyserlingk and Hötzel, 2015).

While some believe that improving animal welfare is guaranteed through state or federal mandates, there are concerns with the "legislative approach," primarily due to challenges of appropriate enforcement that undermines confidence (e.g., Ventura et al., 2016) and the lack of harmonization between different jurisdictions (Fraser et al., 2018). For example, in the UK, a 2014 report found that the 2013 EU regulation 1099/2009 aimed and protecting the welfare of farmed animals was not being implemented (Commission E., 2014). Although the U.S. Humane Methods of Slaughter Act (HMSA) (USDA, 1978) was passed in 1978 to provide protection of animals from undo suffering during slaughter, it was rarely enforced over the first 20 years and research clearly demonstrated that serious issues remained (Grandin, 1997; GAO, 2004).

Industry Led Animal Welfare Standards

In North America, the development of animal-welfare standards was initially driven primarily by the food animal industries, with the first program funded, developed, and administered by the pig industry, Pork Quality Assurance (1989), followed by the Beef Quality Assurance (1991) program and the United Egg Producers (UEP) who initiated the process of publishing animal husbandry guidelines in 1999 (for complete history of the UEP process please, see Mench, 2011). These industries created books, videos, and on-farm training programs aimed at promoting best practices. However, while many of the best practices were related to animal welfare the communication and incentive was largely focused on the quality of the end product, not the quality of life of the animal. This was not surprising given the links animal health has with food safely (CAST, 2012) and the historic emphasis placed by farmers and veterinarian on biological function and health (Te Velde et al., 2002; Lassen et al., 2006; Vanhonacker et al., 2008). These voluntary industry programs have been argued by some industries as ones that improve farm practices that affect carcass quality (Garcia et al., 2008), but some have questioned whether these industry-led programs will effectively improve animal welfare on farms in the long run (von Keyserlingk and Hötzel, 2015; Wolf and Tonsor, 2017). However, there is some evidence that the industries do respond more broadly to increased pressure to improve animal welfare; in 2011, the pig and beef sectors both substantively modified their original programs designed nearly two decades earlier, placing much more emphasis on animal welfare compared to the first versions. The newest version of the UEP guidelines has also been modified and now explicitly considers cage-free systems (UEP, 2017).

Additional examples of industry efforts focused on the welfare of cattle are the Dairy Farmers of Canada (DFC) and the Canadian Cattlemen's Association (CCA) who each partnered with Canada's National Farm Animal Care Committee (NFACC) to work together to create code of practice for the care and handling of dairy cattle (DFC-NFACC, 2009) and the code of practice for the care and handling of beef cattle (CCA-NFACC, 2014). The development of the NFACC codes of practice is a multistep process that begins with a scientists committee preparing a review of the contentious issues that is then sent for peer review. Once the committee has addressed the reviewers comments the document is then distributed to a range of stakeholders, including producers, scientists, government officials, veterinarians, grocery-chain distributors, and representatives of the humane movement that use it to develop a consensus based code of practice (NFACC, 2013). A draft version of the "Code" is posted for public comment; with the comments referred back to the code development committee for consideration before the final version is published. The intent of the National Farm Animal Care Code development process is that they are to be reviewed every 5 years and revised once every 10 years (NFACC, 2013).

Today, the dairy code of practice (DFC-NFACC, 2009) provides the foundation of the Animal Care module of the Canada's Pro-Action: a mandatory animal-welfare assessment that is currently being implemented in Canada and will be required of all dairy farms. As with all science-based standards of animal care the incorporation of new scientific evidence is key to the long-term sustainability of the industry. We see some risk for the Canadian Dairy Industry given that the Code is only scheduled to be revised every 10 years (NFACC, 2014) and thus may not reflect the most recent scientific evidence or societal values.

In the USA, the National Federation of Milk Producers (NMPF) created the F.A.R.M. program (Farmers Assuring Responsible Management) in 2009 (NMPF, 2017). Currently on version 3, the program guidance document is revised every 3 years. Initially drafted by an NMPF-appointed technical writing committee that includes representatives from the NMPF board of directors, various milk cooperatives, the National Cattlemen's Beef Association, an audit company as well as scientific and veterinary advisors, the draft document is then transferred to the NMPF staff who then sends it to select external stakeholders for public comment. The resulting comments are reviewed by NMPF with the final version voted on by the NMPF board of directors (NMPF, 2017).

AUDITS

What Is an Audit?

While many people are familiar with the word "audit," there is often much confusion as to how a first-party audit differs from a second-party audit or a third-party audit, particularly when applied to animal welfare in the context of the supply chain. The International Organization for Standardization (ISO) is a global federation of national standards bodies whose technical committees are responsible for drafting international standards and who we have relied on to provide clarity on this matter (ISO, 2011). Under the heading of *Quality Management and Quality Assurance* (ISO, 2011), this organization provides specific guidelines regarding the process of auditing management systems and in ISO/IEC 17021 they provide guidance on the management of an audit program. It is not our intent to provide a complete review of the ISO standards but to explain briefly some of the basic definitions formally associated with audits and auditing. An audit is a systematic, independent, and documented process for obtaining audit evidence and evaluating it objectively to determine the extent to which the audit criteria are fulfilled (ISO, 2011). Audit evidence includes records, statements of fact or other information, which are relevant and verifiable. An auditor conducts the audit to collect evidence to evaluate whether the system conforms, or fulfills specific audit requirements.

There are six key principles to auditing (ISO, 2011).

1. Integrity
2. Fair presentation—the obligation to report truthfully and accurately
3. Due professional care—the application of diligence and judgment in auditing
4. Confidentiality
5. Independence—the basis for impartiality of the audit and objectivity of the audit conclusions being free from bias and conflict of interest
6. Evidence based—reliable and reproducible audit conclusions rely on verifiable audit evidence. This requires an appropriate sample size and sample methodology.

It is important to recognize that while the first four principles are dependent on the auditors themselves, the last two are dependent on to whom the auditor is accountable and how the audit is written. In Table 11.1, we have outlined the three most common audit approaches, first party, second party, and third party, and where some of the industry audits that address farm animal welfare fall in terms of the type of approach and which principles can be met by each approach. Understanding

Table 11.1   Approaches to Auditing—Audits can be Conducted by First, Second, or Third Parties

Internal Auditing	External Auditing	
"First-party audit"	"Second-party audit"	"Third-party auditing"
Conducted by the organization itself for management to review and inform the need for improvement (e.g., any audit conducted by the herd's veterinarian or an employee of the company/farm)	Conducted by parties having an interest in the organization such as customers or by persons on behalf of customers (i.e., suppliers).	For legal, regulatory, verification, or certification purposes conducted by <u>independent</u> auditing organizations.
	Industry Programs	
Beef Quality Assurance (BQA) Pork Quality Assurance (PQA-Plus) National American Meat Institute (NAMI)-Slaughter Audit* National Milk Producers Federation (NMPF) Farmers for the Assurance of Responsible management (F.A.R.M.)—when conducted by employees of the farmer Cooperative Canadian Feedlot* when done by an employee of the farm	NAMI—Slaughter Audit* F.A.R.M.—When conducted on behalf of the milk processor Animal Care module of Pro-Action (Canada) Canadian Feed-Lot Audit*	NAMI—Slaughter Audit* Common Swine Industry Audit* Canadian Feedlot* Various Proprietary audits used for consumer facing label claims*
	Principles of Auditing Met	
Integrity	Integrity	Integrity
Fair	Fair	Fair
Professionalism	Professionalism	Professionalism
Confidentiality	Confidentiality	Confidentiality
Evidence based*	Evidence based*	Evidence based*
		Independence

Programs highlighted by an * indicate audits that are written in such a way that reported outcomes are verifiable and use appropriate sample size and methodology (see ISO 19011 for more details).

the six key principles and identifying when they are not met allows us to understand the roles, the advantages and the limitations of some of the industry-led programs. As outlined previously, we see roles for first-, second-, and third-party audits in improving animal welfare and assuring that farm animals under our care are provided a reasonably good life.

What Makes a Good Audit? A Case Study: Development of the North American Meat Institute (NAMI) Audit

It was not until after McDonalds Inc. voiced concerns regarding the potential tarnishing of their "brand" and in turn their profitability if their customers were made aware of what appeared to be systemic failures in regulatory oversight within the supply chain that, at the request of McDonalds in 1991, Dr. Temple Grandin developed what we believe became the first North American focused third-party audit. This audit focused on assessing if individual slaughter houses met the minimum standards set out in the 1978 Humane Methods of Livestock Slaughter Act (HMSA). Approximately 5 years later this third-party audit was applied to the 800 USDA inspected slaughter facilities that collectively slaughter approximately 95% of the 140 million head of livestock slaughtered each year (Grandin, 1977). To the shock of the numerous stakeholders working in the beef cattle industry 64% of cattle facilities failed to meet the requirements set out in the HMSA legislation regarding effective stunning of cattle (Grandin, 1997). It became clear that legislation, at least in terms of the slaughter process, presumably reflective of our social consensus ethic, was not enough to protect and promote the welfare of farmed animals. In 2004, a congressional report by the U.S. General Accounting Office found that the USDA failed to provide consistent criteria for enforcement of the HMSA including incomplete or inconsistent records, enforcement, and reporting (GAO, 2004). A follow-up report in 2010 by the US General Accounting Office in 2010 again stated that USDA inspectors were still not taking consistent action to enforce the HMSA (GAO, 2010).

Given the apparent failure of enforcing HMSA in the years since its inception the question arises as to whether the welfare of animals during slaughter has improved? The original 1991 audit commissioned by McDonalds Inc. is now known as the North American Meat Institute (NAMI, 2013) audit and since 2001 the major US chain restaurants have implemented mandatory third-party audits of all slaughter houses in their supply chain—a market-driven demand that was effective in reducing the failure rate of effective stunning to from 64% in 1996 to 3.6% in 2011 (Grandin, 2011).

To understand how this success was achieved, we must understand how market mandates work and in the case of farm animal welfare what motivates corporations to look for assurances that the standards of animal welfare are sufficient and above all, what makes a good third-party audit.

Audit Certification

In 2016, the National Cattle Feeders Association of Canada developed the Canadian feedlot Animal Care Assessment program with the input of feedlot producers, processors, retailers, feedlot veterinarians, the SPCA, and animal-welfare specialists from government and industry. Unlike the DFC stewarded Animal Care module of Pro-action and the US based F.A.R.M. programs, the Canadian Feedlot program is certified by the Professional Animal Auditors Certification Organization (PAACO). PAACO is a nonprofit organization that certifies animal-welfare audits based on established criteria and trains auditors in an effort to establish a qualified and consistent auditing resource. The key attributes of a PAACO-certified audit is that the audit requirements must be auditable (measurable and verifiable), with the intent that the audit results in improved animal welfare. Prospective audits submitted to PAACO are sent for peer review with recommendations sent back for consideration prior to the final document being reviewed by the PAACO board. Any audit that is PAACO approved must also provide sample methodology sufficient to provide an adequate representation of the animals on the operation.

Who Sets the Standards That Are Used in the Audit?

Regardless of the audit or industry-led initiative most programs should be written with the input of scientists who have an established expertise in the welfare of the species of interest. However, while science can describe what is possible on farms it does not determine what society believes what farms "ought" to do—standards that resonate with societal values will no doubt be more sustainable in the long run (Weary et al., 2016).

As food animal agriculture has not historically required adherence to animal-welfare standards (Fraser, 2008), corporations in the food service and food retailer space have become increasingly involved in mandating improvements in animal welfare (Brown and Hollingsworth, 2005). They have done so by including animal-welfare-specific requirements into their supplier agreements and more and more frequently auditing against a standard (Sorensen and Fraser, 2010). When it comes to business or nongovernmental mandates focused on animal welfare the fundamental motivations, each of which are not necessarily mutually exclusive, can be described as being approached from the following perspectives: risk mitigation, potential for developing possible market advantages and improving animal welfare.

At a minimum, business and corporations are wanting to mitigate risk by providing their consumers assurance that animals raised for food that enter their supply chain are treated humanely, as the risk of being associated with a supplier farm accused of animal cruelty, neglect or poor

Table 11.2   The Case of Tail Docking in Dairy Cattle in North America—1992 to 2017

1992	Tail docking of dairy cattle introduced into North America on the basis that it would improve cleanliness and udder health at the American Association of Bovine Practitioners annual meeting (Johnson, 1992)
2001–2002	Three studies show no benefits to tail docking (Eicher et al., 2001; Tucker et al., 2001; Schreiner and Ruegg, 2002)
2004	American Veterinary Medical Association (AVMA) officially announces that they oppose tail docking of cattle
2005	Canadian Veterinary Medical Association (CVMA) officially announces that they oppose tail docking of cattle
2007	Forty percent of cows on US dairy farms had docked tails (USDA, 2007)
2006–2008	Two more studies show no benefit to tail docking (Eicher et al., 2006; Fulwider et al., 2008)
2009	California bans tail docking in cattle (California Penal Code, Section 597n)
2009	Canada's code of practice for the care and handling of dairy cattle (NFACC, 2009; p. 34) specifies that "Dairy cattle must not be tail docked unless medically necessary"
2010	Evidence arises that farms that dock tails actually have dirtier cows than do farms that keep tails intact (Lombard et al, 2010)
2010	American Association of Bovine Practitioners (AABP) officially announces that it opposes the routine tail docking of cattle
2011	Study published indicating that the most stakeholders, including farmers and the public, do not support tail docking in dairy cattle (Weary et al., 2011)
2011	National Mastitis Council (NMC) announces that it opposes the routine tail docking of dairy cattle
2015	Saputo announces Animal Welfare Policy[a]— specifically stating that farms within their supply chain will no longer be allowed to tail dock cows
2016	National Milk Producers Federation announces that they will no longer allow tail docking in farms participating in the FARM program effective January 1, 2017
2017	Thirty-three percent of cows on US dairy farms had docked tails when surveyed in 2014 (USDA, 2017)

[a] www.saputo.com/-/media/Ecosystem/Divisions/Corporate-Services/Sites/Saputo-Com/Saputo-Com-Documents/Our-Promise/Responsible-Sourcing/Animal-Care/2017_Saputo_Animal-Welfare-Policy_English_FINAL.ashx?la=en.

husbandry can have tremendous reputational and economical consequences. Animal-welfare audits or requirements motivated primarily by risk mitigation have historically focused on specific practices that have been the target of criticism from special interest groups. Gestation crates and battery cages are the two most familiar examples with numerous companies now publicly stating that they are transitioning away from these types of housing systems (Sullivan et al., 2017) (see also von Keyserlingk and Hötzel, 2015 for additional discussion). From the perspective of this chapter, there is merit in reviewing the topic of tail docking in dairy cattle in North America. This is an interesting example given that despite a suite of scientific studies on this issue all failing to show any benefits, the Canadian and US dairy industries differed in how they managed this practice. Unlike the DFC who announced in 2009 that this practice was no longer acceptable (DFC-NFACC, 2009), dairy farmers in the US were reluctant to give up this practice, which resulted in other stakeholders, including a major milk processor, driving change (Table 11.2). This provides a clear example where a market mandate is required to overcome industry reluctance to change.

Given the discussion around the need, and arguably the desire, to provide assurance about farm animal welfare some industries have created labels depicting some level of animal-welfare standard (AWI, 2016). Examples included the UEP certified (eggs) (Mench et al., 2011), and Red Tractor in the United Kingdom (2017). The impetuses for these types of certification programs are no doubt driven by the industry to show that they have set standards concerning animal care and those farms within their sector follow these. The reasons for introducing these are likely also to some degree motivated by an aspect of risk mitigation.

However, some North American nongovernmental organizations (NGOs) have also begun offering "certification" of animal-welfare practices resulting in "certification" audits such as SPCA Certified (http://spca.bc.ca/programs-services/certifications-accreditation/spca-certified/), Certified Humane (http://certifiedhumane.org/), American Humane Certified™ www.humaneheartland. org/our-farm-programs/american-humane-certified), Global Animal Partnership (https://global-animalpartnership.org/), and Animal Welfare Approved (https://animalwelfareapproved.us/). The impetus from these humane organizations to engage in developing these certification programs is no doubt to improve farm animal welfare. However, those farmers that set out to be "certified" are likely doing so with the added goal of obtaining a market advantage. Farms seeking certification by one or more of these types of labels typically must pay for a third-party auditor to come and verify whether the farm is compliant in regard to the "label claims." To the best of our knowledge, the only audit that has been reviewed and shown to indeed improve the welfare of animals is the NAMI audit (see Grandin, 2011).

With the growing demand for assurance that animal-welfare standards are being met on farms, there is a need to also determine how best to motivate farmers to adopt best practices on farms. These include a number of human factors that pay a role in verifying cattle welfare (Fraser and Koralesky, 2017), two of which include benchmarking and considering a professional model of animal production.

The Role of Benchmarking and Professionalism

There is a growing body of research indicating that the failure to adopt proven practices that improve animal welfare on farms can be explained, in part, by lack of farmer awareness of the problem. The existence of the gap between proven research and changed practices on commercial farms work is worrisome. The failure to get research implemented into practice can be due to a number of factors. Maybe the initial research question was not relevant to farmers? Or maybe the scientist has just not worked hard enough to create spaces where they can dialogue with farmers about their work and how it might be useful? Or maybe we need to develop new ways of closing the gap between science and changed practice on farms? Some work has begun to address this issue. For example, work done in the United Kingdom on lameness showed that this malady was greater on farms where the farmer underestimated the extent of the problem within their herd (Leach et al., 2010a) and that working toward reducing this problem requires a thorough understanding of the farmers motivations (Leach et al., 2010b).

One potentially effective approach is "benchmarking" welfare relevant measures as a way of getting farmers interested in the problems. Described by Fong et al. (1998) as the process of measuring performance using specific indicators, and then comparing the performance of one peer with other peers, benchmarking provides farmers with information on their farms but also in relation to how they compare to their peer group (von Keyserlingk et al., 2012). Ideally this process identifies areas of poor performance, initiates conversations, and ultimately drives improvements. Although most commonly used as a driver of economic efficiency (Anderson and McAdam, 2004), it has also been successfully used to motivate changes in other non-economic outcomes (Magd and Curry, 2003) such as improving the quality of patient care in human medicine (Woodhouse et al., 2009).

There is some evidence that this approach will be effective in the cattle industries, at least in the case of some farmers, as a vehicle to motivate them to strive for changed practices resulting in better lives for the animals on their farms (lameness, Chapinal et al., 2014b; calf feeding, Atkinson et al., 2017; Sumner et al., 2018). As stated earlier benchmarking works because it provides the farmer with real data from their own farm but also places this into context by comparing their results with their farmer peer group (thus allowing the farmer to decide if the problem is worth addressing) (von Keyserlingk et al., 2012). This approach can be done at the local, regional, or national level through first- and second-party audits that use widely accepted standards (see Sorensen and Fraser, 2010 for more discussion). Equally important is that these types of audits should encourage new conversations between the farmer and the professional team they work with, including their veterinarian, the nutritionist, and other consultants working with the farmers. In this way, benchmarking can provide trusted advisors to the farmers a new way to get their clients interested in specific issues, and to help their clients develop tailored solutions to the problems (von Keyserlingk et al., 2012). Thus, benchmarking may also provide advisors a means of developing their practices beyond the current services provided (Atkinson et al., 2017).

The issue of formally addressing professionalism through the adoption of a professional model has also received some interest within the academic literature. Fraser (2014) states that given the recent debates surrounding the need for increased food production, animal production systems will be hopefully seen as an important service to society. He also states that the issue of certification of farms could then be incorporated into a professional model that could aid in maintaining public trust. A professional model of animal agriculture would serve the industry, the animals, and consumers in fostering an environment where farmers are accustomed to being held accountable to an agreed upon standard. As long as the standards were set to drive improvement, accountability is likely to improve animal welfare while gaining consumer trust.

The promotion of a culture of self-governance (i.e., professionalism) within the industry (Fraser, 2014) has other advantages when it used as a vehicle to provide farmers with "confidential" access to critical feedback that allows them access to information that can help identify areas of concern. There are tremendous opportunities for rigorous confidential first-party audits that align with animal-welfare standards that mirror third party audits as a method for preparing farmers to meet standards (Weary and von Keyserlingk, 2017). This confidential assessment of the farms' animal care and management practices allows for dialogue between the farmer and the trusted party of stakeholders (i.e., veterinarians).

FUTURE CHALLENGES

Clearly expectations of what it means for an animal to have a good life have changed over time and will no doubt continue to change (Weary et al., 2016; von Keyserlingk and Weary, 2017). Historically farmers focused primarily on providing assurances that animals are in good health and producing well (von Keyserlingk et al., 2009). Not surprisingly, many animal-welfare audits already require information on the prevalence of lameness and injuries, and more work will be needed to curb these maladies on farm as many farms in the US and Canada continue to be plagued with high rates of lameness (von Keyserlingk et al., 2012; Cook et al., 2017). Disease rates during the transition period in dairy cattle and in the feedlot are also worrisome and will no doubt be included in future audits. The traditional practice of underfeeding the milk fed calf, the fact that individual housing during the milk feeding period remains the norm in many countries, and the fate of the bull calf will no doubt be areas where standards will be discussed and debated in the future.

There is also a growing body of evidence that for most lay citizens animal welfare is more than just good health and production. People want assurances that the animal is feeling well, free of pain, and able to experience pleasure (Fraser et al., 1997). Animal-welfare standards that ensure pain mitigation when doing routine painful procedures will likely not be optional for all programs in both North American beef and dairy cattle production systems in the very near future. This is already a reality for many European countries, for example, in Sweden, Denmark, and The Netherlands pain relief is legally required when disbudding/dehorning regardless of age (ALCASDE, 2009; Robbins et al., 2015). We see broad adoption of standards that require pain mitigation, for all painful procedures including castration and branding in both beef and dairy cattle in the near future.

The issue of naturalness is extremely important for the public (see Cardoso et al., 2016) and will likely cause the cattle industries the most difficulty (Weary et al., 2016). Given the known challenges associated with restriction of movement associated with hens housed in cages (Mench et al., 2008) and pigs in gestation stalls (Sato et al., 2017), we see tie stall housing for dairy cattle and tethering of any kind (e.g., veal calves, the milk fed calf) becoming problematic in the future (Spooner et al., 2014; von Keyserlingk and Weary, 2017). The public also places great emphasis on pasture access (Cardoso et al., 2016) which will require dairy producers and beef producers raising cattle in strict confinement and possibly feedlot operators to address this issue in the near future. One particularly contentious issue will be cow—calf separation in dairy (Ventura et al., 2015).

FINAL THOUGHTS

In order to achieve a good life for farm animals care givers of animals will need to continually reflect critically on current practices. Research has and will continue to play a key role in identifying best practices that result in improved welfare and ways and means of facilitating timely adoption of these best practices must become a focus by the members of the cattle industries. The development of standards and audits that reflect societal and industry concerns based on understanding of farm animal-welfare requirements and recommended practices will also be crucial. These standards reflecting the latest scientific evidence will ideally be developed and executed in a transparent manner—key if the industries are to be sustainable in maintaining public trust now and in the future. To maintain public trust third-party audits are needed. We also see need for additional efforts to focus on clearly showing that audits result in improved animal welfare. It is here where rigorous first- and second-party audits done to ensure that farms meet the standards on a daily basis will play an important role in preparing farmers to succeed when third-party audits are completed. Lastly, throughout this dynamic process, efforts must focus on producer engagement, such as benchmarking, to improve the adoption of implementing proven welfare solutions on farm.

ACKNOWLEDGMENTS

We thank Katie Koralesky (UBC Animal Welfare Program) for her comments on an earlier version of this manuscript. MvK is supported by supported by Canada's Natural Sciences and Engineering Research Council (NSERC) Industrial Research Chair Program with contributions from the Dairy Farmers of Canada (Ottawa, ON, Canada), British Columbia Dairy Association (Burnaby, BC Canada), Westgen Endowment Fund (Milner, BC, Canada), Intervet Canada Corporation (Kirkland, QC, Canada), Novus International Inc. (Oakville, ON, Canada), Zoetis (Kirkland, QC, Canada), BC Cattle Industry Development Fund (Kamloops, BC, Canada), Alberta Milk (Edmonton, AB, Canada), Valacta (St. Anne-de-Bellevue, QC, Canada), and CanWest DHI (Guelph, ON, Canada).

REFERENCES

ALCASDE. 2009. Report on dehorning practices across EU member states. www.vuzv.sk/DB-Welfare/telata/calves_alcasde_D-2-1-1.pdf. Accessed October 8, 2017.

Anderson, K., and R. McAdam. 2004. A critique of benchmarking and performance measurement: lead or lag? *Benchmarking: An International Journal* 11:1463–1483. doi:10.1108/14635770410557708.

Animal Welfare Act. 1999. Parliamentary counsel office of New Zealand 1999. www.legislation.govt.nz/act/public/1999/0142/56.0/DLM49664.html. Accessed October 8, 2017.

Animal Welfare Institute (AWI). 2016. A consumer's guide to food labels and animal welfare. https://awionline.org/content/consumers-guide-food-labels-and-animal-welfare. Accessed October 7, 2017.

Atkinson, D. J., M. A. G. von Keyserlingk, and D. M. Weary. 2017. Benchmarking passive transfer of immunity and growth in dairy calves. *Journal of Dairy Science* 100:3773–3782. doi:10.3168/jds.2016-11800.

Becker, J., M. G. Doherr, R. M. Bruckmaier, M. Bodmer, P. Zanolari, and A. Steiner. 2012. Acute and chronic pain in calves after different methods of rubber-ring castration. *Journal of Veterinary Science* 194:380–385.

Bergstra, T., J., Bart Gremmen, and Elsbeth N. Stassen. 2015. Moral values and attitudes toward Dutch sow husbandry. *Journal of Agricultural and Environmental Ethics* 28:375–401. doi:org/10.1007/s10806-015-9539-x.

Brambell Committee. 1965. Report of the technical committee to enquire into the welfare of livestock kept under intensive conditions; Command Paper 2836; Her Majesty's Stationery Office: London, UK.

Brown, K. H., and Hollingsworth, J., 2005. The food marketing institute and the National Council of Chain Restaurants: animal welfare and the retail food industry in the United States of America. *Revue scientifique et technique (International Office of Epizootics)* 24:655–663.

Canadian Cattlemen's Association and National Farm Animal Care Council (CCA-NFACC). 2014. Code of practice for the care and handling of beef cattle. www.nfacc.ca/pdfs/codes/beef_code_of_practice.pdf. Accessed October 8, 2017.

Cantrell, R., Lubben, B., and Reese, D. 2013. Perceptions of food animal welfare in extension: results of a two-state survey. *Journal of Extension* 51(2), 2FEA7.

Cardoso, C. S., M. J. Hötzel, D. M. Weary, J. A. Robbins, and M. A. G. von Keyserlingk. 2016. Imagining the ideal dairy farm. *Journal of Dairy Science* 99:1663–1671. doi:10.3168/jds.2015-9925.

Chapinal, N., A. Barrientos, M. A. G. von Keyserlingk, E. Galo, and D. M. Weary. 2013. Herd-level risk factors for lameness in freestall farms in North Eastern US and California. *Journal of Dairy Science* 96:318–328. doi:10.3168/jds.2012-5940.

Chapinal, N., Y. Liang, D. M. Weary, Y. Wang, and M. A. G. von Keyserlingk. 2014a. Risk factors for lameness and hock injuries in Holstein herds in China. *Journal of Dairy Science* 97:4309–4316. doi:10.3168/jds.2014-8089.

Chapinal, N., D. M. Weary, L. Collings, and M. A. G. von Keyserlingk. 2014b. Short communication: lameness and hock injuries improve on farms participating in an assessment program. *The Veterinary Journal* 202:646–648. doi:10.1016/j.tvjl.2014.09.018.

Clark, B., G. B. Stewart, L. A. PanzoneI Kyriazakis and L. J. Frewe. 2016. A systematic review of public attitudes, perceptions and behaviours towards production diseases associated with farm animal welfare. *Journal of Agricultural and Environmental Ethics* 29:455–478. doi:10.1007/s10806-016-9615-x.

Commission E., 2014. Final report of an audit carried out in the United Kingdom from 29 April to 09 May 2014 in order to evaluate the animal welfare controls in place at slaughter and during related operations. In: Office EC-FaV, ed, 2014:21.

Cook, NB. 2017. A comparison of measurable objective outcomes used to assess cattle welfare. In: Tucker, C. B. (ed). *Advances in cattle welfare*. Elsevier: Amsterdam.

Council for Agricultural Science and Technology (CAST). 2012. The direct relationship between animal health and food safety outcomes. CAST Commentary QTA2012-1. CAST, Ames, Iowa.

Dairy Farmers of Canada and National Farm Animal Care Council (DFC-NFACC). 2009. Code of practice for the care and handling of dairy cattle. www.nfacc.ca/pdfs/codes/dairy_code_of_practice.pdf. Accessed October 8, 2017.

Duggan, Tara. 2017. New ballot initiative could increase California farm animal welfare standards. *San Francisco Chronicle*, published August 27, 2017. www.sfchronicle.com/food/article/New-ballot-initiative-could-increase-California-12159349.php. Accessed October 6, 2017.

Eicher, S. D., H. W. Cheng, A. D. Sorrells, and M. M. Schutz. 2006. Short communication: behavioral and physiological indicators of sensitivity or chronic pain following tail docking. *Journal of Dairy Science* 89:3047–3051. doi:10.3168/jds.S0022-0302(06)72578-4.

FAWC (Farm Animal Welfare Council). 1992. FAWC updates the five freedoms. *The Veterinary Record* 131:357.

Flint, H. E., K. S. Schwartzkopf-Genswein, K. G. Bateman, and D. B. Haley. 2014. Characteristics of loads of cattle stopping for feed, water and rest during long-distance transport in Canada. *Animals* 4, 62–81. doi:10.3390/ani4010062.

Fong, S. W., E. W. L. Cheng, and D. C. K. Ho. 1998. Benchmarking: a general reading for management. *Management Decision* 36:407–418. doi:10.1108/00251749810223646.

Fraser, D. 2001. The "new perception" of animal agriculture: legless cows, featherless chickens, and a need for genuine analysis. *Journal of Animal Science* 79:634–641. doi:10.2527/2001.793634x.

Fraser, D. 2006. Animal welfare assurance programmes in food production: a framework for assessing the option. *Animal Welfare* 15:93–104.

Fraser, D. 2008. *Understanding animal welfare: the science in its cultural context.* Wiley-Blackwell, Oxford.

Fraser, D, and K. Koralesky. 2017. Assuring and verifying dairy cattle welfare. *Large dairy herd management e-book.* 3rd Edition, pp 993–1004. doi:10.3168/ldhm.1171.

Fraser, D., K. E. Koralesky, and G. Urton. In press. Toward a harmonized approach to animal welfare law in Canada. *Canadian Veterinary Journal* 59(3):293–302.

Fraser, D., D. M. Weary, E. A. Pajor, and B. N. Milligan. 1997. A scientific conception of animal welfare. *Animal Welfare* 6:187–205.

Fulwider, W. K., T. Grandin, B. E. Rollin, T. E. Engle, N. L. Dalsted, and W. D. Lamm. 2008. Survey of dairy management practices on one hundred thirteen north central and northeastern United States dairies. *Journal of Dairy Science* 91:1686–1692. doi:10.3168/jds.2007-0631.

GAO. 2004. Humane Methods of Slaughter Act: USDA has addressed some but still faces enforcement challenges. In: Office USGA, ed. Office of Public Affairs. 40, https://www.gao.gov/products/GAO-04-247.

GAO. 2010. Humane Methods of Slaughter Act: weakness in USDA Enforcement. In: Office USGA, ed. Office of Public Affairs. 10, https://www.gao.gov/products/GAO-10-487T.

Garcia, L. G., K. L. Nicholson, T. W. Hoffman, T. E. Lawrence, D. S. Hale, D. B. Griffin, J. W. Savell, D. L. VanOverbeke, J. B. Morgan, K. E. Belk, T. G. Field, J. A. Scanga, J. D. Tatum, and G. C. Smith. 2008. National beef quality audit–2005: survey of targeted cattle and carcass characteristics related to quality, quantity, and value of fed steers and heifers. *Journal of Animal Science* 86:3533–3543. doi:10.2527/jas.2007-0782.

Grandin, T. 1997. Survey of stunning and handling in federally inspected beef, veal, pork, and sheep slaughter plants. In: Agricultural Research Service/United States Department of Agriculture-3602-20-00 Project Number -3602-32000-002-08G. http://www.grandin.com/survey/usdarpt.html.

Grandin, T. 2011 Restaurant animal welfare and humane slaughter audits in U.S. federally inspected beef and pork slaughter plants. grandin.com.

Haley D. B., Bailey D. W., and Stookey J. M. 2005. The effects of weaning beef calves in two stages on their behavior and growth rate. *Journal of Animal Science* 83:2205–2214.

Hoeschele I. 1990. Potential gain from insertion of major genes into dairy cattle. *Journal of Dairy Science* 73:2601–2618. doi:10.3168/jds.S0022-0302(90)78947-3.

ISO. 2011. European committee for standardization. EN ISO 19011. Guidelines for auditing management systems (ISO 19011:2011).

Johnson, A. P. 1992. Mastitis control without a slap in the face. Page 146 in *Proc. 24th Annu. Conv. Am. Assoc. Bov. Pract.* 1991, Orlando, FL.

Lassen, J., P. Sandøe, and B. Forkman. 2006. Happy pigs are dirty! – conflicting perspectives on animal welfare. *Livestock Science* 103:221–230. doi:10.1016/j.livsci.2006.05.008.

Leach, K. A., H. R. Whay, C. M. Maggs, Z. E. Barker, E. S. Paul, A. K. Bell, and D. C. J. Main. 2010a. Working towards a reduction in cattle lameness: 1. understanding barriers to lameness control on dairy farms. *Research in Veterinary Science* 89:311–317. doi:10.1016/j.rvsc.2010.02.014.

Leach, K. A., H. R. Whay, C. M. Maggs, Z. E. Barker, E. S. Paul, A. K. Bell, and D. C. J. Main. 2010b. Working towards a reduction in cattle lameness: 2. understanding dairy farmers' motivations. *Research in Veterinary Science* 89:318–323. doi:10.1016/j.rvsc.2010.02.017.

Lombard, J. E., C. B. Tucker, M. A. G. von Keyserlingk, C. A. Kopral, and D. M. Weary. 2010. Associations between cow hygiene, hock injuries, and free stall usage on US dairy farms. *Journal of Dairy Science* 93:4668–4676. doi:10.3168/jds.2010-3225.

Long C. R., and K. E. Gregory. 1978. Inheritance of the horned, scurred and polled condition in cattle. *Journal of Heredity* 69:395–400. doi:10.1093/oxfordjournals.jhered.a108980.

Magd, H., and A. Curry. 2003. Benchmarking: achieving best value in public-sector organizations. *Benchmarking: An International Journal* 10:261–286. doi:10.1108/14635770310477780.

Marti, S., A. Velarde, J. L. de la Torre, A. Bach, A. Aris, A. Serrano, X. Manteca, and M. Devant. 2010. Effects of ring castration with local anesthesia and analgesia in Holstein calves at 3 months of age on welfare indicators. *Journal of Animal Science* 88:2789–2796. doi:10.2527/jas.2009-2408.

Mench, J. A. 2008. Farm animal welfare in the U.S.A.: farming practices, research, education, regulation, and assurance programs. *Applied Animal Behaviour Science* 113:298–312. doi:10.1016/j.applanim.2008.01.009.

Mench, J. A., D. A. Sumner, and J. T. Rosen-Molina. 2011. Sustainability of egg production in the United States—the policy and market context. *Poultry Science* 90:229–240. doi:10.3382/ps.2010-00844.

Miller, Donna. 2014. Great lakes cheese is "outraged" over mistreatment of cows. www.cleveland.com/metro/index.ssf/2014/11/great_lakes_cheese_is_outraged.html. Accessed October 8, 2017.

National Farm Animal Care Council of Canada (NFACC). 2013. Implementing codes of practice - Canada's framework for developing animal care assessment programs. www.nfacc.ca/resources/assessment/Animal_Care_Assessment_Framework_Oct2013.pdf. Accessed October 8, 2017.

NMPF. 2017. National milk producers federation. Farm animal care. www.nationaldairyfarm.com/about-farm. Accessed October 7, 2017.

North American Meat Institue Foundation (NAMI). 2013. Recommended animal handling guide and audit guide: a systematic approach to animal welfare. http://animalhandling.org/ht/a/GetDocumentAction/i/93003.

Olynk, N. J., G. T. Tonsor, and A. W. Christopher. 2010. Consumer willingness to pay for livestock credence attribute claim verification. *Journal of Agricultural and Resource Economics* 35:261–280.

Paul, J. 2015. Colorado authorities investigating dairy cow abuse video; workers fired. www.denverpost.com/2015/06/11/colorado-authorities-investigating-dairy-cow-abuse-video-workers-fired/. Accessed October 8, 2017.

Prickett, R. W., Norwood, F. B., and Lusk, J. L. 2010. Consumer preferences for farm animal welfare: results from a telephone survey of US households. *Animal Welfare* 19:335–347. ISSN 0962-7286.

Red Tractor. (2017). Assured food standards. https://assurance.redtractor.org.uk/. Accessed October 7, 2017.

Robbins, J. A., D. M. Weary, C. A. Schuppli, and M. A. G. von Keyserlingk. (2015). Stakeholder views on treating pain due to dehorning dairy calves. *Animal Welfare* 24:399–406. doi:10.7120/09627286.24.4.399.

Robbins, J.A., B. Franks, D.M. Weary and M.A.G. von Keyserlingk. (2016). Awareness of ag-gag laws erodes trust in farmers and increases support for animal welfare regulations. Food Policy 61:121–125. http://dx.doi.org/10.1016/j.foodpol.2016.02.008.

Schecter, A., and B. Ross. 2010. Got milk? Got ethics? Animal rights v. U.S. dairy industry. http://abcnews.go.com/Blotter/animal-rights-us-dairy-industry/story?id=9658866. Accessed October 8, 2017.

Schreiner, D. A., and P. L. Ruegg. 2002. Effects of tail docking on milk quality and cow cleanliness. *Journal of Dairy Science* 85(10):2503–2511.

Schwartzkopf-Genswein, K. S., J. M. Stookey, T. G. Crowe, and B. M. A. Genswein. 1998. Comparison of image analysis, exertion force and behavior measurements for use in the assessment of beef cattle responses to hot-iron and freeze branding. *Journal of Animal Science* 76:972–979.

Sorensen, J. T., and D. Fraser. 2010. On-farm welfare assessment for regulatory purposes: issues and possible solutions. *Livestock Science* 131:1–7. doi:10.1016/j.livsci.2010.02.025.

Spooner, J. M., C. A. Schuppli, and D. Fraser. 2014. Attitudes of Canadian pig producers toward animal welfare. *Journal of Agricultural and Environmental Ethics* 27:569–589. doi:10.1007/s10806-013-9477-4.

Stafford, K. J., and D. J. Mellor. 2011. Addressing the pain associated with disbudding and dehorning in cattle. *Applied Animal Behaviour Science* 135:226–231. doi:10.1016/j.applanim.2011.10.018.

Step D. L., Krehbiel C. R., DePra H. A., Cranston J. J., Fulton R. W., Kirkpatrick J. G., Gill D. R., Payton M. E., Montelongo M. A., and Confer A. W. 2008. Effects of commingling beef calves from different sources and weaning protocols during a forty-two-day receiving period on performance and bovine respiratory disease. *Journal of Animal Science* 86:3146–3158. doi:10.2527/jas.2008-0883.

Sullivan, R., N. Amos, and H. A. van de Weerd. 2017. Corporate reporting on farm animal welfare: an evaluation of global food companies' discourse and disclosures on farm animal welfare. *Animals* 7(3):17. doi:10.3390/ani7030017.

Sumner, Christine L., Marina A.G. von Keyserlingk and Daniel M. Weary (2018). How benchmarking motivates farmers to improve dairy calf management. Journal of Dairy Science 101:3323–3333. https://doi.org/10.3168/jds.2017-13596.

Te Velde, H., N. Aarts, and C. Van Woerkum. 2002. Dealing with ambivalence: farmers' and consumers' perceptions of animal welfare in livestock breeding. *Journal of Agricultural and Environmental Ethics* 15:203–219. doi:10.1023/A:1015012403331.

Thuer, S., S. Mellema, M. G. Doherr, B. Wechsler, K. Nuss, and A. Steiner. 2007. Effect of local anaesthesia on short- and long-term pain induced by two bloodless castration methods in calves. *Journal of Veterinary Science* 173:333–342. doi:10.1016/j.tvjl.2005.08.031.

Tucker, C. B., D. Fraser, and D. M. Weary. 2001. Tail docking dairy cattle: effects on cow cleanliness and udder health. *Journal of Dairy Science* 84:84–87. doi:10.3168/jds.S0022-0302(01)74455-4.

UEP (United Egg Producers). http://uepcertified.com/wp-content/uploads/2015/08/2017-UEP-Animal-Welfare-Complete-Guidelines-6.2017-FINAL.pdf. Accessed October 6, 2017.

USDA. 1978. Humane Methods of Slaughter Act-92 STAT.1069. In: Congress t, ed. Public law.

USDA 2008a Beef 2007–08, Part III: changes in the U.S. beef cow-calf industry, 1993–2008. www.aphis.usda.gov/animal_health/nahms/beefcowcalf/downloads/beef0708/Beef0708_dr_PartIII.pdf.

USDA. 2008b. Beef 2007–08, Part IV: reference of beef cow-calf management practices in the United States, 2007–08. USDA-APHIS-VS, CEAH. Fort Collins, CO. #N523-0210.

USDA. 2009. Dairy 2007, Part IV: reference of dairy cattle health and management practices in the United States, 2007. www.aphis.usda.gov/animal_health/nahms/dairy/downloads/dairy07/Dairy07_dr_PartIV.pdf.

USDA. 2017. Dairy 2014, "Health and Management Practices on U.S. Dairy Operations, 2014". USDA-APHIS-VS-CEAH-NAHMS. Fort Collins, CO. #696.1017.

Vanhonacker, F., W. Verbeke, E. Van Poucke, and F. A. M. Tuyttens. 2008. Do citizens and farmers interpret the concept of farm animal welfare differently? *Livestock Science* 116(1–3):126–136. doi:10.1016/j.livsci.2007.09.017.

Ventura, B. A., M. A. G. von Keyserlingk, C. A. Schuppli, and D. M. Weary. 2013. Views on contentious practices in dairy farming: the case of early cow-calf separation. *Journal of Dairy Science* 96:6105–6116. doi:10.3168/jds.2012-6040.

Ventura, B.A. D.M. Weary, A.S. Giovanetti, and M.A.G. von Keyserlingk. 2016. Veterinary perspectives on cattle welfare challenges and solutions. *Livestock Science* 193:95–102. doi:10.1016/j.livsci.2016.10.004.

Verbeke W., F. J. A. Pérez-Cueto, M. D. de Barcellos, A. Krystallis, and K. G. Grunert. 2010. European citizen and consumer attitudes and preferences regarding beef and pork. *Meat Science* 84:284–292. doi:10.1016/j.meatsci.2009.05.001.

von Keyserlingk, M. A. G., and M. J. Hötzel. 2015. The ticking clock: addressing farm animal welfare in emerging countries. *Journal of Agricultural and Environmental Ethics* 28:179–195. doi:10.1007/s10806-014-9518-7.

von Keyserlingk, M. A. G., and D. M. Weary. 2017. INVITED REVIEW: animal welfare in the journal of dairy science – the first 100 years. *Journal of Dairy Science* doi:10.3168/jds.2017-13298.

von Keyserlingk, M. A. G., A. Barrientos, K. Ito, E. Galo, and D. M. Weary. 2012. Benchmarking cow comfort on North American freestall dairies: lameness, leg injuries, lying time, facility design, and management for high-producing Holstein dairy cows. *Journal of Dairy Science* 95:7399–7408. doi:10.3168/jds.2012-5807.

von Keyserlingk, M. A. G., J. Rushen, A. M. B. de Passillé, and D. M. Weary. 2009. Invited review: the welfare of dairy cattle – key concepts and the role of science. *Journal of Dairy Science* 92:4101–4111. doi:10.3168/jds.2009-2326.

Weary D. M., and von Keyserlingk M. A. G. 2017. Public concerns about dairy-cow welfare: how should the industry respond? *Animal Production Science* 57:1201–1209. doi:10.1071/AN16680.

Weary, D. M., C. A Schuppli, and M. A. G. von Keyserlingk. 2011. Tail docking dairy cattle: responses from an online engagement. *Journal of Animal Science* 89:3831–3837. doi:10.2527/jas.2011-3858.

Weary, D. M., B. A. Ventura, and M. A. G. von Keyserlingk. 2016. Invited review: societal views and animal welfare science: understanding why the modified cage may fail and other stories. *Animal* 10:309–317. doi:10.1017/S1751731115001160.

Wolf, C. A., and G. T. Tonsor. 2017. Cow welfare in the U.S. dairy industry: willingness-to-pay and willingness-to-supply. *Journal of Agricultural and Resource Economics* 42:164–179. ISSN 1068–5502.

Woodhouse, D., M. Berg, J. van der Putten, and J. Houtepen. 2009. Will benchmarking ICUs improve outcomes? *Current Opinion on Critical Care* 15:450–455. doi:10.1097/MCC.0b013e32833079fb.

CHAPTER 12

Transportation and Slaughter of Beef and Dairy Cattle

Kurl D. Vogel
University of Wisconsin

CONTENTS

INTRODUCTION

Overview of the Size and Scope of North American Beef Slaughter

According to the United States Department of Agriculture's (USDA) Agricultural Marketing Service (AMS) and National Agricultural Statistics Service (NASS), the total number of cattle slaughtered under federal inspection in 2015 in the United States was nearly 28.7 million. The majority (80.2%) of those animals were heifers (7.5 million, 26.04% of total cattle slaughter) and steers (15.5 million, 54.13% of total cattle slaughter). Within the mature cattle population, dairy cows contributed 2.9 million (10.31% of total cattle slaughter) and beef cows contributed 2.3 million (7.92% of total cattle slaughter) toward the total of 5.2 million cows (18.22% of total cattle slaughter) that were slaughtered in 2015. Bulls and stags contributed slightly more than 467,000 animals (1.62% of total cattle slaughter).

In 2015, the USDA reported approximately 7.5% of the mature beef cow population and 31% of the mature dairy cow population were slaughtered. Mature beef and dairy cows are culled from their respective herds for a variety of reasons. In general, the reasons that cows are culled are divided between two categories: voluntary culling and involuntary culling. Voluntary culling involves the removal of an animal from the herd by the choice of the farmer or rancher. Often, voluntary culling is associated with reduced productivity and the availability of replacement animals that are younger with greater production potential. Involuntary culling typically consists of a "forced decision" for the producer. Animals are involuntarily culled due to health and welfare disorders that are often chronic, untreatable, or life threatening in severe cases. Hadley et al. (2006) reported that 79.5% of dairy cows in the Upper Midwest and Northeast U.S. that were culled from dairy herds enrolled in the Dairy Herd Improvement program were culled due to health reasons. Those reasons included reproduction, injury, death, mastitis, feet and legs, disease, and udder problems. All of the previously listed conditions are classified as involuntary culling reasons. In the same study, the remaining 21.5% of dairy cows were culled due to low production (12.8%) or sold for dairy purposes on other farms (7.7%). It is important to recognize that mature cows that arrive at slaughter facilities were not raised and managed with the primary objective of meat production like beef heifers and beef and dairy steers. As a result, the mature cow population will typically manifest health and welfare issues in greater frequency and severity.

Dairy cows are most often sold through terminal markets and auctions in the U.S. In a 2007 survey, the USDA National Animal Health Monitoring System (NAHMS) reported that 76.2% of dairy cows that were removed from their respective farms were sold at auctions or terminal markets and 26.5% of cows that were permanently removed from dairy herds were sold directly to a packer or processor (NAHMS, 2007).

MODES OF TRANSPORTATION

Prevalence of Transport Methods

Prior to the development of the U.S. Interstate Highway System, long-distance transportation of live cattle often occurred on hoof or by rail. As the infrastructure developed to support the movement of goods by truck, straight-trucks and tractor-trailers became the predominant modes of cattle transportation to slaughter.

The 2016 Market Cow and Bull National Beef Quality Audit (NCBA, 2016a) reported that the average travel time for mature cows and bulls prior to slaughter was 6.7 ± 6.4 hours (NCBA, 2016a). The minimum transportation time was 0.2 hours and the maximum was 39.5 hours. The mean distance traveled within that population was 283.2 ± 273.9 miles. The minimum distance traveled was 2 miles and the maximum was 1412.9 miles.

The travel times and distances for steers and heifers were less than for mature cows and bulls (NCBA, 2016b). The average travel time prior to slaughter for that population of cattle was 2.7 ± 2.4 hours. The average travel distance was 135.8 ± 132.5 miles.

PRE-TRANSPORT HANDLING

A wide variety of handling facilities can be observed at the farms, ranches, terminal markets, and collection points throughout North America. At smaller farms and ranches, it is commonplace to observe the use of purpose-built cattle trailers that may be attached at the rear bumper or within the cargo bed of pickup trucks for moving cattle directly to slaughter or to a terminal market or collection point. In locations where greater volumes of cattle will be loaded for transport, the handling facilities will typically include a loading ramp or a raised elevation throughout the handling facility in relation to a large semi-trailer to facilitate the movement of animals into the trailer.

CONTROL OF TRANSPORT CONTAINER ENVIRONMENT

The environment inside a cattle trailer is greatly impacted by ambient temperature and humidity, speed of trailer movement, and stocking density. In addition, each compartment within a livestock trailer has its own microclimate, which often differs from the other compartments within the trailer.

The primary means of controlling the temperature on cattle trucks during transport includes movement of the truck. During periods of hot weather, slaughter establishments often advise livestock transporters to not park their trucks and wait to unload. Rather, transporters are often encouraged to continue driving until they can be unloaded to promote airflow through the trailer while it is in motion. Some transport trailers are equipped with plumbing that allows sprinklers or misters to be used to assist with heat mitigation. These types of systems are generally used in parked transport trailers at locations where a water supply and drainage are available. The availability of water lines and sufficient pressure usually limits the number of mister-equipped trailers that can operate at the same time once the transporters have reached their destination. Some slaughter establishments have installed fan banks with the capability of blowing air through the vented sides of trailers. The space requirement to provide a fan bank for multiple trucks that are waiting to unload limits the utilization of this approach. As a result, careful scheduling of transporter unload times and timely unloading are necessary and commonly used to prevent the buildup of excessive heat and humidity in parked trailers.

ANIMAL WELFARE MONITORING DURING TRANSPORT

Relevant Laws

The primary law in the United States regarding the transport of farm animals is the 28-hour law (9 U.S.C. 80502). Originally adopted in 1873, the law was focused on the transportation of animals by rail and ship. In 2005, The Humane Society of the United States, Farm Sanctuary, Compassion over Killing, and Animals' Angels—all animal advocacy groups—petitioned the USDA to interpret the 28-hour law to include motor vehicle carriers due to nearly complete transition of livestock transportation away from rail cars to trucks and trailers. The USDA responded that the appropriate interpretation of the law included livestock trailers. The USDA is responsible for enforcement of the 28-hour law through a joint effort by the Animal and Plant Health Inspection Service (APHIS) and the Food Safety and Inspection Service (FSIS). Specifically, the 28-hour law mandates that "a rail carrier, express carrier, or common carrier (except by air or water), a receiver, trustee, or lessee of one of those carriers, or an owner or master of a vessel transporting animals from a place in a State, the District of Columbia, or a territory or possession of the United States through or to a place in another State, the District of Columbia, or a territory or possession, may not confine animals in a vehicle or vessel for more than 28 consecutive hours without unloading the animals for feeding, water, and rest." The 28-hour law also requires that, "Animals being transported shall be unloaded in a humane way into pens equipped for feeding, water, and rest for at least 5 consecutive hours."

TRANSPORTATION CHALLENGES

Most cattle have limited experience with the transportation process. As a result, loading facilities, transportation containers, and receiving facilities must be designed and maintained to support a safe and minimally stressful experience for the animals. The physical condition of each animal, the ambient weather conditions, and the quality of the driving of the transport container are all substantial factors in the overall transport experience by farm animals.

In general, the most challenging population of cattle to transport is mature former breeding stock with preexisting health and welfare deficiencies. This population of cattle requires additional care and caution during all phases of handling prior to slaughter. The first step that many cattle dealers and transporters take is sorting the injured and high-risk cattle from the rest of the group. Those high-risk animals are the last animals that are loaded onto semi-trailers. They will occupy the compartment within the trailer that is closest to the door. This compartment is commonly referred to as the "tail." The tail of the trailer is often bedded with wood shavings or straw to provide traction, absorb urine, and promote rest by lying down. The stocking density in the tail is often reduced to allow compromised cattle space to lie down comfortably and lunge to get up.

It is not customary or common for transporters to apply bedding to compartments other than the tail. For most transport distances, the objective is to keep cattle standing due to the potential for animals to be trampled and injured if they lay down while the majority of others are standing.

POST-TRANSPORTATION HANDLING

The process of loading livestock trailers is typically performed by the truck driver with assistance by farm or ranch personnel. At the point of unloading, the truck driver may not be as involved with movement of animals off of the trailer. Several U.S. slaughter establishments have developed policies that deter transporters from handling animals on their premise to reduce the risk of inhumane handling by untrained drivers. The FSIS clarified the way in which enforcement action should be initiated and provided specific examples of the types of behavior that would result in agency action in Directive 6900.2 Revision 2. Ultimately, the decision to allow truck drivers to participate in the loading and unloading of their trucks is under the discretion of the shippers and receivers of the loads of animals. Some slaughter establishments have invested in training the truck drivers that bring cattle to them and require training for all new truck drivers as well.

LAIRAGE

Lairage refers to the holding pens where animals are kept prior to slaughter. This step in the pre-slaughter handling process is important to allow animals to recover from transportation stress. The

Humane Slaughter Act requires access to water in all occupied cattle holding pens any time animals are in them. It is very common for indoor holding pens to be bedded to control the accumulation of liquid where animals are likely to lie down. Outdoor pens are not commonly bedded because the bedding is difficult to maintain without protection from precipitation and wind. All federal and state-inspected slaughter establishments in the United States are required to have a designated space for animals that require extra attention during the inspection process. This space is commonly referred to as a "Suspect Pen" because the animals are labeled as U.S. Suspect until their carcasses are inspected during the slaughter process. Since animals that are likely to be identified as U.S. Suspect commonly display health conditions that reduce their typical resistance to changes in temperature and exposure to sun and rain, it is a legal mandate that the suspect pen must be covered by a roof.

Typically, animals that are destined for slaughter spend less than one day in lairage if they arrive at a slaughter establishment during the work week and the plant is operating normally. If a plant chooses to receive animals during the weekend with the intent of slaughtering them during the work week or an equipment failure or other stoppage of plant operation occurs, it may be necessary to hold animals in lairage for longer periods of time. There is no requirement regarding the maximum amount of time that animals may be held in lairage, but all animals must be fed if they are held for longer than 24 hours. The regulations do not specify the type of diet the animals must receive as the primary objective of feeding animals that are held in slaughter facilities is to provide gut fill to curb hunger.

PRESLAUGHTER HANDLING

Handling Principles

Calm and quiet actions serve as catalysts for successful handling after cattle arrive at the slaughter establishment. In plants that slaughter steers and heifers that were raised for the purpose of becoming beef, the same handling principles that are commonly employed at the feed yard and throughout the transportation process apply. Mature cattle, particularly dairy cows, require additional time and patience to handle in some cases. Lameness is a substantial contributing factor in dairy culling decisions. Lame cows experience changes in their gait that slow their movement during handling. In addition, many plants that slaughter mature dairy cows have dealt with challenges in getting animals to move smoothly and willingly through their single file chutes and restrainer entrances. This is a multifactorial problem to solve. First, dairy cattle are used to being handled closely by people. As a result, pressure to the flight zone during handling may not be as effective as we observe in fed cattle and beef cows. In addition, most dairy cattle have not experienced facilities that resemble those in a large slaughter establishment. The novelty of a new facility to navigate presents handling challenges for any type of cattle, but it is compounded by the lack of familiarity in design afforded to beef cattle. Lameness also contributes to the challenge of moving mature dairy cows into the restraint device prior to stunning.

For most facilities that slaughter mature dairy cows, the key to safe, efficient, and humane handling lies in the selection and training of patient and skilled handlers. Many plants struggle to identify viable options for the redesign of cull dairy cow handling facilities that remain conducive to safety and efficiency. However, the quality of handling and skillset of the handlers can provide substantial support for less-than-ideal facilities.

Effects of Handling on Meat Quality

The acute beef quality issues that occur as a result of handling are largely the result of physical trauma during preslaughter handling and distress shortly before slaughter. Physical trauma appears to be the most consistent and substantial category of causative factors that impact beef quality during the final days and hours of cattle's lives. The 2016 National Beef Quality Audit reported that 23% of steer and heifer carcasses displayed bruising that resulted in greater than one pound of trim loss (NCBA, 2016b). Bruising can occur at any stage in the handling and transport process that brings an animal to slaughter. Unfortunately, bruising is generally not identifiable until an animal has been slaughtered and the hide removed. When cattle are purchased on a live weight basis, bruising results in a loss for the purchaser. This type of purchasing agreement does not hold parties that are selling cattle accountable for carcass weight loss due to bruising. The 2016 National Beef Quality Audit also reported that 1.9% of steer and heifer carcasses displayed dark cutting beef. This was the lowest incidence of the condition since the audit was started in 1991 (NCBA, 2016b).

The U.S. beef industry has conducted audits at approximately 5-year intervals since 1991 to monitor the occurrence of defects that ultimately result in lost value for the farmers and ranchers that sell animals on a carcass weight basis and the slaughter establishments that purchase animals on a live weight basis. Many of those defects are rooted in animal health and welfare problems. The basis upon which an animal is sold presents the potential for unscrupulous farmers, ranchers, cattle dealers, livestock markets, and other intermediate handlers to avoid financial repercussion for defects that they caused. When cattle are sold on a live weight basis, it is impossible for the buyer of that animal's carcass to know the amount of carcass loss that may occur due to bruising, arthritic joints, and improperly placed injection site lesions because the sale price of the animal is determined before the animal is slaughtered and defects are discovered. Sale of cattle on a carcass weight basis allows the buyer to only pay for the useable carcass after the animal is slaughtered. In turn, the seller is held accountable for handling and husbandry-related carcass defects. Research has suggested that mature cows that were marketed with average to moderate body condition scores were most favorable for cattle producers and slaughter establishments based on value (Apple, 1999). Ultimately, the type of cattle that a slaughter establishment favors depends on the market for that beef they sell. If a plant specializes in lean beef production, a cow with lower body condition score may be more favorable.

RESTRAINT METHODS

Center-track Restrainers

In North America, the center-track conveyor restrainer is commonly used to restrain cattle prior to stunning or ritual slaughter in larger slaughter establishments. A center-track restrainer consists of a conveyor that fits between the legs and under the brisket and abdomen and supports the entire weight of the animal as their feet are not allowed to make contact with the floor or any other solid surface. The sides of the restrainer are adjustable, typically through the actuation of hydraulic cylinders. To enter the center-track restrainer, cattle typically walk along a downward-sloping cleated ramp that allows them to walk until their weight is supported by the conveyor. The type of animal seems to have an impact on willingness to enter the restrainer. In general, dairy cows present greater challenges when entering the center-track restrainer due to the occurrence of large inflamed udders, lameness, and an overall lack of experience in moving through facilities that are unfamiliar.

Stun Boxes

Prior to the introduction of conveyor-type restrainers for cattle, individual stalls that have the ability to fully restrain an animal prior to stunning were the predominant means of restraining cattle prior to slaughter. In larger, high-throughput slaughter establishments, stun boxes are less common than smaller facilities with smaller capacity. However, stun boxes are very common in small and very small slaughter establishments and are very effective if they are appropriately designed and maintained. As general principles, restraint boxes must be equipped with nonslip flooring and maintain a width that does not allow cattle to turn around—either completely or in part. In addition, the sides and front of the restraint box must be tall enough to prevent cattle from trying to jump out. Most restraint boxes are equipped with solid sides made from impact resistant materials such as concrete, steel, or wood. The solid sides prevent the occurrence of injury to the legs of animals that become scared or restless during the time they are in the restraint box. A variety of configurations exist for the front of cattle restraint boxes.

STUNNING

Mechanical Stunning Devices

Nearly all conventionally slaughtered cattle in North America are stunned with mechanical stunning devices. Mechanical stunning devices include penetrating captive bolt stunners, nonpenetrating captive bolt stunners, and firearms. Of these three types of mechanical stunning devices, the penetrating captive bolt stunner is most common.

The Humane Methods of Slaughter Act (7 U.S.C. §1901) requires all animals that it covers to be rendered unconscious and insensible to pain by a method that is rapid and immediate before being shackled, cast, cut, or thrown. Although mechanical stunning devices are highly effective in general, stunning problems are consistently cited as the most common reason for regulatory action regarding humane handling and stunning. Common causes of mechanical stunning noncompliance have been outlined by Grandin (2002) and include inappropriate device placement, poor maintenance, insufficient air supply for pneumatic stunners, exposure of gunpowder charges to humidity, insufficient or ineffective restraint, and poor ergonomic design of stunners and stunning platforms.

ANIMAL WELFARE MONITORING AT SLAUGHTER ESTABLISHMENTS

Relevant Laws

The Humane Methods of Slaughter Act, the primary animal protection law for cattle at U.S. slaughter establishments, was signed into law in 1958 and originally enacted in 1960. The Act initially served as the basis for the regulatory framework used by the USDA's FSIS in federally inspected slaughter establishments that sold product to the federal government. Those regulations that enforce the Humane Methods of Slaughter Act can be found in 9 C.F.R. §313. In its original form, the Act did not grant provisions to food safety inspectors to stop production if infractions were observed. As a result, minimal enforcement action occurred in the early years of the Act. In 1978, the Act was amended to apply to all federally inspected slaughter establishments, regardless of vendor status with the U.S. Government. By default, all state-inspected slaughter establishments were required to abide by the same law starting in 1978 as well.

The purposes of the Humane Methods of Slaughter Act are as follows: (1) to prevent needless suffering; (2) to improve safety and conditions for workers; (3) To improve meat products; and (4) to increase economy of slaughter operations. Since the inception of the Act, lawmakers recognized the potential to improve the welfare of *both* animals and workers, as well as the quality of meat products and the economy of the slaughter process *simultaneously*.

The Humane Methods of Slaughter Act prescribes two acceptable means of approaching the process of stunning animals prior to beginning other slaughter procedures. The first acceptable class of methods is generally focused on conventional slaughter. The second method is focused on ritual slaughter procedures. The conventional method requires that "All animals are rendered insensible to pain by a single blow or gunshot or an electrical, chemical, or other means that is rapid and effective before being shackled, cast, thrown, or cut." Ritual slaughter procedures that omit preslaughter stunning are allowed by the Humane Methods of Slaughter Act provided they are, "By slaughtering in accordance with the Jewish faith or any other faith… where by the animal loses consciousness by simultaneous and instantaneous severance of the carotid arteries."

In other areas of the world, ritual slaughter without stunning has become a highly contentious topic as it is a point of intersection between animal welfare and religious freedom. Animal welfare advocates often view the allowance of slaughter without stunning as a substantial contributor to animal suffering while members of the religious groups that utilize slaughter procedures without stunning view opposition to slaughter without stunning as an affront to their right to practice their religion. Iceland, Norway, Switzerland, and Sweden have banned the slaughter of animals without stunning. Many large-scale slaughter facilities in the U.S. that produce Kosher meat apply captive bolt stunning immediately after the ritual cut is complete. This approach helps to quickly render the animal unconscious to prevent suffering and allows higher capacity slaughter facilities to keep their production lines full and moving. Although the use of post-cut captive bolt stunning is one answer to the question of controlling the potential for an animal to experience suffering during ritual slaughter procedures, it does not directly address the potential for the animal to experience pain, distress, and suffering as a result of the cut while the animal is still conscious. In addition, such an approach suggests that the following of a religious prescription to kill animals for food without stunning prior to bleeding is more important than the careful, logical, and data informed consideration of the welfare of the animals that experience the procedure. There is a legitimate need for the religious groups that prescribe slaughter procedures that do not begin with stunning to open-mindedly consider if methods are available today that can ensure a positive welfare state for animals in their final moments of life. It is plausible that the severance of the carotid arteries and jugular veins with a very sharp blade was—at one time—the most humane method of slaughtering animals. However, consistently effective preslaughter stunning was not a viable option at that time. It is today.

The regulations that enforce the Humane Methods of Slaughter Act also prescribe standards for facility condition and layout, handling tools, handling of nonambulatory animals, and feed and water requirements in lairage. The facility layout and condition requirements are generally vague, requiring that pens are maintained to prevent injury, nonslip flooring is provided in handling areas, the provision of a roof over the suspect pen(s), and the minimization of sharp corners and direction reversals.

Assessment Standards

NAMI Standards

The North American Meat Institute (NAMI) released the first iteration of *Recommended Animal Handling Guidelines for Meat Packers* in 1991. The document was written by Dr. Temple Grandin and was the first of its kind in animal agriculture. The document was focused on providing practical guidance to slaughter establishments to reduce or prevent animal suffering. In 1997, a second document was developed by Dr. Grandin that included the general audit framework. When the initial standards within the first iteration of the audit were established, virtually no scientific data were available to help Dr. Grandin determine the appropriate performance standards for audit criteria such as stunning efficacy, electric prod usage, and vocalization. As a result, the initial standards were set by using survey data and practical intuition. The focus of the NAMI animal handling guidelines has been on quantifying performance in a limited number of areas within slaughter establishments that can provide maximum information on the overall performance regarding humane animal handling and stunning.

The NAMI auditing guidelines consist of two primary sections: transportation and slaughter. The slaughter guidelines were established first. The core criteria of the slaughter audit guidelines are the most crucial standards for an audited slaughter establishment to pass. Failure on any one of the core criteria will result in complete failure of the audit. The following are considered core criteria within the slaughter audit: Effective Stunning, Bleed Rail Insensibility, Falls, Vocalizations, Electric Prod Usage, and Willful Acts of Abuse. Willful acts of abuse are rare, but they result in immediate audit failure if they are observed. The NAMI Animal Care and Handling Guide defines a willful act of abuse as:

... includ(ing) but not limited to: (1) Dragging a conscious, nonambulatory animal; (2) intentionally applying prods to sensitive parts of the animal such as the eyes, ears, nose, anus, or testicles; (3) deliberate slamming of gates on livestock; (4) malicious driving of ambulatory livestock on top of one another either manually or with direct contact with motorized equipment (this excludes loading a nonambulatory animal for transport); (5) purposefully driving livestock off high ledges, platforms or off a truck without a ramp (driving market weight or adult animals off a low stock trailer is acceptable); (6) hitting or beating an animal; or (7) animals frozen to the sides or floor of the trailer.

The performance standards for each of the core criteria, with the exception of willful acts of abuse, are based on the occurrence of events associated with each core criterion. For Effective Stunning, 96% or more of a minimum of 100 animals must be stunned with a single application of the stunning device. For Bleed Rail Insensibility, 100% of the animals must be insensible while suspended on the bleed rail. It is completely unacceptable to begin slaughter procedures on animals that are not insensible to pain. It is also required that 100% of animals remain insensible on the bleed rail in 9 C.F.R. §313.

For falls, 1% or less of the animals are allowed to experience a fall during handling. In early iterations of the NAMI Animal Care and Welfare Guidelines, both slips and falls were included as core criteria; however, substantial debate regarding the appropriate definition of a slip and the minimal impact of slipping on animal welfare so long as it did not result in falling prompted the NAMI Animal Welfare Committee to remove slipping from the core criteria and more clearly define a fall. The definition of fall that is used in the NAMI Animal Care and Welfare Guidelines is:

A fall occurs when an animal loses an upright position suddenly in which a part of the body other than the limbs touches the ground. All falls that occur in a stun box or restrainer before stunning or religious slaughter are counted as falls. Equipment that is designed to cause falling before stunning or religious slaughter should not be used.

Vocalizations are counted for cattle in the crowd pen, lead-up chute, stun box, or restrainer at slaughter facilities. The maximum allowable vocalization occurrence is 3% of the animals that pass through this area of the facility. The maximum for facilities that use head restraint to securely restrain the heads of cattle during ritual slaughter and/or stunning is 5% vocalization. The rationale between the difference in standards is that cattle with restrained heads tend to vocalize in response to the stress associated with securing the head of the animal. Head restraint is considered to be good for animal welfare as long as the restraint is not maintained for a length of time that causes animals to become distressed. Head restraint helps to ensure accurate placement of captive bolt stunning devices and facilitates the placement of the cut during ritual slaughter. The NAMI Guide recommends that auditors do not attempt to score vocalization in lairage because cattle tend to vocalize to communicate with others when they are not overtly distracted by stressors.

For all types of cattle, electric prod use is limited to 25% or less of all animals audited for prod use. In the 1990s, when this criterion was set, some plants struggled to comply with the standard because electric prods had become their primary handling tool. Increased attention on electric prod use as a result of this standard caused beef slaughter establishments to become more attentive to facility and handling issues that were detrimental to animal movement. In the early 2000s a slaughter establishment in the Great Lakes region developed a vibrating prod from a modified engraving tool powered by compressed air. Although these tools have helped to greatly reduce the use of electric prods, they must be used carefully and with discretion. If the tip of the tool is too sharp, the author has observed instances in which the tip of the tool was burrowed into the skin of cattle due to abuse by an operator.

Assessment Methods

In-person

There are three parties that are potentially involved in the animal welfare auditing process. They are simply known as first, second, and third parties. First-party auditors are employed by the slaughter facility, typically in the quality assurance and food safety departments. In general, first-party audits are considered to possess the least inherent credibility because they are analogous to grading one's own work. Although first-party audit results often have a high level of fidelity, they are not considered to be highly trustworthy outside of plant walls due to the obvious conflict of interest. Second-party audits are conducted by personnel that work for customers of a slaughter establishment. These audits have a greater level of credibility than first-party audits but still contain an implicit conflict of interest because a financial relationship exists between supplier and customer in which the customer has a vested interest in the success of the supplier due to their contractual relationship. Third-party audits are considered to provide the greatest level of inherent credibility because an entity that is not directly linked to the supplier or the customer performs the audit. Most third-party animal welfare audits in the U.S. Beef industry are conducted by food safety and quality auditing companies. The third-party audit company has a financial relationship with the entity that pays for the audit, but this relationship is not practically avoidable in a fee-for-service arrangement. Third-party animal welfare audits were initially offered in in-person format, where an auditor would travel to the slaughter establishment and spend half a day or longer conducting their audit. With improvements in video and data transfer technologies, remote video auditing has become increasingly common.

The regular assessment of animal handling and welfare began in the late 1990s and early 2000s in cattle and pig slaughter establishments. These types of assessments were initially conducted by second-party food safety auditors of major quick service restaurant companies. As slaughter facilities began to understand the value of the data that were collected during the animal handling and welfare assessments, first-party auditing was widely adopted by the industry.

Video

In the mid-2000s, reports began to emerge among in-plant quality assurance personnel that first-party audit data were not reflective of actual animal handling performance in slaughter facilities. To a great extent, the discrepancy could be attributed to human nature as people typically put forth their best effort when they are being monitored and may relax their individual performance standards when they do not feel that they are being directly supervised. Remote video-auditing technology had been developed for hospitals and daycare centers by the mid-2000s and remote video auditing was developed and piloted for slaughter establishments by the end of 2006. The release of the undercover expose at Hallmark Westland Packing in Chino, CA, in the beginning of 2007 served as a catalyst for adoption of remote video-auditing technology by North American slaughter establishments.

Remote video-auditing systems consist of a few standard components. They include the on-site video recording system, Internet connection with appropriate bandwidth to support video data transfer, and the off-site auditing center. The general concept of operating a remote video-auditing system is based on the premise that the audit is conducted from outside of the location that is being audited. The ability to monitor human performance on tasks such as animal handling from an off-site location provides the opportunity to observe an accurate representation of employee behavior. The physical presence of an auditor during an in-person audit promotes employee behaviors that will improve performance on the audit.

PRESLAUGHTER HANDLING AND STUNNING CHALLENGES

Animal Condition

The condition of a portion of mature cattle presented for slaughter has been a long-term challenge for the beef industry. Although it is important to remember that mature cows are slaughtered for reasons that include common health and welfare issues, the reason for culling does not permit a producer to neglect animals and allow them to reach severe states of those conditions. A 2010 report on the condition of cows that were sold for slaughter in the western U.S. revealed that 34.8% of culled dairy cows and 10.4% of culled beef cows displayed body condition scores less than 2.0 (Ahola et al., 2010). The same study identified 2% of dairy cows as severely lame. A large national study reported that 2.7% of dairy cows displayed severe lameness at slaughter facilities (Nicholson et al., 2013). In contrast, 0.2% of beef cows were identified as severely lame in the same study. The 2016 National Beef Quality Audit reported that 0.1% of steers and heifers displayed a mobility score of 3 on a 4 point scale. A score of 3 was defined as "Exhibit(ing) obvious stiffness, difficulty taking steps, obvious limp, obvious discomfort, lags behind normal cattle." All other cattle in the study were able to keep up with other cattle during movement and the vast majority (96.8%) displayed no apparent lameness (NCBA, 2016b). When animal welfare and quality issues, such as poor body condition, poor udder condition, and severe lameness are identified, it is important for feedback regarding those animals to reach the producers that introduced animals with such conditions to the marketing chain. In the most recent iteration of the National Dairy FARM Program, materials have been incorporated to assist dairy farmers in making good choices regarding culling practices and drug administration. Some university dairy science programs have begun to incorporate beef quality and culling decision materials in locations within their curricula that all students are likely to encounter. The appropriate and careful use of education is a powerful tool in addressing and fixing animal welfare and quality issues that are chronic in nature with multiple causative factors.

Identification of Nonambulatory Animals in Lairage

In 2007, an undercover expose of abusive and inhumane handling practices was released from Hallmark Westland Packing in Chino, CA. The video was captured and released by the Humane Society of the United States. The release of the video to national media outlets resulted in increased public concern regarding the handling of cattle at slaughter facilities. In addition, further investigation by the USDA revealed that nonambulatory cattle were being slaughtered for human consumption at Hallmark Westland without passing FSIS antemortem inspection. Due to the potential for food safety implications as a result of allowing meat from nonambulatory cattle into the food supply, the Hallmark Westland incident resulted in the largest beef recall in U.S. history at that time. Much of the product included in the recall had already been consumed, but the company filed for bankruptcy protection in the wake of the financial loss associated with the event.

Prior to the Hallmark Westland incident, it was illegal to slaughter nonambulatory cattle for human consumption with the exception of cattle that had passed antemortem inspection by FSIS and became nonambulatory immediately before the restraint system during preslaughter handling. In those cases, plant personnel were allowed to stun nonambulatory cattle in their location and drag the carcass into the production process. After the beef recall associated with Hallmark Westland Packing, this exception was no longer allowed. If cattle become nonambulatory at any time after antemortem inspection by FSIS, they are required to be humanely euthanized and have their carcasses disposed appropriately.

The Humane Methods of Slaughter Act and 9 C.F.R §313 clearly require that all cattle in lairage must have access to water at all times. Recently, slaughter establishments have begun to grapple with the challenge of identifying cattle in sternal recumbence that are nonambulatory among other cattle that are in sternal recumbence but able to rise and walk, in a timely manner. Since both types of cattle have the same appearance while laying down, it is difficult to determine which animals are not able to access the water trough in lairage. Cattle are capable of withstanding water withdrawal for 18 hours in ambient temperatures near freezing (Vogel et al. 2011) before showing physiological changes associated with dehydration. However, warmer ambient temperatures are likely to increase water turnover within cattle and reduce the interval between drinking events. At this time, a common solution to the challenge of identifying nonambulatory cattle in lairage is to have plant personnel check the animals in lairage to be sure they can stand up and walk twice per day during the days that the slaughter process is not operating and for cattle that are held in lairage overnight. Although this solution is not perfect because it is unknown exactly when animals reach nonambulatory status, it begins to address the issue. Future innovation regarding this challenge may include the expansion of video monitoring capability or the adaption of other technologies, such as thermal imaging, to reduce the labor associated with identifying nonambulatory cattle in lairage and allow timely identification and response to those animals.

LAYOUT AND ERGONOMICS OF STUNNING PLATFORM

Multiple factors have been identified that contribute to the occurrence of ineffective stunning in beef slaughter facilities (Grandin, 2002). However, the layout and ergonomic profile of the stunning platform is often overlooked when slaughter establishments are evaluating causes for stunning problems. It is important to make sure that stunning equipment is fully operational, well maintained, and stunner operators are well-trained in stunner placement and operation. In addition, the position that the stunner operator is required to maintain during the stunning process must be evaluated. The stunner operator must have a clear view of the heads of cattle that will be stunned. The best vantage points are directly in front of the animal or to the front and side of the animal. Standing directly adjacent to, or behind cattle that are about to be stunned impedes the ability of the stunner operator to clearly see the appropriate landmarks to effectively stun the animal. Requiring stunner operators to operate stunning equipment in physical positions that are unnatural and require substantial reaching or movement of heavy equipment are also likely to result in the development of operator fatigue. As workspaces in the stunning and restraint area of beef slaughter facilities are remodeled or newly constructed, substantial planning should be allocated to assuring that the work flow and ergonomic design of the space and equipment helps to alleviate fatigue.

REFERENCES

Ahola, J. K., H. A. Foster, D. L. VanOverbeke, K. S. Jensen, R. L. Wilson, J. B. Glaze Jr., T. E. Fife, C. W. Gray, S. A. Nash, R. R. Panting, and N. R. Rimbey. 2010. Survey of quality defects in market beef and dairy cows and bulls sold through livestock auction markets in the Western United States: I. Incidence rates. J. Anim. Sci. 89: 1474–1483.

Apple J. K. 1999. Influence of body condition score on live and carcass value of cull beef cows. J Anim. Sci. 77: 2610–2620.

Grandin, T. 2002. Return-to-sensibility problems after penetrating captive bolt stunning of cattle in commercial beef slaughter plants. J. Am. Vet. Med. Assoc. 221: 1258–1261.

Hadley, G. L., Wolf, C. A., and Harsh, S. B. 2006. Dairy cattle culling patterns, explanations, and implications. J. Dairy Sci. 89: 2286–2296.

NAHMS. 2007. Dairy 2007, Part 1: Reference of dairy cattle health and management practices in the United States, 2007. USDA-APHIS-VS, CEAH, National Animal Health Monitoring System. Fort Collins, CO. www.aphis.usda.gov/dairy/dairy96/Dairy96_dr_Part1.pdf. Accessed August 21, 2010.

NCBA. 2016a. 2016 Market Cow and Bull National Beef Quality Audit Executive Summary. National Cattlemen's Beef Association. Centennial, CO. www.bqa.org/Media/BQA/Docs/nbqa-exec-summary_cowbull_final.pdf. Accessed October 5, 2017.

NCBA. 2016b. 2016 National Beef Quality Audit Executive Summary. www.bqa.org/Media/BQA/Docs/2016nbqa_es.pdf. Accessed October 6, 2017.

Nicholson, J. D. W., K. L. Nicholson, L. L. Frenzel, R. J. Maddock, R. J. Delmore Jr., T. E. Lawrence, W. R. Henning, T. D. Pringle, D. D. Johnson, J. C. Paschal, R. J. Gill, J. J. Cleere, B. B. Carpenter, R. V. Machen, J. P. Banta, D. S. Hale, D. B. Griffin, and J. W. Savell. 2013. Survey of transportation procedures, management practices, and health assessment related to quality, quantity, and value for market beef and dairy cows and bulls. J. Anim. Sci. 91: 5026–5036.

Vogel, K. D., J. R. Claus, T. Grandin, G. R. Oetzel, and D. M. Schaefer. 2011. Water access and health status effects on blood components, body weight loss percentage, and fresh meat characteristics of Holstein slaughter cows. J. Anim. Sci. 89: 538–548.

CHAPTER 13

Optimal Human Animal Interactions for Improving Cattle Handling

Temple Grandin
Colorado State University

CONTENTS

To have good animal welfare during handling for vaccinations, truck loading, milking, moving between pastures or movement through a slaughter plant REQUIRES management who is committed to maintaining high standards. Too often people want the new state-of-the-art facility but they do not have the management commitment to make it work right. I have a saying, "People want the thing more than they want the management." Equipment is only half the equation and the attitude of either the producer who owns a ranch or the manager of a large operation will determine how animals are treated during handling. If you had a choice between two different cattle operations, which one would you choose? The first ranch has state-of-the-art equipment, and a poor management or a ranch with older adequate facilities and excellent management. I will choose the one with the excellent manager.

IMPROVEMENTS IN HANDLING AND NUMERICAL SCORING

During a career that spans over 40 years I have seen many improvements in cattle handling at both slaughter plants and in large feedlots. A major driver of this change has been pressure from consumers and large meat buyers such as McDonald's Corporation and Wendy's International (Grandin, 2000b, 2006).

A numerical scoring system developed by the author (Grandin, 1998a) made it possible to put hard numbers on animal-handling practices. This enables people to determine if their practices are getting better or becoming worse. Numerical scoring is now routinely used in slaughter plants, feedlots, and ranches to evaluate handling practices (Bourquet et al., 2011; Dunn, 1990; Edge et al., 2005; Edge and Barnett 2009; Grandin, 1998a, 1998b, 2010; NCBA, 2009; Simon et al., 2016; Welfare Quality, 2009; Woiwode et al., 2016a). Some of the variables that are measured for beef cattle are as follows:

- Percentage of cattle that fall during handling
- Percentage of cattle that slip or stumble
- Percentage that vocalize in a restraint device (squeeze chute)
- Percentage moved with an electric prod
- Percentage that run during handling
- Percentage that balk or turn back
- Percentage miscaught in a squeeze chute headgate

Numerical scoring of handling has also been used to evaluate handling of dairy cattle. The following measures are used (Bertenshaw et al., 2008; Dotzi et al., 2011; Rouse et al., 2004; Waiblinger et al. 2003).

- Percentage of cows kicking during milker attachment
- Percentage moved with aggressive handling such as slapping
- Flight zone or avoidance distance of the dairy cows from people
- Percentage of cows flinch stepping during attachment of the milker
- Percentage of cows yelled at.

TRACKING IMPROVEMENTS IN HANDLING

When baseline scores were first collected at slaughter plants, the usage of electric prods was high. In some places, each animal was poked more than once with an electric prod. Recent surveys of large feedlots in Kansas, Colorado, and Nebraska indicate that the average electric prod use is on 5% or less of the cattle (Barnhardt et al., 2015; Woiwode et al., 2016a). People manage the things that they measure. They are also motivated to change when a large customer requires it. Another motivator is benchmarking where feedlot or meat plant managers can determine how they compare to others. This has been effective for reducing lameness in dairy cows (VonKeyserlingk et al., 2012). People are often motivated to be better than other producers.

Vocalization Scoring During Cattle Handling and Restraint—A high percentage of cattle vocalizing that occurs while they are either moving through a chute (race) or held in squeeze chute is an indicator of a severe welfare issue that must be corrected. Since this is a measure of handling, vocalizations associated with a painful procedure such as branding, are not scored. The vocalization score is recorded after the animal is fully restrained, before procedures are started. Vocalization scoring is effective for identifying problems that are associated with electric prod use, excessive pressure from a restraint device or sharp edges (Grandin, 1998a, 2001; Bourquet et al., 2011). When handling and restraint at a packing plant or a feedlot is done well, the average vocalization score will be 3% or less at the plant and under 2% at a feedlot (Barnhardt et al., 2015; Woiwode et al., 2016a; Grandin, 2000a, 2005). The range in the feedlot survey was 0 to 6% of the cattle vocalizing (Woiwode et al., 2016a). When problems start occurring, such as excessive pressure from a restraint device or increased electric prod use, vocalization averages can rise to 23% (Grandin, 2001), 25%

Table 13.1   Maximum Acceptable and Excellent Scores for Handling on Published Industry Guidelines on Cattle Handling in Slaughter Plants and Beef Feedlots

	Falls	Slip/Stumble	Vocalization	Electric Prod Use
NAMI Slaughter Plants				
Excellent	0%	3%	1%	5%
Acceptable	1%		5%	25%
BQA/NCBA Feedlot and Ranch				
Excellent			No Excellent Score	
Acceptable	2%	10%	10%	10%
National Cattle Feeders Association Canadian Feedlot Assessment				
Excellent			No Excellent Score	
Target Less Than	1%	5%	15%	10%

(Bourquet et al., 2011), and 57% (Hayes et al., 2015). Woiwode et al. (2016a) clearly showed that feedlots could easily achieve the published cutoff points for vocalization, electric prod use, falling and stumbling. The Beef Quality Assurance Feedlot Assessment Guide allows 2% falls, 10% electric prod use, and 5% vocalization. Simon et al. (2016) conducted a survey of 25 ranches in California. They did numerical scoring of cattle being moved through chutes into squeeze chutes. Average scores for vocalization and falling were similar to the feedlot surveys. Vocalization in the squeeze chute ranged from 0 to 20% with an average score of 5.2% (Simon et al., 2016). One area where the ranchers need to improve is reducing electric prod use. It averaged 25% of the cattle. Cattle that vocalize during restraint have higher cortisol levels and lower average daily gain (Dunn, 1990; Hemsworth et al., 2011; Woiwode et al., 2016b) (Table 13.1).

U.S. and Canadian slaughter plants can easily comply with these limits (Grandin, 2005, 2006). Many European plants can also comply. Plants in Mexico and Columbia need to work on improving. One plant in Colombia had a 42% electric prod score, 11% vocalization, and 8% falls (Romero et al., 2017). The falls were due to a poorly designed 45° unloading ramp. In the stunning area of a plant in Mexico, the vocalization score was 10%. Sixty-seven percent of the cattle were moved with an electric prod (Minanda et al., 2012).

A kosher slaughter plant that was not audited by major customers had 47% of the cattle vocalizing (Hayes et al., 2015). This was probably due to both restraint equipment problems and electric prod use. Well-run kosher plants can achieve 5% or less vocalization scores (Grandin, 2012). Taken together, these figures show that when an abattoir has problems, the vocalization scores will clearly reflect them.

Positive Attitude is Important—The use of numerical scoring can prevent bad welfare. To achieve the highest level of animal welfare requires stock people who have positive attitudes and really like animals. Coleman et al. (2014), Fukasawa et al. (2016), and Noffsinger and Locatelli (2004) have all clearly shown that people who like animals have more productive animals. If cattle are afraid of people, and have a large flight zone, they will have lower milk production, higher somatic cell count or lower weight gain (Fulwider et al., 2007; Rushen et al., 1999; Waiblinger et al., 2002). Breuer et al. (2000) found that negative handling by a stock person who yelled at cows and hit them lowered milk production. The science is clear, good stockmanship improves both welfare and animal performance. There is a need in animal agriculture to put the "husbandry" back into animal science.

W.D. Mumford wrote A Tribute to the Stockman (Upson and Garrigus, 1995). A major building at University of Illinois Animal Science Department was named after him. Dr. Mumford's words emphasize the art of stockmanship when he writes, "Behold the Stockman, Artist, and Artisan."

Mumford's words also make it obvious that the animals like the stock person and the stock person likes the animals. Mumford writes further about the animal's response to the stockperson.

"His approaching footsteps call forth the affectionate whinny of recognition." His calm well-modulated voice inspires confidence and wins affection."

W.D. Hoard, the founder of the Hoard's Dairyman magazine wrote, "The rule to be observed in the stable at all times, toward the cattle, young and old, is that of patience and kindness. A man's usefulness to the herd ceases at once when he loses his temper or bestows rough usage."

DO NOT UNDERSTAFF AND OVERWORK

For a stockperson to have a positive attitude, it is essential that they are not worked to exhaustion where they get too tired to care about doing things right. Unpublished data in the broiler chicken and pork industry indicate that after a person has worked for over 6 hours loading either chickens or pigs onto trucks, they start to get tired and death losses go up. Another problem is understaffing. It is impossible for a stockperson to have a positive attitude if they are doing a job that two people should be doing. Specialists in low-stress cattle handling emphasize the importance of people getting to know the unique behaviors of individual animals. This enables the animals to develop trust with the stockpersons (Smith-Thomas, 2016). To do this, the stock person must not be overworked. In large operations, each stock person should be given a group of animals he/she can take ownership of and care for. In a big feedlot, the pens of cattle could be divided up between the different pen riders.

IMPROVE THE PROFESSIONAL STATUS OF STOCK PEOPLE

There is a need to increase the professional status of the people who handle cattle, doing jobs such as vaccinating the animals or riding the feedlot pens to look for sick cattle. There are many people who would be really good at handling cattle on either large ranches or feedlots. Low pay is one reason it is difficult to get good stock people. I had a student tell me that she loved working with pigs in the farrowing barn, but she was not getting sufficient pay to live in a decent house. She quit the job because she did not want to live in a trailer. Managers need to place a higher value on people who either already have, or are willing to develop really good stockmanship skills. Fortunately, there are now many classes for training stock people, because managers are recognizing the importance of stockmanship. Progressive managers are enrolling employees in stockmanship training.

ANIMALS REMEMBER BAD EXPERIENCES AND HAVE SPECIFIC MEMORIES

Cattle have long memories and they do not forget bad experiences. In one of my early research projects, I learned that cattle could remember being accidentally banged on the head with the head-gate of the squeeze chute. Thirty days later, the next time they entered the chute, they were more likely to balk (Grandin, 1993). A study with sheep clearly showed that they remembered an aversive experience a year later (Hutson, 1980). When electric prods were used, cattle were more likely to balk and more force was required to move them through a chute in the future (Goodewardene et al., 1999). Dairy cows have specific things they are afraid of Lindahl et al. (2013) found that a cow's heart rate was higher and they balked more when being moved to the hoof-trimming chute, compared to being moved to the milking parlor. Animal memories are specific because animal memories are sensory based. For example, if a horse is trained to tolerate blue and white umbrellas, that does not transfer to orange tarp (Leiner and Fendt, 2011). Umbrellas and tarps look totally different. Cooke (2014) also found that getting cattle accustomed to being fed from a person on a truck does not transfer to being handled in the corrals. In both of these cases, the tarp and the corrals were new, novel, and scary. This same problem can happen with extensively raised cattle that have become accustomed to riders on horseback. The rider on horseback is perceived as familiar and safe and a person on foot is new, novel, and frightening. The author has observed that the flight zone between a horse and rider may be 5 ft (2 m) but it may expand to 20 ft (6 m) to 30 ft (9 m) when the animals first see a person on foot. This can be very dangerous when the cattle are being moved into a small pen. To get the handler out of their flight zone, the cattle may suddenly turn back and rebound out of a pen and run over a person. It is now clear that an animal's previous experiences will have an effect on how it will react to handling in the future (Grandin, 1997; Grandin and Shivley, 2015).

ACCLIMATE ANIMALS TO HANDLING

Progressive ranchers take the time to acclimate young heifers to having people walk through them, and to moving them through the corrals (Cooke, 2014; Cooke et al., 2009). It is important to use this method on young heifers. It is less likely to be effective on older cows (Cooke, 2014). Hutson (1985) improved the movement of sheep through a handling system by giving them rewards of a small amount of tasty grain. During the first few passes through the handling system, the heifers should be allowed to walk through it. All the gates should be open and backstops tied up. To keep the heifers calm, the acclimation sessions should be limited to one per day. Doing several sessions in rapid succession can make the animals stressed. Each time they passed through the handling facility they became increasingly agitated. They had no time to calm down between the rapidly repeated acclimation sessions.

Frequent moving of beef cattle between different pastures improves the animal's temperament and produces calmer animal (Ceballos et al., 2016). Animal temperament can interact with acclimation. Stock people must be careful to not frighten temperamentally reactive cattle during acclimation. Acclimating low reactive dairy heifers to the milking facilities before freshening, reduced flight distances and kicking during milking machine attachment. Unfortunately, high reactive heifers became worse (Sutherland and Huddart, 2012). The author speculates that the poor effects of acclimation may be partly due to the lack of feed rewards given in the milking parlor. An earlier study showed that longer term acclimation of dairy heifers to positive interactions with people before calving greatly reduced kicking and other agitated behavior when they first went in the milking parlor (Bertenshaw et al., 2008).

UNDERSTANDING BEHAVIORAL PRINCIPLES

It is essential that people handling cattle understand some basic principles of animal handling. The first principle is that calm animals are easier to handle them agitated fearful animals. Agitated behaviors that occur during handling are usually motivated by either fear or pain. Fear is a proper scientific word. The presence of fear circuits in the animal's brain is well documented (Panksepp et al., 2011). Australian researchers have studied the interactions between animals and people and they emphasized the importance of reducing fear (Hemsworth et al., 2011).

- Signs of Fear in Cattle—Listed in order of appearance.
  - Tail swishing back and forth when no flies are present (Grandin, 2017)
  - Defecation
  - Eye white shows (Core et al., 2014; Sandem et al., 2006)
  - Rearing or agitated behavior when confined in a chute
- Handlers Should Be Quiet—Loud noise and yelling and screaming at cattle is very stressful (Waynert et al., 1999). Lima et al. (2016) found that replacing rough cattle handling with calm, quiet handling reduced Cortisol levels.

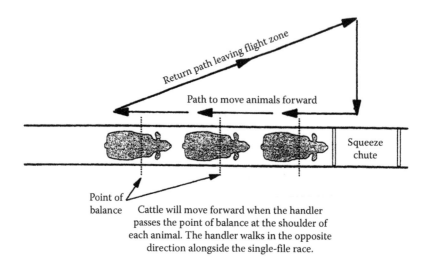

Figure 13.1 shows the text: "Return path leaving flight zone", "Path to move animals forward", "Squeeze chute", "Point of balance"

Point of
balance   Cattle will move forward when the handler
           passes the point of balance at the shoulder of
           each animal. The handler walks in the opposite
           direction alongside the single-file race.

Figure 13.1 Handler movement pattern to encourage an animal to move forward in a single-file race (chute).

- <u>Understand Flight Zone Principles</u>—Unless cattle are completely tame and trained to lead, they will have a flight zone. The four factors that determine the size of the flight zone are (1) genetics, (2) duration of contact with people, (3) the quality of the contact (Grandin, 2000a), and (4) the speed and angle of the handler's approach (Patrick Tobola, personal communication, 2017). Animals that seldom see people or they have been handled in a rough, aggressive manner will have larger flight zones than cattle that have had regular gentle contact. A common mistake made by novice handlers is to stand in the flight zone when there is no way for the cattle to move away. This is most likely to occur when cattle are held in a single-file chute. If an animal rears, becomes agitated, or starts to defecate, the handler should back up and remove him/herself from the flight zone.

TWO ZONES AROUND CATTLE

There are two zones around cattle where a person has an effect on moving an animal. The innermost zone is the flight zone. When a person enters the flight zone, an animal on an open pasture will move away. The outermost zone, which is outside the flight zone, is called the zone of awareness or the pressure zone. When a person enters this zone, the cattle will turn and look at you. There is much more information on these zones in Grandin and Deesing (2008) and Grandin (2017).

<u>Point of Balance</u>—When people are moving cattle through a single-file chute, they need to learn to never stand at the head of the animal and poke it on the butt. This gives it conflicting signals. A counterintuitive method to urge an animal to move forward in a single-file chute is to quickly walk back past it in the opposite direction of desired movement (Grandin, 2000a) (Figure 13.1). When the handler crosses the point of balance at the shoulder, it will move forward. Cattle-handling specialist Tom Noffsinger recommends stroking the animal from withers to tail to encourage it to go forward (Personal communication, 2015).

NEVER USE AN ELECTRIC PROD AS THE PRIMARY DRIVING TOOL

A flag, plastic paddle or no driving aid should be used for moving cattle. People often get overly aggressive with driving aids. Proponents of low-stress cattle handling emphasize learning how to move animals by using very small body movements. Some welfare advocates want to completely ban electric prods. The author recommends banning constant carrying of electric prods. It should only be picked up when a stubborn animal refuses to enter the squeeze chute. These cattle are most likely to be animals that have had a bad experience with the squeeze chute in the past. A single zap with an electric prod is preferable to hard tail twisting or hitting the animal.

PRINCIPLES OF PRESSURE AND RELEASE

A basic principle of animal handling is that when an animal does something you want, you reward it by giving it relief. When working the flight zone, you back off when it moves where you want it to go. If you have twisted a tail, let it go the instant the animal starts to cooperate. Letting go rewards the cow for moving.

Prevent Acts of Abuse—Abusive behavior by people during cattle handling is never acceptable. Numerous guidelines for handling cattle have prohibited behaviors toward animals that are never acceptable. One of the big drivers of stopping acts of abuse has been undercover video shown on the internet.

The OIE World Organization of Animal Health Transportation Guideline (2016), forbids the use of abusive methods such as sticks with pointed ends, pressure on sensitive areas of the animal such as eye or genitals. Lifting by tail or horns is also prohibited. During truck loading or handling at a slaughter plant, animals should not be thrown, dragged, or dropped. Other guidelines written by industry organizations have strong prohibitions against abuse. The author has observed that there are some people who do not have the right temperament for working with animals. These people will need to be removed.

STEPS FOR MANAGERS TO TAKE TO IMPROVE WELFARE DURING CATTLE HANDLING

Animal welfare concerns range from preventing abuse to giving animals a life worth living. Many consumers are concerned that farm animals do not lead a natural life. Beef cattle have fewer welfare issues compared to other species because in extensive operations, the cows and bulls are on pasture. Improving cattle welfare during handling ranges from the basics such as preventing abuse, to developing high levels of stockmanship where both the stock people and the cattle trust each other.

Step 1. Prevent Abuse and Neglect—Prevent bad welfare and pain and distress by preventing acts of abuse, such as beating animals or dragging conscious downed non-ambulatory cattle. Do not understaff stockmanship positions, or overwork people until they are too exhausted to care.

Step 2. Numerical Scoring—Implement the use of numerical scoring to achieve an acceptable level of animal handling. This will bring handling practices up to an acceptable level.

Step 3. Training Employees—Provide employees with training classes and support so they can learn the art of stockmanship and develop a herd of mother cows that trusts them (Grandin, 2017). When cattle learn to trust the handler, they will be more willing to drive straight and not turn around to look at the person (Cote, 2003). Ranchers on extensive western ranches have become increasingly interested in methods to place cattle on pasture and rotate them without fencing. This requires a high degree of stockmanship skill. Barnes and Hibbard (2016) have a paper that describes these methods. Stephenson et al. (2016a, 2016b) found that a combination of Bud William's low-stress handling methods and supplements was effective for placing cattle. The author has been on operations where the numerical scores were better than the BQA guidelines, but further changes in procedures resulted in calmer cattle handling. At one feedlot visited by the author changing the position of one employee and stopping him from constantly waving a flag, caused the cattle to stand much more quietly in a single animal scale.

Step 4. Learn the Art of Stockmanship—Both management and the employees handling cattle owe it to the animals we use for food to give them a life worth living. Bud Williams, the originator of many concepts of low-stress handling, emphasizes the importance of having the right attitude (Smith-Thomas, 2016). Some people are more intuitive with animals than others. They have a feel for them. Managers need to find these natural stock people employees and pay them decently. In the pork industry, stockmanship has become more important as the industry transitions from gestation stalls to group housing To teach young pigs to successfully use electronic feeders requires a calm patient person to be their Hog Whisperer (Coleman, 2016). To find this right person was not easy, and many people tried and failed to become the successful "Hog Whisperer." To find these excellent stockpeople, the managers MUST care. Changing attitudes starts at the top of the organization.

REFERENCES

Barnhardt, T.T., Thomson, D.U., Terrell, S.P., Rezac, D.J., Frese, D.A., and Reinhardt, C.D. (2015) Implementation of industry oriented animal welfare and quality assurance assessment in Kansas feeding operations, Bovine Pract., 45: 81–87.

Barnes, M. and Hibbard, W. (2016) Strategic grazing management using low stress herding and night penning for animal impact, Stockmanship Journal, 5(2): 57–65.

Bertenshaw, C., Rowlingson, P., Edge, H., Douglas, S., and Shiel, R. (2008) The effect of different degrees of positive human-animal interaction during rearing and welfare and subsequent production of commercial dairy heifers, Appl. Anim. Behav. Sci., 114: 65–75.

Bourquet, C., Deiss, V., Tannugi, C.C., and Terlouw, E.M.C. (2011) Behavioural and physiological reactions of cattle in a commercial abattoir: Relationships with organizational aspects of the abattoir and animal characteristics, Meat Sci., 88: 158–168.

Breuer, K., Hemsworth, P.H., Barnett, J.L., Matthews, L.R., and Coleman, G.J. (2000) Behavioral response to humans and the productivity of commercial dairy cows, Appl. Anim. Behav. Sci., 66: 273–288.

Ceballos, M.C., Gois, K.C.R., Sant Anna, A.C., and Paranos de Costa, M.J.R. (2016) Frequent handling of grazing beef cattle maintained under rotational stocking method improves temperament over time, Anim. Prod. Sci. doi: 10.1071/AN16025.

Coleman, L.L. (2016) Achieving high productivity in group housed sows, Advances in Pork Production, Vol. 27, University of Alberta, Banff Pork Seminar Proceedings, pp. 95–102.

Coleman, G.J. and Hemsworth, P.H. (2014) Training to improve stockperson beliefs and behavior towards livestock enhances welfare and productivity, Rev. Sci. Technol., 33: 131–137.

Cooke, R.F. (2014) Bill E. Kunkle Interdisciplinary Beef Symposium: Temperament and acclimation to human handling influence growth health and reproductive performance in Bos Taurus and Bos indicus cattle, J. Anim. Sci., 92: 5325–5333.

Cooke, R.F., Arthington, J.D., Austin, B.R., and Yelich, J.V. (2009) Effects of acclimation to handling on performance, reproductive and physiological responses of Brahman-crossbred heifers, J. Anim. Sci., 87: 3403–3412.

Cote, S. (2003) Stockmanship: A powerful tool for grazing lands management, USDA Natural Resources.

Core, S., Widowski, T., and Mason, G. (2014) Eye white percentage as a predictor of temperament in beef cattle, J. Anim. Sci., 87: 2168–2174.

Dotzi, M.S. and Muchenje, V. (2011) Avoidance related behavioral variables and their relationship to milk yield in pasture based dairy cows, Appl. Anim. Behav. Sci., 133: 11–17.

Dunn, C. (1990) Stress reactions of cattle undergoing ritual slaughter using two methods of restraint, Vet. Rec., 126: 522–525.

Edge, M.K., Maguire, T., Dorian, J., and Barnett, J.L. (2005) National Animal Welfare Standards for Livestock Processing Establishments Preparing Meat for Human Consumption, Part 1: The Standards Anim. Welf. Sci Centre, Werribee, Victoria, Australia: Australian Meat Ind. Counc., Crows Nest, NSW, Australia.

Edge, M. and Barnett, J. (2009) Development of animal welfare standards for the livestock transport industry: Process, challenges, and implementation, J. Vet. Behav. Clin. Appl. Res., 4: 187–192.

Fukasawa, M., Kawahata, M., Higashiyama, Y., and Komatsu, T. (2016) Relationship between the stockperson's attitudes and dairy productivity in Japan, Anim. Sci. J., (88:394–400).

Fulwider, W.K., Grandin, T., Garrick, D.J., Engle, T.E., Lamm, W.D., Dalsted, N.L., and Rollin, B.E. (2007) Influence of stall base on tarsal joint lesions and hygiene in dairy cows, J. Dairy Sci., 90: 3559–3566.

Goodewardene, L.A., Price, M.A., Okine, E., and Berg, R. (1999) Behavioral response to handling and restraint in dehorned and polled cattle, Appl. Anim. Behav. Sci., 64: 159–167.

Grandin, T. (1993) Behavioral agitation during handling of cattle is persistent over time, Appl. Anim. Behav. Sci., 36: 1–9.

Grandin, T. (1997) Survey of Stunning and Handling in Federally Inspected Beef, Veal, Pork, and Sheep Slaughter Plants, United States Department of Agriculture, Agricultural Research Service Project 3602-32000-08G, USDA, Beltsville, MD.

Grandin, T. (1998a) Objective scoring of animal handling and stunning practices at slaughter plants, J. Am. Vet. Med. Assoc., 212: 36–39.

Grandin, T. (1998b) The feasibility of using vocalization scoring as an indicator of poor welfare during cattle slaughter, Appl. Anim. Behav. Sci., 56: 121–128.

Grandin, T. (2000a) Livestock Handling and Transport, CABI International, Wallingford, Oxfordshire, UK.

Grandin, T. (2000b) Effect of animal welfare audits of slaughter plants by a major fast food company on cattle handling and stunning practices, J. Am. Vet. Med. Assoc., 216: 848–851.

Grandin, T. (2001) Cattle vocalizations are associated with handling and equipment problems at beef slaughter plants, Appl. Anim. Behav. Sci., 7: 191–201.

Grandin, T. (2005) Maintenance of good animal welfare standards in beef slaughter plants by use of auditing programs, J. AM. Vet. Med. Assoc., 226: 370–373.

Grandin, T. (2006) Progress and challenges in animal handling and slaughter in the U.S., Appl. Anim. Behav. Sci., 100: 126–139.

Grandin, T. (2010) Auditing animal welfare at slaughter plants, Meat Sci., 86: 56–65.

Grandin, T. (2012) Developing measures to audit welfare of cattle and pigs at slaughter, Anim. Welf., 21: 351–356.

Grandin, T. (2017) Temple Grandin's Guide for Working with Farm Animals, Storey Publishing, North Adams, MA, USA.

Grandin, T. and Deesing, D.M. (2008) Humane Livestock Handling, Storey Publishing, North Adams, MA, USA.

Grandin, T. and Shivley, C. (2015) How do farm animals react and perceive stressful situations such as handling, restraint, and transport, Animals, 5(4): 1233–1251.

Hayes, N.S., Schartz, C.A., and Maddock, R.J. (2015) The relationship between pre-harvest stress and carcass characteristics of beef heifers that qualified for kosher designation, Meat Sci., 100: 134–138.

Hemsworth, P.H., Rice, M., Karlen, M.G., Calleja, L., Barnett, J.L., Nash, J., and Coleman, G.J. (2011) Human-animal interactions at abattoirs: Relationships between handling and animal stress in sheep and cattle, Appl. Anim. Behav. Sci., 135: 24–33.

Hutson, G.D. (1980) The effect of previous experience on sheep movement through yards, Appl. Anim. Ethol., 6: 233–240.

Hutson, G.D. (1985) The influence of barley food rewards on sheep movement through a handling system, Appl. Anim. Behav. Sci., 14: 263–273.

Leiner, L. and Fendt, M. (2011) Behaviour fear and heart rate responses of horses after exposure to novel objects: Effect of habituation, Appl. Anim. Behav. Sci., 13: 104–109.

Lima, M.L.P., Negrao, J.A., Paz, C.C.F., and Grandin, T. (2016) Effect of corral modification for humane live-stock handling on cattle behavior and cortisol release, J. Anim. Sci., 94: 31–32 (Abstract).

Lindahl, C., Lundquist, P., Hagevoort, R., Lunner Kolstrup, C., Douphrate, D., and Pinzke, S. (2013) Occupational health and safety aspects of animal handling and dairy production, Journal of Agromedicine, 18(3), 274–283.

Minanda, G.C., Serrano, I.G.L., Sanchez-Lopez, C.P., Maria, G.A., and Saavedra, F.F. (2012) Assessment of cattle welfare at a commercial plant in the Northwest of Mexico, Trop. Anim. Health Prod., 44: 497–504.

NCBA (2009) Beef Quality Assurance Feedyard Assessment Assessor's Guide, National Cattlemen's Beef Association, Centennial, CO, www.bqa.org (Accessed Sep. 20, 2016).

National Cattle Feeders Association (2015) Canadian Feedlot Animal Care Assessment Program.

Noffsinger, T. and Locatelli, L. (2004) Low stress handling and overlooked dimension of management, Proceedings Meeting Academy of Veterinary (accessed Janury 15, 2017).

Panksepp, J. (2011) The basic emotional circuits of mammalian brains: Do animals have affective lives? Neurosci. Biobehav. Rev., 35: 1791–1804.

Romero, M.H., Uribe-Valasquez, L.F., and Miranda de la Lima, G.C. (2017) Conventional versus modern abattoirs in Columbia: Impacts on welfare indicators and risk factors of high muscle pH in commercial zebu young bulls, Meat Sci., 123: 173–181.

Rouse, T., Bonde, M., Bandsburg, J.H., and Sorensen, J.T. (2004) Stepping and kicking behavior during milking in relation to response in human animal interaction test and clinical health in loose housed dairy cows Livest. Prod. Sci., 88: 1–8.

Rushen, J., dePassille, A.M.B., and Munksgaard, L. (1999) Fear of people by sows and effects on milk yield, behavior, and heart rate at milking, J. Dairy Sci., 82: 720–727.

Sandem, A.I., Janczak, A.M., Salte, R., and Braastad, B.O. (2006) The use of diazepam as pharmacological validation of eye white as an indicator of emotional state in dairy cows, Appl. Anim. Behav. Sci., 97: 177–183.

Simon, G.E., Hoar, B.A., and Tucker, C.B. (2016) Assessing cow-calf welfare, Part 2, Risk factors for beef cow health and behavior and stockperson handling, J. Anim. Sci., 94: 3488–3500.

Smith-Thomas, H.S. (2016) Training pen riders in low stress cattle handling, Feedlot, Dec, p. 8–9.

Stephenson, M.B., Bailey, D.B., Howery, L.D. and Henderson, L. (2016a) Efficacy of low stress herding and low moisture block to target grazing locations of New Mexico rangelands, J. Arid. Environ., 130: 84–93.

Stephenson, M.B., Bailey, D.W., Bruegger, R.A. and Howry, L.D. (2016b) Factors affecting efficacy of low stress herding and supplement placement to target cattle grazing locations. Rangeland Ecology and Management, (70:202–209).

Sutherland, A. and Huddart, F.J. (2012) The effect of training first lactation heifers to the milking parlor on the behavioral reactivity to humans and the physiological and behavioral responses to milking and productivity, J. Dairy Sci., 95: 6983–6993.

Upson, S. and Garrigus, O.T. (1995) Herbert Winsor Mumford, 1871–1938: A brief Biography, J. Anim. Sci., 73: 3499–3502.

VonKeyserlingk, M.A.C., Barrientos, A., Ito, K., Gallo, E. and Weary, D.M. (2012) Benchmarking cow comfort on North American freestall dairies, Lameness leg injuries, lying time, facility design, and management for high producing Holstein dairy cows, J. Dairy Sci., 95: 7399–7408.

Waiblinger, S., Menke, C., and Coleman, G. (2002) The relationship between attitudes, personal characteristics and behavior of stock people and subsequent behavior and production of dairy cows, Appl. Aim. Behav. Sci., 79: 195–219.

Waiblinger, S., Menke, C., and Folsch, D.W. (2003) Influence on the avoidance and approach behavior of dairy cows towards humans on 35 farms, Appl. Aim. Behav. Sci., 84: 23–39.

Waynert, D.F., Stookey, J.M., Schartzkopf-Genswein, K.S., Watts, J.M., and Waltz, C.A. (1999) The response of beef cattle to noise during handling, Appl. Anim. Behav. Sci., 62: 27–42.

Welfare Quality (2009) Assessment protocol for cattle, welfarequalitynetwork.net (accessed Jan. 4, 2017).

Woiwode, R., Grandin, T., Kirch, B., and Patterson, J. (2016a) Compliance of large feedyards in the Northern high plains with Beef Quality Assurance Feedyard Assessment, Prof. Anim. Sci., 32: 750–757.

Woiwode, T., Grandin, T., Kirch, B., and Paterson, J. (2016b) Effects of initial handling practices on behavior and average daily gain of fed steers, Int. J. Livest. Prod., 7: 12–18.

CHAPTER 14

Cattle Welfare and Principles of Handling Facility Design

Temple Grandin
Colorado State University

CONTENTS

There are many different designs for milking parlors and beef cattle handling facilities. Personal preference often determines the type of facility that is chosen from a wide variety of good designs. During a long career, the author has observed that there are certain design mistakes that people keep making over and over. There are three kinds of mistakes. The first kind is that the people designing the facility did not take the time to look up design information. This results in layout mistakes, which cause balking and animals fail to move easily through the facility. The second type of mistakes is cutting costs on construction to the point where welfare may be severely compromised. Common cost-cutting methods are overcrowding dairy cows in a free stall barn or flimsy construction that is more likely to fail. The third area where many mistakes are made is flooring surfaces.

AVOID THESE COMMON MISTAKES IN FACILITY DESIGN

The most common mistake is a hard troweled glass smooth floor installed in a facility for handling cattle. This results in animals slipping and falling. In 2016, the author visited a university Experiment Station that had a new beef cattle handling facility with glass smooth floors. The professors knew it was wrong, but it happened because nobody told the mason who was finishing the floor that it should not be totally smooth.

Ramps for loading or unloading trucks that are too steep are another common mistake. Romero et al. (2017) compared cattle handling in two abattoirs in the country of Colombia. One was an old abattoir and the other one was a modern new abattoir designed to improve both animal welfare and beef quality. Unfortunately, the new plant had a very steep unloading ramp on a 45° angle and the old plant had a ramp that was properly designed with a 20° angle. Slipping was 3% on the properly designed ramp and 7% on the steep ramp. The modern plant also had more slipping in the stun box, 17.6% vs. 31.7% (Romero et al., 2017). The author has observed many new stun boxes in South America with floors that were too slick. Scoring of slips and falls during handling can be used to evaluate both flooring and cattle handling practices (Grandin, 1998; Velarde et al., 2012; Welfare Quality, 2009).

Since 2015, the author has observed several cases of substandard cheap construction. On one farm outside the U.S., the concrete work for supporting a rotary milking parlor was done so poorly that it collapsed during construction. To insure reliable rotation of the platform, it MUST be installed on a high-quality concrete foundation that will remain level under the tremendous weight of all the cows. The author has also visited dairies where substandard construction was used for the free stall barn roof. There was evidence of the roof steel work starting to bend under a heavy snow load. About every 10 years, the state where the dairy is located will get 2 feet (0.75 m) of wet snow. This building with its almost flat roof and flimsy steel structure is at a high risk of collapsing onto the cows.

CONFLICTING GOALS ON DAIRY FACILITY DESIGN

When I search through the scientific literature, I find conflicting reports on dairy housing and management. Several papers state very clearly that to maintain high-producing cows and prevent lameness, each cow needs a dry soft place to lie down (Cook et al., 2016; deVries, et al., 2015). On a well-managed free stall dairy, the percentage of lame high-producing cows is very low—2.87% (Cook et al., 2016). There is a big difference between well-managed and poorly managed dairies. A poorly managed dairy had 36% lame cows (Cook et al., 2016). There is additional information on lameness and lameness scoring in Flower et al. (2008), Gibbons et al. (2012), March et al. (2007), Chapinal et al. (2014), Schlageter-Tlelo et al., (2014), and Von Keyserlingk et al. (2012). It is also important to insure that the cows are not competing for feed. If a high-producing cow gets pushed away from the feed, she will lose body condition. Then another paper in the same year from the same scientific journal will tell the dairyman that he will have an economic advantage if he over-stocks the free stalls (deVries, et al., 2016). In this situation, the individual cow may suffer but the entire house makes more money because construction costs are reduced.

When free stall housing was first invented, most barns had open sides and high-pitched roofs with a wide ridge vent. This design provided good natural ventilation and supplemental electric fan ventilation was seldom needed. The air quality in these buildings was usually excellent and they stayed cool in the summer because the combination of the high roof and wide ridge vent allowed heat to escape. There is a trend today to build free stall barns with a low roof, which are completely dependent on mechanical ventilation. Many of these buildings have no skylights to admit natural light. It is possible to maintain good air quality in these buildings but my big concern is they will not satisfy the public concern about providing the cow with access to natural elements such as sunlight.

There is a reason why building contractors promote the mechanically ventilated building with low roofs. They require much less labor to build. A large crane is not required to build the low roof. Contractors promote building designs where they can make money. They make more money on the low-profile construction and selling and servicing all the fans that are required to ventilate it.

NEVER SKIMP ON SITEWORK GRADING IN OPEN DIRT LOTS FOR BEEF OR DAIRY CATTLE

For beef and dairy cattle that will be living in open dirt pens, the SINGLE MOST IMPORTANT part of construction is sloping the pens so they will drain (Grandin, 2016). This MUST be done before the pens are built. Cutting costs on dirt work and grading will result in a muddy feedlot that does not drain. The pens should have a 2% to 3% slope away from the feedbunk (Mader and Griffin, 2015).

BEEF CATTLE FED ON CONCRETE GET SWOLLEN JOINTS

In the 1970s numerous companies built indoor beef cattle feedlots with floors constructed from concrete slats. Many of these facilities are now abandoned because the cattle got swollen joints from laying on the concrete slats. The author observed that if the cattle gain more than 400 lbs. on the concrete floor, their joints became swollen. Today there are some producers who are going back to these same types of facilities. Today one of the reasons people are going back to slatted floor barns is to bring the cattle closer to large sources of wet distiller's grain from corn ethanol plants. Compared to dry kernel corn, this product is more expensive to ship to feedlots located in the drier high plains. The problems the author observed with swollen joints in the 1970s are now appearing on cattle housed in new concrete slatted floor facilities. Wagner (2016) has documented problems with swollen joints. Installation of rubber mats slows down the development of swollen joints, but it does not eliminate them. In the Midwest, some producers are housing beef cattle raised indoors on a bedded pack. When this is done correctly, it results in clean cattle with no leg injuries. The main problem that may occur with a bedded pack is failure to use sufficient straw, chopped cornstalks, or other bedding materials to keep the animals clean.

ANIMAL-BASED OUTCOME MEASURE TO MONITOR PROBLEMS WITH HOUSING OR MANAGEMENT

To maintain standards for acceptable welfare, animals should be scored on the following critical control points. Each bovine is scored with a numerical scale on the following variables: (1) lameness (NAMI, 2015; Von Keyserlingk, et al., 2012; Zinpro, 2016), (2) body condition (Elanco, 2009; Wildman et al., 1982, Ferguson et al., 1994), (3) cleanliness (hygiene) (McKeith et al., 2015), (4) swollen joints, and hock lesions (Fulwider et al., 2007), (5) coat condition, and (6) signs of heat stress, such as open mouth panting (Gaughan et al., 2008; Mader and Griffin, 2015). Assessment tools for scoring each one of these critical control points are readily available (Grandin, 2015; Welfare Quality, 2009). There is often a big difference between the best and the worst producers on lameness and hock lesions. For example, in high-producing dairies, the best producers had 2.8% lame cows and the worst was 36% (Cook et al., 2016). A big advantage of scoring is that it enables a producer to determine if they are getting better or becoming worse. There is also evidence that benchmarking can help motivate producers to improve (von Keyserlingk, et al. 2012). When they see how they rank compared to others, they are motivated to improve.

Providing a soft surface to lie on will help prevent lameness in dairy cows (Devries et al., 2015). Fulwider et al. (2007) found that good management of the bedding surface helped reduce swollen hocks. Dairy cattle kept in dirty free stalls are more likely to have hock lesions. Dirty cows also have higher somatic cell count (Reneau et al, 2005).

DESIGN OF BEEF CATTLE HANDLING FACILITIES

There are two approaches to the design of facilities for handling beef cattle for veterinary work. A facility can be built that is really economical, but it will require higher stockmanship skill to handle cattle both effectively and safely. The alternative is more elaborate and expensive facility

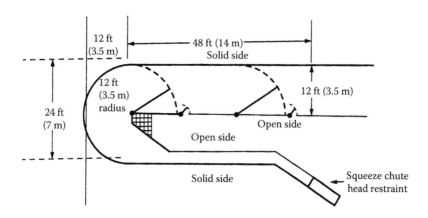

Figure 14.1 Cattle handling facility where the handler stands on the hatched area and works cattle from the crowd gate pivot. This takes advantage of the natural behavior of going back to where they came from.

that requires less stockmanship skill. For beef cattle handling facilities, the above principles are evident when comparing the Bud Box design (Gill and Machen, 2014; Stookey and Watts, 2014) to circular designs (Grandin, 2017; Grandin and Deesing, 2008). A Bud Box is simple to construct from portable panels but the stock person has to be in with the cattle. The handler has to know how to stand in the correct position so that the cattle will flow around him. Circular facilities are easier for people to learn how to use but they are more expensive to build. Both designs, when they are laid out correctly, rely on the behavioral principle of cattle going back to where they came from. In the survey of 25 feedlots in Colorado, Nebraska, and Kansas by Woiwode et al. (2016), 11% of the feedlots had Bud Boxes and all 89% had circular designs. To reduce costs, there are now less-expensive circular designs where all the catwalks are eliminated and the handler works at the pivot point of the round crowd pen (Grandin, 2014, 2015) (Figure 14.1). Drawings for both types of designs are in Grandin and Deesing (2008) and Grandin (2017). Working at the pivot of the crowd gate allows the cattle to circle around the handler.

MISTAKES THAT MAKE HANDLING FACILITIES WORK POORLY

When evaluating different designs, one must differentiate between poor performances caused by the basic design being wrong, or other factors not related to design that will ruin performance. Below is a list of problems that may cause cattle to balk or turn back. They may occur in ANY TYPE of design. A good stockperson needs to watch carefully and see where cattle stop and hesitate. It is especially important to observe this with cattle that are not familiar with the facility. An old experienced dairy cow will walk right by something that a new heifer will balk at. A calm animal will stop and look at a thing in the facility that is bothering them. If you rush the cattle up quickly, they will balk and turn back. A calm animal will look at distractions that need correcting and show them to you.

Lighting Problems—Lighting problems can cause cattle to balk and turn back in the best well-designed facilities. This is especially a problem with new animals that are not familiar with the facility such as young dairy heifers or newly arrived beef calves that are entering a feedlot. The worst problem is the dark movie theater effect. This occurs when there is a building over a handling facility and the cattle refuse to enter the dark building. This problem is mostly like to occur on bright sunny days. There are two ways to fix this problem. Cattle will enter more easily if they can see daylight through the building. In some facilities, cattle entry can be improved by installing white translucent skylights in the building to admit more shadow-free daylight. Cattle do not like going into dark places. At night, lights can be installed in the building to attract the animals into it. Unfortunately, electric lights are not bright enough to attract cattle into dark buildings on a really bright, sunny day.

Remove Distractions—When cattle are being handled for veterinary work at auctions or at meat plants, they are entering a novel facility where distractions are more likely to impede movement (Grandin, 1996). Some of the things that should be removed are dangling chains, coats on fences, paper cups, and hoses lying across the floor. Cattle notice little things that people tend to not notice. Parked vehicles alongside a cattle handling facility may cause cattle to balk.

Avoid High Contrasts—Animals will often stop if there is a change in flooring. Common balk points are changing from a dirt floor to either metal or concrete. Contrast can be reduced by scuffing dirt onto the concrete so that the line between the two types of materials becomes less distinct. Sprinkling hay or straw on the floor can also make the change less distinct. In premanufactured portable facilities, animals may refuse to cross the metal pipes or struts on the floor. These struts are required to hold a freestanding portable facility together. Covering the struts with dirt, straw, or shavings may help facilitate movement. Facilities should be painted all one color. Cattle also tend to stop where the color of the fence panels change.

Recent Tall Cattle Problems—In older facilities, vertical slide gates or top mounted anti-back up gates may be mounted too low to allow tall Holsteins to easily pass under them. Cattle will balk if they start to bump their backs. The pivot point of a backstop may need to be raised.

Balking at the Single-File Chute (Race) Entrance—There are four main problems that cause animals to refuse to enter a single-file chute from the crowd pen. They are (1) people standing next to it, (2) laid out wrong (3) backstop gate is closed, or (4) the chute is full. From a design standpoint, the biggest problem are layout mistakes where the single-file chute joins the crowd pen. To facilitate entering the single-file chute, an animal standing at the entrance must be able to see two body lengths up ahead. A common layout mistake is to bend the single-file chute too sharply where it joins the crowd pen. More information on layout can be found in Grandin and Deesing (2008) and Grandin (2017). If there is a backstop at the single-file chute entrance, it may cause cattle to balk and refuse to enter. Equipping it with a remote control rope so it can be easily held open will facilitate cattle entry.

Nonslip Flooring Essential—Animals get scared when they start to slip on the floor. A slick floor is one of the main causes of cattle becoming agitated in the stun box, single animal scale, or squeeze chute. Flooring in facilities for holding animals for veterinary procedures and other occasional treatments usually needs to be rougher than flooring used in places where the cattle walk through it every day, such as a milking parlor. A smooth, hard troweled surface must never be used. Rough broom finishes also work poorly because they wear out too quickly. Flooring that is too rough in a milking parlor can damage feet because the cows walk on it every day. In feedlots, a woven tire mat made from old tire treads is very effective for preventing slips and falls when cattle exit a squeeze chute. Woiwode et al. (2016) reported that 78% of the feedlots in Kansas, Nebraska, and Colorado she surveyed had these mats. More information on flooring can be found in Grandin and Deesing (2008) and Grandin (2017), for beef cattle handling facilities. These recommendations are for facilities where cattle do not walk on them every day.

Solid vs. Open Sides on Chutes—For the last 20 years, beef cattle have been selected for a calmer temperament. Many studies show that cattle that race quickly out of squeeze chutes (high exit speed) or get behaviorally agitated in the chute have lower weight gains (Café et al., 2014; Voisinet et al., 1997). When wild cattle are handled by unskilled people, solid sides on chutes help keep cattle calmer. Today some specialists in low stress handling recommend removing solid sides. Which is correct? It depends on the situation.

The most important part of chute to cover up is the sides along the outer perimeter. This blocks the animal's reaction to vehicles and other distractions that occur outside the facility. Feedlots that have a facility with open sides usually have it located within a building, which effectively puts solids side around the outer perimeter.

If a chute has open sides, stock people MUST stay away from it, except when they walk toward it to move the animals. If they continually stand too close to the animal in an open sided chute, it may become agitated. Agitation occurs because the person is inside the flight zone and the animal cannot move away. With the exception of halter broke tame cattle, a people-free space MUST be maintained alongside a chute with open sides. It can be visualized as a "force field" that extends outside the chute. The size of the force field will be determined by the wildness or tameness of the cattle. A skilled stock person can easily work the animal's point of balance and flight zone by walking on the ground. All catwalks are eliminated. If the animals are handled by less-skilled people or in a place with a lot of employee turnover, solid sides may be the best choice.

Restraint Device Design and Operation—The first rule is NEVER leave an animal unattended in any restraint device. If cattle vocalize in direct response to being held or caught in a squeeze chute or other restraint device, there is a problem that needs to be corrected. Excessive pressure or sharp edges will cause cattle to vocalize (moo or bellow) in a restraint device such as a neck stanchion (Bourquet, et al., 2012; Grandin, 2001). On California cattle ranches, Simon et al. (2016) found that vocalization was greater in hydraulic squeeze chutes compared to manually operated chutes. This is likely to be due to excessive pressure. A hydraulic squeeze chute should be adjusted so that it will automatically stop applying pressure BEFORE the animal vocalizes. In both the Woiwode et al. (2016) and Barnhardt et al. (2015) feedlot handling surveys, vocalization scores were low. The average vocalization score was 1.4% of the cattle (Woiwode et al., 2016). All the feed yards in the Woiwode et al. (2016) survey had hydraulic squeeze chutes. Greater awareness of Beef Quality Assurance principles may have made feedlot managers more aware of the need to limit the maximum amount of pressure.

Some squeeze chutes are equipped with additional hydraulic devices to hold the animal's head still. These must be set at a much lighter pressure than the body squeeze. To prevent injuries, these devices must be used carefully. If use of these devices causes vocalization or straining, they are applying excessive pressure.

Cattle will remain calmer in a squeeze chute if they can stand in a balanced position. The best squeeze chutes have sides that squeeze in evenly from both sides. If a chute has only a single movable side, it must be carefully adjusted to prevent throwing the animal off balance.

There are two types of design for the squeeze sides. Hinged at the bottom to form a V or the sides remain vertical during squeezing. Sides that remain vertical are best for mature dairy cows, big bulls, and very wide, fat cattle. The V-shaped chute often works better for smaller younger animals. There are two basic types of headgates. They can have either straight vertical neck bars or curved bars that provide better control of the head. A major advantage of a straight bar stanchion is that a bovine can safely lie down in it, without choking. If a curved bar stanchion is used, the chute must be designed to prevent the cow from lying down. This can be achieved with either a brisket bar in a chute with straight sides or V-shaped squeeze sides. There are a wide variety of designs that can be used. The ultimate determinants of how appropriate a certain design of a squeeze chute or other restraint device are the scores obtained from animal-based measures.

RESTRAINT DEVICE EVALUATION

- Percentage of cattle vocalizing in direct response to being held by the restraint device (Grandin, 1998, 2001, 2010, 2010). Should be 5% or less of the cattle. Score before procedures such as eartagging or injections (Barnhardt et al., 2015; Simon et al., 2016; Woiwode et al., 2016).
- Percentage of animals that balk at the entrance (Welfare Quality, 2009). Balking may be caused by the chute being too narrow at the bottom. Vocalizations caused by use of an electric prod are counted on the vocalization score.
- Struggling and agitation while being restrained (Grandin, 1993). This is often caused by slipping on the floor or chute that throws the animal off balance.

- Percentage of cattle where an electric prod is used to move them into the squeeze chute.
- Percentage that fall or stumble while exiting. Score falling and stumbling as separate variables.

To maintain high welfare standards during handling requires a combination of both good stockmanship skills and well-designed facilities. In housing facilities, there may be a conflict on what is best for the individual animal and economies. Skimping on space requirements may improve the economics of a dairy but the individual cow may suffer. Each cow should have her own free stall.

REFERENCES

Barnhardt, T.T., Thomson, D.U., Terrell, S.P., Rezac, D.J., Frese, D.A., and Reinhardt, C.D. (2015) Implementation of industry-oriented animal welfare and quality assessments in Kansas feeding operations, Bovine Practice, 45:81–87.

Bourquet, C., Deiss, V., Tannugi, C.C., and Terlouw, E.M. (2012) Behavioral and physiological reactions of cattle in a commercial abattoir, relationships with organizational aspects of the abattoir and animal characteristics, Meat Science, 68:158–168.

Café, L.M., Robinson, D.L., Ferguson, D.M., McIntyre, B.L., Geesink, G.H., and Greenwood, L. (2014) Cattle temperament: Persistence of assessments and associations with productivity, efficiency, carcass and meat quality traits, Journal Animal Science, 89:1452–1465.

Chapinal, N., Weary, D.M., Collings, L., and von Keyserlingk, M.A.G. (2014) Lameness and hock injuries improve on-farms participating in an assessment program, The Veterinary Journal, 202:646–648.

Cook, N.B., Hess, J.P., Foy, T.B., Bennett, T.B., and Bratzman, R.L. (2016) Management characteristics, lameness, and body injuries of dairy cattle housed in high performance dairy herds in Wisconsin, Journal of Dairy Science, 99:5879–5891.

deVries, M., Bukkers, E.A.M., van Reenen, C.G., Engel, B., van Schaik, G., Dijkstra, T., and de Boev, I.J.M. (2015) Housing and management factors associated with indicators of dairy cow welfare, Preventive Veterinary Medicine, 110:80–92.

deVries, A., Dechassa, H., and Hogeveen, H. (2016) Economic evaluation of stall stocking density of lactating dairy cows, Journal of Dairy Science, 99:3848–3857.

Elanco (2009) The 5 Point Body Condition Scoring System, Elanco Animal Health, Greenfield, Indiana, Accessed December 25, 2016.

Ferguson, J.O., Galligan, D.T., and Thomsen, N. (1994) Principle descriptors of body condition score in Holstein cows, Journal of Dairy Science, 77:2695–2703.

Flower, F.C., Sedlbauer, M., Carter, E., von Keyslerlingk, M.A.G., Sanderson, D.J., and Weary, D.M. (2008) Analgesics improve the gait of lame dairy cattle, Journal of Dairy Science, 91:3010–3014.

Fulwider, W., Grandin, T., Garrick, D.J., Engle, T.E., and Rollin, B.E. (2007) Influence of freestall base on tarsal joint lesions and hygiene in dairy cows, Journal of Dairy Science, 90:3559–3566.

Gaughan, J.B., Mader, T.L., Holt, S.M., and Lisle, A. (2008) A new heat load index for feedlot cattle, Journal of Animal Science, 86:226–234.

Gibbons, J., Vasseur, E., Rushen, J., and dePasille, A.M. (2012) A training program to ensure high repeatability of injury scoring of dairy cows, Animal Welfare, 21:379–388.

Gill, R. and Machen, R. (2014) Designing a Bud Box, Accessed January 29, 2017. http://agrilifeedn.tamu.edu/beeffinto/files/2014/o1/designing-a-bud-box.pdf.

Grandin, T. (1993) Behavioral agitation during handling of cattle is persistent over time, Applied Animal Behaviour Science, 36:1–9.

Grandin, T. (1996) Factors that impede animal movement at slaughter plants, Journal of the American Veterinary Medical Association, 209:757–759.

Grandin, T. (1998) Objective scoring of animal handling and stunning practices at slaughter plants, Journal American Veterinary Medical Association, 212:36–39.

Grandin, T. (2001) Cattle vocalizations are associated with handling and equipment problems at beef slaughter plants, Applied Animal Behavior Science, 71:191–201.

Grandin, T. (2010) Auditing animal welfare at slaughter plants, Meat Science, 86:56–65.

Grandin, T. (2014) Livestock Handling and Transport, 4th Edition, CABI International, Wallingford, Oxfordshire, UK.

Grandin, T. (2015) Improving Animal Welfare: A Practical Approach, 2nd Edition, CABI International, Wallingford, Oxfordshire, UK.

Grandin, T. (2016) Evaluation of the welfare of cattle housed in outdoor feedlot pens, Veterinary and Animal Science, 1:23–28.

Grandin, T. (2017) Temple Grandin's Guide for Working with Farm Animals, Storey Publishing, North Adams, MA, USA.

Grandin, T. and Deesing, M. (2008) Humane Livestock Handling, Storey Publishing, North Adams, MA, USA.

Mader, T.L. and Griffin, D. (2015) Management of cattle exposed to adverse environmental conditions, Veterinary Clinic of North America, Food Animal Practice, 31:247–258.

March, S., Brinkman, J., and Winkler, C. (2007) Effect of training on the inter-observer reliability of lameness coring in dairy cattle, Animal Welfare, 16:131–133.

McKeith, R.O., Gray, G.D., Hale, D.S., Karth, C.R., Griffin, D.B., Savell, J.W. et al. (2015) National Beef Quality Audit 2011: Harvest floor assessments of targeted characteristics that effect quality and value of cattle carcasses and byproducts, Journal Animal Sciences, 90:5135–5142.

NAMI (2015) Mobility Scoring of Cattle, Video handling.org, Accessed December 26, 2016.

Reneau, J.K., Seykova, A.J., Heins, B.J., Endres, M.I., Farnsworth, R.J., and Bey, R.F. (2005) Association between hygiene scores and somatic cell scores in dairy cattle, Journal American Veterinary Association, 227:1297–1301.

Romero, M.H., Uribe-Velasquez, L.F., and Miranda de la Lima, G.C. (2017) Conventional versus modern abattoirs in Columbia: Impacts on welfare indicators and risk factors of high muscle pH in commercial Zebu young bulls, Meat Science, 123:173–181.

Schlageter-Tlelo, A., Bokkers, E.A.M., Grootkoerkamp, P.W.G., Van Hertern, T., Viazzi, S., and Romanini, C.B. et al. (2014) Effect of merging levels of locomotion scores for dairycows on intra- and inter-rates reliability and agreement, Journal of Dairy Science 97:5533–5542.

Simon, G.E., Hoar, B.R., and Tucker, C.B. (2016) Assessing cow-calf welfare, Part 2: Risk factors for beef cow health and behavior and stockperson handling, Journal of Animal Sciences, 94:3488–3500.

Stookey, J.M. and Watts, J.M. (2014) Low stress restraint and handling and sorting of cattle, In: T. Grandin (Editor) Livestock Handling and Transport, 4th edition, CABI International, Wallingford, Oxfordshire, UK, pp. 65–76.

Velarde, A. and Dalmau, A. (2012) Animal welfare assessments at slaughter: Moving from inputs to outputs, Meat Science, 92:244–251.

Voisinet, B.D., Grandin, T., Tatum, J.D., O'Connor, S.F., and Struthers, J.J. (1997) Feedlot cattle with calm temperaments have higher average daily gain than cattle with excitable temperaments, Journal of Animal Science 75:892–896.

Von Keyserlingk, M.A.G., Barrientos, A., Ito, K., Gallo, E. and Weary, D.M. (2012) Benchmarking cow comfort on North American freestall dairies: Lameness leg injuries, lying time, facility design and management for high producing Holstein dairy cows, Journal of Dairy Science, 95:7399–7408.

Wagner, D. (2016) Behavioral analysis and performance response of feedlot steers on concrete slots versus rubber slats (Abstract) American Society of Animal Science, Salt Lake City, Utah.

Welfare Quality Network (2009) Assessment protocol for cattle, Accessed, December 11, 2016. www.welfarequalitynetwork.net.

Wildman, E.E., Jones, G.M., Wagner, P.E., Boman, R.L., and Troutt, H.F. et al. (1982) A dairy cow body condition scoring system and its relationship to selected production characteristics, Journal of Dairy Science, 65:495–501.

Woiwode, R. (2014) Survey of BQA cattle handling practices that occurred during processing of feedlot cattle, ADSA, ASAS, CSAS Joint Annual Meeting, Kansas City (Abstract).

Woiwode, R., Grandin, T., Kirch, B., and Patterson, J. (2016) Compliance of large feedyards in the Northern high plains with Beef Quality Assurance Feedyard Assessment, The Professional Animal Scientist, 32:750–757.

Zinpro (2016) Firststep Dairy Locomotion Scoring Videos, Accessed December 26, 2016. www.zinpro.com/video-library/dairy-locomotion-videos#/videos/list.

CHAPTER 15

Pain Mitigation in Cattle

Johann F. Coetzee and Michael D. Kleinhenz
Kansas State University

CONTENTS

INTRODUCTION

It has been well-documented and well-established that cattle of all ages are capable of experiencing pain. The pain may be the result of procedures performed to alter animals for the betterment of the herd and the end product of beef. Cattle also experience pain as a result of disease conditions that either themselves are painful, as in the case of lameness; or as a result of a medical procedure to treat a disease condition, such as laparotomies for correction of left displaced abomasum.

Regardless of the origin of pain, animal owners, caretakers, and veterinarians have an obligation to provide the best animal welfare and well-being as possible. It is also important to utilize science-based treatment and prevention strategies as these diseases and procedures are under the scrutiny of the public and thus consumers of the products these animals provide. In this chapter, we review the pharmacology of analgesic drugs in specific disease conditions and elective procedures. Additionally, where applicable, we briefly explore alternatives and preventive strategies to minimize the pain an animal experiences or remove the need for a procedure.

PHARMACOLOGY OF ANALGESIC DRUGS

The use of pharmaceuticals as analgesic is an important factor in the treatment and well-being of animals. However, it is important to recognize the challenges and factors that must go into utilizing these agents for pain mitigation, especially in food producing animals. First is the regulatory component of using these drugs. As of July 2017, there is only one compound with a label for pain mitigation in the United States. Furthermore, this compound, flunixin meglumine transdermal, is only approved for the treatment of pain associated with foot rot in cattle. Beyond the one drug for one disease condition, there are no other drug approvals for pain mitigation in cattle.[1] The Animal Medicinal Drug Use Clarification Act (AMDUCA) of 1994 permits extra-label drug use (ELDU) in order to relieve suffering in cattle.[2] As such, analgesics would be permitted under AMDUCA given the criteria for ELDU are followed. This requires a valid veterinary–client–patient relationship (VCPR) and the prescribing veterinarian must ensure the ELDU does not result in violative drug residues in meat or milk.

The basic pharmacology of the commonly used and researched analgesic compounds will be discussed in the remainder of this section before discussion of individual procedures or disease conditions.

Local Anesthesia

Local anesthetics are the most commonly used analgesics in cattle. They function by blocking sodium channels within nerve cells and preventing the conduction and transmission of the pain signal.[3] By blocking the nerve transmissions, the regions innervated by the nerves are devoid of sensation. Additionally, nerve fibers responsible for pain and temperature are blocked before those fibers are involved with touch, pressure, and motor activity.[4] The addition of vasoconstrictors (e.g., epinephrine) to the local block increase contact time to the nerve and prolong the length of desensitization.[4] Onset of activity for lidocaine, a common local anesthetic agent, is rapid, with desensitization lasting about 90 minutes.[4] Other local anesthetics in veterinary medicine include bupivacaine, which has a longer duration but also is slower to onset of action.[5]

Non-steroidal Anti-inflammatory Drugs (NSAIDs)

The mode of action for NSAIDs is the inhibition of cyclo-oxygenase (COX) isoenzymes and subsequent reduction in the production of prostaglandins (PGs) from arachidonic acid.[6] PGs contribute to the inflammatory response through recruitment of inflammatory cells, vasodilation, and working in concert with other cytokines and neuropeptides. All these actions result in the reduction of the action potential threshold in nociceptors and propagation of the pain signal.[7] Peripherally, PG production causes local hyperalgesia or peripheral sensitization; centrally excess PG lead to central sensitization and chronic pain.[8] Both COX-1 and COX-2 isoenzymes are thought to be responsible for the inflammatory response, with the initial effects a result of COX-1 derived PG and the delayed effects due to upregulation of COX-2 expression.[7] For immediate and prolonged pain control, an NSAID targeting both isoenzymes may be advantageous as inhibition of COX-1 is associated with renal and gastrointestinal effects.[6]

Flunixin Meglumine

Flunixin meglumine is derived from nicotinic acid in the anthranilic acid NSAID class. Flunixin is the only FDA-approved NSAID for cattle in the United States and is the only NSAID with a label approval for the treatment of pain. There are two approved formulations of flunixin in the

United States—an injectable formulation and a pour-on formulation for transdermal absorption. Currently, injectable flunixin is indicated for the control of fever associated with respiratory disease or mastitis, and fever and inflammation associated with endotoxemia. The transdermal flunixin is indicated for the control of fever associated with respiratory disease and the control of pain associated with foot rot. Flunixin has been shown to have age-related pharmacokinetics for both the IV injectable and transdermal routes.[8]

Ketoprofen

Ketoprofen is an NSAID of the propionic acid class. It has approval as an adjunctive therapy for fever, pain, and inflammation associated with mastitis and inflammatory and painful conditions of bones and joints lass of NSAIDs in the European Union and Canada.[9] Ketoprofen is administered as a racemic mixture (50:50) with chiral RS± enantiomers. Interestingly, the R(−) enantiomer will undergo chiral inversion to S(+) which is clinically relevant since the S(+) enantiomer is a more potent PGE_2 inhibitor.[10] Ketoprofen has a relatively short half-life due to rapid metabolism and elimination, efforts to sustain analgesia may require multiple doses. Nevertheless, many studies have investigated the analgesic potential of ketoprofen largely due to its approvals in the EU and Canada.

Carprofen

Carprofen, like ketoprofen, is an NSAID in the propionic acid class.[11] Carprofen is also administered as a racemic (50:50) mixture of RS± enantiomers. However, carprofen is not known to undergo chiral inversion.[12] For cattle in the European Union, carprofen is indicated as an adjunct to antimicrobial therapy associated with respiratory disease and mastitis.

Unique pharmacokinetic properties of carprofen in cattle include a prolonged half-life, slow clearance, and possibly biliary drug secretion as observed in dogs.[13] Like flunixin meglumine, the pharmacokinetics of carprofen are age-dependent, with a prolonged half-life in younger animals (<10 weeks), most likely due to the decreased clearance common to neonates.[12]

Meloxicam

Meloxicam is a member of the oxicam class of NSAIDs. It has approval for use in the European Union and Canada for adjunctive therapy of acute respiratory disease, diarrhea, acute mastitis, and as an analgesic to relieve pain following dehorning in calves. In the United States, meloxicam is approved to control pain associated with osteoarthritis in humans, dogs, and cats. The pharmacokinetics of meloxicam in cattle indicates a prolonged half-life and a high bioavailability when administered orally.[14] Due to these favorable properties for providing practical analgesia in cattle, many studies have recently investigated oral meloxicam as an analgesic.

Firocoxib

Firocoxib is an NSAID of the coxib class. This is a newer group of NSAIDs demonstrating COX-2 selectivity in dogs and horses, thereby potentially limiting adverse effects caused by COX-1 inhibition.[15] Currently, firocoxib is indicated for the treatment of pain and inflammation associated with osteoarthritis for dogs and horses in the United States. Limited information is available about firocoxib in cattle, with only one study conducted in preweaned calves.[16] Unique pharmacokinetic properties in preweaned calves include high oral bioavailability, prolonged terminal half-life, and an extensive tissue distribution (high volume of distribution).

Gabapentin

Gabapentin is a γ-aminobutyric acid (GABA) analog historically used as an anti-seizure medication, but it has been recognized for its improvement in chronic, neuropathic pain. Gabapentin works by decreasing excitatory neurotransmitter release as a result of modulation of voltage-gated calcium channels.[17] Additionally, analgesic activity is synergistically enhanced when used in conjunction with an NSAID.[18,19]

CASTRATION

Castration is a common husbandry practice in the United States and Canada to minimize the danger associated with intact males, unwanted pregnancies in feedlot situations, and improve carcass quality. The benefits of castration are widely known and deemed important to the cattle industry. It is known that castration causes pain, distress, and other neuroendocrine changes to the animal regardless of the technique used and age of animal.[20]

The use of sexed semen is a practice to reduce the number of males born, especially in the dairy industry. The use of sexed semen does come at a higher cost per unit of semen and lower conception rates, but these cost may be offset by higher milk production in the lactation following carrying a female calf.[21]

The use of an immunocastration vaccine against gonadotropin releasing hormone (GnRH) has been demonstrated to be an effective alternative for the physical castration of cattle.[22] Although attractive in principle, this technique has not gained widespread adaptation due to the need for repeated injections, lack of approved products for cattle in the United States, and human health implications of self-injection of the vaccine antigens. Therefore, physical castration methods are primarily used for castration of cattle.

Pain and distress caused by castration is widely known, resulting in research looking into ways to mitigate pain using pharmacologic agents. Commonly used drug classes include local anesthetics such as lidocaine and NSAIDs such as flunixin meglumine, ketoprofen, and meloxicam.

Local anesthetics are beneficial in reduction of the acute pain associated with castration by blocking the transmission of pain impulses to the central nervous system. A review of 15 castration studies evaluating maximal cortisol concentrations indicated a 25.8% reduction of cortisol in cattle receiving a local anesthetic compared to controls.[23] Local anesthetics alone have minimal effects on the overall feed consumption, average daily gain, and inflammatory mediators.[24–27]

Investigations into the analgesic effects of flunixin indicate pain-relief during the initial period post-castration. Flunixin administration is associated with pain-relief, although, in general, these reported changes are not reported to persist beyond 8 hours. Cortisol concentrations were lower in calves treated with flunixin following burdizzo clamp castrations and surgical castration compared to castrated controls.[28,29] Furthermore, stride length increases when treated with flunixin and a lidocaine or xylazine epidural.[30,31]

In both surgically and non-surgically castrated cattle, preoperatively administered ketoprofen demonstrated reduced cortisol concentrations compared with castrated controls.[25–27,32] This effect was potentiated when combined with a local anesthetic.[23] Additionally, average daily gain is improved by the administration of ketoprofen combined with a lidocaine local anesthetic for surgical[25] but not non-surgical castration.[26] Taken together with the cortisol data, when using ketoprofen, multimodal therapy is needed for maximizing pain-relief during castration.

Calves administered carprofen following non-surgical clamp castration had lower cortical concentrations compared to untreated controls.[28] Calves undergoing surgical castration had reductions in cortisol concentrations, but these effects were not significant.[33] Meloxicam has also been evaluated in surgically and non-surgically castrated cattle. Administration of meloxicam to cattle before surgical castration following arrival to the feedlot resulted in lower pull rates and respiratory disease treatment rates.[34] Furthermore, calves receiving meloxicam following surgical castration had a reduced inflammatory response compared to placebo controls.[35]

DEHORNING

Dehorning occurs on more than 90% of the dairy farms in the United States.[36] In the beef cattle sector, it occurs less than half the time due to the use of polled genetics.[37] The purpose of dehorning is to lower risk of injury to humans and other animals, and to minimize carcass bruising. Additionally, cattle without horns require less bunk space compared to their horned counterparts. The removal of horns can be accomplished one of three ways: disbudding prior to the horn tissue attaching to the skull; removal of the horn after horn tissue has adhered to the skull; and the use of polled genes, in which cattle are born without horns. Both disbudding and dehorning are associated with pain and activation of the neuroendocrine system.

The use of polled genes is an alternative to dehorning cattle. In cattle, the polled gene is dominant, meaning only one chromosome must carry the gene for the phenotypic expression. Use of polled genetics is prevalent in the beef cattle industry due to the use of Angus genetic lines that typically carry the polled gene. Utilization of polled genetics has not gained significant ground in the dairy industry due to its negative relationship to milk production.[38] It is estimated that the use of polled genetics may reduce the herd's genetic merit and predicted net merit.[38] The use of polled genetics in the Jersey breed may be less than those seen in Holsteins.[38]

There are differences in invasiveness of the various dehorning methods, but all methods are considered painful. The American Veterinary Medical Association (AVMA) and American Association of Bovine Practitioners (AABP) both recommend disbudding or dehorning to be accomplished as early in life as possible. The use of caustic paste in calves younger than 1 week of age has been shown to be less invasive of a procedure compared to cautery dehorning.[39] Although less invasive, caustic dehorning does result in pain and discomfort to the animal. This discomfort can be attenuated using local anesthetic blocks.[40]

The effects of local anesthesia in dehorning or disbudding studies are similar to castration. The acute pain and distress associated with dehorning is attenuated compared to untreated controls as determined by cortisol concentrations.[41,42] This effect is primarily observed for the duration of the local anesthetic activity for up to 5 hours post-dehorning.[42] Moreover, evaluation of the autonomic nervous system through heart rate variability and ocular temperatures indicate an imbalance 2–3 hours post-dehorning with lidocaine administration, coinciding with the time associated with loss of lidocaine activity.[43]

The use of NSAID therapy in conjunction with local anesthetics has been shown to have a positive impact on pain biomarkers, but this practice is not widely used by veterinarians and producers.[44,45] Flunixin meglumine is a commonly studied NSAID. Flunixin administration provided benefits to calves during the acute distress and painful phase following dehorning. Flunixin meglumine administered to calves following a cornual nerve block reduced cortisol concentrations compared to untreated controls in calves undergoing chemical[46] and amputation dehorning,[29] respectively.

Recently, a transdermal flunixin meglumine was released in the European Union and approved in the United States. Due to its novel pharmacokinetic properties, it has the potential to be a needleless method of providing analgesia to calves. When administered at the time of dehorning, calves treated with the transdermal flunixin meglumine had significantly lower cortisol at 1.5 hours post-dehorn.[47] Cortisol levels remained lower in treated calves for up to 12 hours post-dehorning, but these levels were not significant.[47] The transdermal flunixin had no effect on substance P concentrations or MNT at the dehorning site.[47] However, there was a significant improvement in the MNT at a control site, indicating improvement in central sensitization. Further research on how this formulation can fit into a multimodal analgesic plan is needed as this formulation of flunixin has an FDA label for pain.

Ketoprofen administration in combination with a local anesthetic resulted in the amelioration of the acute cortisol response, with effects lasting up to 5 hours compared to untreated controls.[48–51] When administered without local anesthesia, the typically observed cortisol plateau was attenuated; however, peak cortisol concentrations were only mildly reduced.[48] Carprofen has been shown to lower cortisol levels following cautery dehorning.[52,53] Additionally, Stock et al. found when given either orally or subcutaneously, carprofen-treated calves tended to tolerate more pressure around the horn bud over the course of the study.[53]

There are several studies supporting the use of meloxicam in cattle at the time of disbudding or dehorning. When used following caustic paste, meloxicam alone did not provide substantial analgesia as compared to calves receiving a local anesthetic block.[40] Meloxicam given by the intravenous route did not show the same effects as intramuscularly injected meloxicam.[54,55]

Oral (PO) meloxicam has been evaluated in 8- to 10-week-old calves at the time of cautery dehorning demonstrating reduced cortisol and substance P compared to placebo-treated controls.[56] Furthermore, 6-month-old calves undergoing amputation dehorning demonstrated improved ADG when treated with oral meloxicam compared with placebo-treated controls.[57] Meloxicam had no effect on pain sensitivity, ocular temperature, or haptoglobin.[56,57] When meloxicam was combined with gabapentin in a dehorning trial, researchers found an increase in average daily gain but saw no improvement in physiologic responses.[56]

A clinical trial evaluating oral firocoxib administered in combination with a lidocaine corneal nerve block administered 10 minutes before cautery dehorning was conducted on 4- to 6-week-old calves.[58] Although the acute effects of cautery dehorning as determined by physiologic and nociception changes were unaffected by treatment, firocoxib calves had an overall reduced integrated cortisol response compared to placebo-treated controls suggesting a role for firocoxib in the control of the delayed distress response.

BRANDING

Branding, by either hot iron or freeze, is practiced on 44% of the beef operations, with distinct regional differences.[37] In western Canada, branding is practiced on 55% of farms, but analgesia is provided only 4% of the time.[45] In the dairy industry, branding is done on 4% of the farms, accounting for 13% of all dairy cattle.[36] Freeze branding appears to cause less pain than hot-iron branding, but both procedures cause pain and result in long-lasting pain afterwards. In fact, pain can last up to 71 days when a hot iron is used.[59–61] Behavioral indicators of the pain associated with branding include tail flicking, kicking and vocalization.[59,60]

There are few studies investigating pharmaceutical intervention and analgesic strategies for branding. Tucker et al. described the healing of hot-iron brands and found that a single dose of flunixin did not provide substantial analgesia.[61] This is likely due to the duration of healing and relatively short half-life of flunixin given IV to cattle. The use of cooling gels has been proposed, but their use has not proven beneficial.[62]

TAIL DOCKING

Tail docking, in which the tail is removed, is an elective procedure practiced in the dairy industry. This practice gained prominence due to the convenience of dairy employees, especially milkers. The switch, at the end of the tail, would often be a nuisance due to its penchant for picking up dirt and fecal matter; furthermore cows were perceived to be cleaner without tails. This perceived cleanliness has not been fully supported by the literature, as two papers demonstrated no benefit to udder hygiene or udder health in cows with docked tails.[63,64] Furthermore, a national survey of 491 United States dairy farms revealed hygiene was actually better on farms that did not dock tails when compared to farms that practiced tail docking, indicating environmental management plays a more important role than tail docking for cow hygiene.[65]

Tail docking also carries with it a strong negative perception by the general public.[66] This is highlighted by the National Milk Producers Federation, the AVMA and the AABP publically stating they do not support the practice. The National Milk Producers Federation's Farmers Assuring Responsible Management (FARM) program mandates cessation of routine tail docking as of January 1, 2017. To further support this mandate, milk processors have also mandated a no tail docking standard as part of agreements to purchase milk from producers.

However, there are circumstances in which tail docking is necessitated. These circumstances are limited to injuries of the tail due to trauma. In confined beef cattle, housed on slatted floors, tail injuries are common and tail docking is performed for therapeutic purposes. However, the use of a lidocaine epidural and injection of flunixin meglumine did not provide adequate analgesia to cattle undergoing tail docking compared to untreated controls undergoing the same procedure.[67]

INVASIVE SURGERIES

In cattle production, there is often a need to perform invasive surgeries. Furthermore, these surgeries are not typically elective in nature but rather a result of serious medical conditions that need correction. Cesarean sections and laparotomies (for correction of displaced abomasum) are common surgical procedures performed in cattle. These surgeries are often performed without general anesthesia, but with only local anesthesia with a provision of sedation in certain cases. However, a survey of practitioners revealed 99% give analgesia to cattle undergoing Cesarean sections.[68]

Lidocaine is the typical local anesthetic used. There are numerous methods to achieve anesthesia of the surgical site. These methods are practitioner and case-dependent and reach beyond the scope of this chapter. However, the application of local anesthetic is important for the welfare of the patient. It is also for the safety for the practitioner performing the procedure, as kicking is the first response to inappropriate or insufficient anesthesia.

Controlled clinical studies and experimental models investigating the use of analgesics in cattle following surgeries are sparse. Ketoprofen has been studied in a clinical trial following left displacement of the abomasum and was given at the time of surgery and the following day. Cattle receiving ketoprofen had no differences in heart rate, respiratory rate, β-hydroxybuterate, and milk production, between cattle receiving ketoprofen or saline.[69]

LAMENESS

Lameness is a highly prevalent disease in dairy herds, with prevalence reported as high as 33.7%. However, reported lameness prevalence as reported by owners is about 10%.[70–72] In the beef cattle sector, lameness is a common reason for cows and bulls to leave the herd.[73] A survey of cows and bulls arriving at slaughter facilities indicated that 26.6% and 36.3% of beef cows and bulls, respectively, were evaluated to be lame on inspection. This same survey had dairy cow lameness at 39.2% of dairy cows arriving to slaughter facilities as being lame.[74] It should be noted that lameness is usually underreported by producers compared with independent observers potentially due to a decreased sensitivity in detecting lame cattle.[75,76] There are economic and production losses that are associate with lameness as well as the obvious pain and distress.[77] The evaluation and prevalence of lame cattle is included in third-party welfare audits, including the FARM program, the New York State Cattle Health Assurance Program (NYSCHAP) and Validus.[78–80]

Lameness, unlike the previously discussed painful procedures, does not have a known start point. Lameness is found by producers and animal caretakers. The time from when the animal starts to experience pain to the time when the lameness is treated is not known. This has been the impetus for the early detection and treatment of lameness. Locomotion scoring systems have been developed for routine detection of lame cattle. Early detection and treatment of lameness is obviously suggested but may also prevent central sensitization.[81] Central sensitization is responsible for the observed pain-related behavioral changes through increased sensitivity of pain (hyperalgesia) and pain from non-painful stimuli (allodynia).[81] Analgesic treatment difficulties in chronic lame cattle may be best explained through the aforementioned central sensitization based on current pain models. As a result, preemptive analgesia that is usually advocated is difficult—if not impossible—to implement in lame cattle.[81] Recommendations for the treatment of lameness include a multimodal approach that includes the use of therapeutic trimming, hoof blocks, and pharmaceutical analgesics.[20]

Causes of lameness can be infectious or noninfectious. Infectious causes of lameness include foot rot (interdigital phlegmon) and digital dermatitis. Noninfectious causes include sole ulcers, toe ulcers, white-line disease, and bone fractures. It is important to note that both infectious and noninfectious lameness may be present in any clinical case.

Prevention is the most useful tool for the reduction of lameness on farms. Prevention strategies should be targeted for the control of both infectious and noninfectious lameness causes. Farm and cattle environmental cleanliness is the mainstay of prevention for controlling infectious lameness. This is accomplished though frequent cleaning of fecal material and water in pens and alleys; the use of footbaths containing antibiotics, salts, or antiseptics; and minimizing movement through muddy areas. A regular hoof trimming schedule is a key component of prevention programs. Hoof trimming maintains appropriate weight balance on the foot and allows for early diagnosis of lesions.[20] Nutrition, calving management, and environmental management all factor into prevention of noninfectious causes of lameness, but these strategies are beyond the focus of this chapter.

Pharmacologic management of pain associated with lameness is dependent on the cause of disease and also the duration of the lameness. There are numerous antibiotics in the United States labeled for treatment of foot rot. Although very important in the overall treatment of lameness, antibiotics do not provide any analgesic effects.

In the United States, a novel formulation of the NSAID flunixin meglumine has been approved for the control of pain associated with foot rot. This is the first drug to be given FDA approval for pain mitigation in cattle. This novel formulation has transdermal absorption, thus allowing safe, easy, convenient, and needleless dosing. In documents supporting the approval of the new flunixin meglumine formulation, cattle were experimentally infected with *Fusobacterium necrophorum*, the causative agent of bovine foot rot. Infected calves had improved lameness by 48 hours postinfection.[82]

The use of intravenously injected flunixin meglumine has been studied in both a clinical study and also an amphotericin B lameness model study. In the clinical study, lame cattle were administered flunixin meglumine immediately before corrective hoof trimming and 24 hours following the first treatment.[83] Gait scores as determined by a visual analog scale (VAS) and weight distribution were not significantly affected by the provision of analgesia. Cows treated with flunixin had shorter laying times than untreated controls. A mild significant reduction in variation (SD) of weight distribution of the rear legs was observed following analgesia administration as observed acutely with lidocaine or ketoprofen administration.[83–85]

A study using flunixin meglumine following lameness induction using amphotericin B showed an improvement in lameness. Cattle treated with flunixin meglumine had improved lameness scores, placed more pressure and surface area contact on pressure mats with their affected foot as well as the contralateral claw of the affected foot. Cortisol concentrations were numerically lower in flunixin-treated cattle, but there numbers were not statistically different. Additionally, untreated controls in the study spent more time laying following induction and treatment compared to flunixin-treated cattle, but these results were not seen after the first day.[86]

The NSAID ketoprofen has been evaluated in multiple clinical field trials to determine its analgesic effectiveness on pain mitigation of lameness in cattle.[83,87,88] In all studies, adult lactating dairy cattle were detected as lame using a locomotion scoring technique before enrollment into the study.[89] Following ketoprofen administration of a gradually increasing dose with a maximum dose of 3 mg/kg, mild improvements were observed in gait attributes, including an increased number of symmetrical steps and a more even weight distribution among all four limbs.[88] Moreover, a reduced variation in weight distribution has also been reported following administration of keto-profen analgesia.[83] Ketoprofen was also shown to result in a sound lameness score 35 days after an acute lameness episode when given as part of a trim and block treatment regimen.[90] However, when given as a preemptive analgesic to acutely lame cattle before corrective trimming, ketoprofen failed to alleviate stress results.[91]

Meloxicam has been shown to improve lameness scores when given orally at doses between 0.5 and 1 mg/kg.[92,93] In a trial using a lameness induction model, calves receiving meloxicam once daily for 4 days had improved lameness signs.[94] With an increased step count in meloxicam-treated animals compared to placebo-treated controls, meloxicam concentrations were inversely associ-ated with lameness scores and positively associated with pressure and contact of the ipsilateral limb.[94] Meloxicam administered with gabapentin demonstrated a beneficial analgesic response as evidenced by increased stride length and force applied to the ipsilateral claw compared to placebo-treated controls.[94] This response was greater than gabapentin administered without concurrent meloxicam administration.[94]

Local anesthetics have used for desensitization of the lower limb and foot for the treatment of lameness and/or diagnosis of the condition. However, effects are relatively short-lived after removal of the tourniquet required for use. The training and facilities required to utilize such techniques rel-egate them to use on severely lame cows and surgery of the distal limb. In a clinical study evaluating lameness, cattle given a local anesthetic block using lidocaine had improved gait scores but these improvements were not significant. The differences in the weight bearing of the affected foot and contralateral foot were reduced following local anesthesia. This indicated cattle shifted less weight from the affected foot to the contralateral foot.[95]

REFERENCES

1. Smith GW, Davis JL, Tell LA, et al. FARAD digest—extra label use of nonsteroidal anti-inflammatory drugs in cattle. *J Am Vet Med Assoc* 2008; 232:697–701.
2. Animal Medicinal Drug Use Clarification Act of 1994 (AMDUCA). U.S. Food and Drug Admini-stration website. Available at: www.fda.gov/AnimalVeterinary/GuidanceComplianceEnforcement/ActsRulesRegulations/ucm085377.htm. Accessed 10 September 2017.
3. Catterall WA, Mackie K 2011. Local anesthetics. In *Goodman & Gilman's manual of pharmacology and therapeutics.* (12th Edn.) Eds. LL Brunton, BA Chabner, BC Knollmann. McGraw-Hill Medical. New York, NY, pp. 565–582.
4. Coetzee JF. A review of analgesic compounds used in food animals in the United States. *Vet Clin North Am Food Anim Pract* 2013; 29:11–28.
5. Webb AI, Pablo LS 2009. Injectable anesthetic agents. In *Veterinary pharmacology and therapeutics.* (9th Edn.) Eds. JE Riviere, MG Papich. Wiley-Blackwell. Ames, IA, pp. 381–399.
6. Grosser T, Smyth E, FitzGerald GA. 2011. Anti-inflammatory, antipyretic, and analgesic agents: pharmacotherapy of gout. In *Goodman & Gilman's manual of pharmacology and therapeutics.* (12th Edn.) Eds. LL Brunton, BA Chabner, BC Knollmann. McGraw-Hill Medical. New York, NY, pp. 959–1004.

7. Svensson CI, Yaksh TL. The spinal phospholipase-cyclooxygenase-prostanoid cascade in nociceptive processing. *Annu Rev Pharmacol Toxicol* 2002; 42:553–583.

8. Kleinhenz MD, Van Engen NK, Gorden PJ, et al. Effect of age on the pharmacokinetics of intravenous and transdermal flunixin in Holstein calves. *Am J Vet Res* 2018; 79:568–575.

9. Veterinary Medicine Expert Committee on Drug Information USP. USP veterinary pharmaceutical information monographs—anti-inflammatories. *J Vet Pharmacol Ther* 2004; 27(1):1–110.

10. Aberg G, Ciofalo VB, Pendleton RG, et al. Inversion of (R)- to (S)-ketoprofen in eight animal species. *Chirality* 1995; 7:383–387.

11. Stilwell G, Lima MS, Broom DM. Comparing plasma cortisol and behaviour of calves dehorned with caustic paste after non-steroidal-anti-inflammatory analgesia. *Livest Sci* 2008; 119:63–69.

12. Delatour P, Foot R, Foster AP, et al. Pharmacodynamics and chiral pharmacokinetics of carprofen in calves. *Br Vet J* 1996; 152:183–198.

13. Rubio F, Seawall S, Pocelinko R, et al. Metabolism of carprofen, a nonsteroid anti-inflammatory agent, in rats, dogs, and humans. *J Pharm Sci* 1980; 69:1245–1253.

14. Coetzee JF, KuKanich B, Mosher R, et al. Pharmacokinetics of intravenous and oral meloxicam in ruminant calves. *Vet Ther* 2009; 10.

15. Lees, P 2009. Analgesics, anti-inflammatory, antipyretic drugs. In *Veterinary Pharmacology and Therapeutics*. (9th Edn.) Eds. JE Riviere, MG Papich. Wiley-Blackwell. Ames, IA, pp. 457–492.

16. Stock ML, Gehring R, Barth LA, et al. Pharmacokinetics of firocoxib in preweaned calves after oral and intravenous administration. *J Vet Pharmacol Ther* 2014; 37:457–463.

17. Taylor CP. Mechanisms of analgesia by gabapentin and pregabalin - Calcium channel alpha(2)-delta [Ca-v alpha(2)-delta] ligands. *Pain* 2009; 142:13–16.

18. Hurley RW, Chatterjea D, Feng MHR, et al. Gabapentin and pregabalin can interact synergistically with naproxen to produce antihyperalgesia. *Anesthesiology* 2002; 97:1263–1273.

19. Picazo A, Castaneda-Hernandez G, Ortiz MI. Examination of the interaction between peripheral diclofenac and gabapentin on the 5% formalin test in rats. *Life Sci* 2006; 79:2283–2287.

20. Coetzee JF, Shearer JK, Stock ML, et al. An update on the assessment and management of pain associated with lameness in cattle. *Clin North Am Food Anim Pract* 2017; 33:389–411.

21. Hinde K, Carpenter AJ, Clay JS, Bradford, BJ. Holsteins favor heifers, not bulls: Biased milk production programmed during pregnancy as a function of fetal sex. *PLoS One* 2014; 9(2): e86169. doi:10.1371/journal.pone.0086169.

22. Finnerty M., Enright W, Morrison C, Roche, J. Immunization of bull calves with a GnRH analogue–human serum albumin conjugate: Effect of conjugate dose, type of adjuvant and booster interval on immune, endocrine, testicular and growth responses. *J Reprod Fertil* 1994; 101(2):333–343.

23. Coetzee JF. Assessment and management of pain associated with castration in cattle. *Vet Clin North Am Food Anim Pract* 2013; 29:75–101.

24. Webster HB, Morin D, Jarrell V, et al. Effects of local anesthesia and flunixin meglumine on the acute cortisol response, behavior, and performance of young dairy calves undergoing surgical castration. *J Dairy Sci* 2013; 96:6285–6300.

25. Earley B, Crowe MA. Effects of ketoprofen alone or in combination with local anesthesia during the castration of bull calves on plasma cortisol, immunological, and inflammatory responses. *J Anim Sci* 2002; 80:1044–1052.

26. Ting ST, Earley B, Hughes JM, et al. Effect of ketoprofen, lidocaine local anesthesia, and combined xylazine and lidocaine caudal epidural anesthesia during castration of beef cattle on stress responses, immunity, growth, and behavior. *J Anim Sci* 2003; 81:1281–1293.

27. Ting ST, Earley B, Crowe MA. Effect of repeated ketoprofen administration during surgical castration of bulls on cortisol, immunological function, feed intake, growth, and behavior. *J Anim Sci* 2003; 81:1253–1264.

28. Stilwell G, Lima MS, Broom DM. Effects of nonsteroidal anti-inflammatory drugs on long-term pain in calves castrated by use of an external clamping technique following epidural anesthesia. *Am J Vet Res* 2008; 69:744–750.

29. Ballou MA, Sutherland MA, Brooks TA, et al. Administration of anesthetic and analgesic prevent the suppression of many leukocyte responses following surgical castration and physical dehorning. *Vet Immunol Immunopathol* 2013; 151:285–293.

30. Currah JM, Hendrick SH, Stookey JM. The behavioral assessment and alleviation of pain associated with castration in beef calves treated with flunixin meglumine and caudal lidocaine epidural anesthesia with epinephrine. *Can Vet J* 2009; 50:375–382.

31. Gonzalez LA, Schwartzkopf-Genswein KS, Caulkett NA, et al. Pain mitigation after band castration of beef calves and its effects on performance, behavior, *Escherichia coli*, and salivary cortisol. *J Anim Sci* 2010; 88:802–810.

32. Stafford KJ, Mellor DJ, Todd SE, et al. Effects of local anesthesia or local anesthesia plus a non-steroidal anti-inflammatory drug on the acute cortisol response of calves to five different methods of castration. *Res Vet Sci* 2002; 73:61–70.

33. Pang WY, Earley B, Sweeney T, et al. Effect of carprofen administration during banding or burdizzo castration of bulls on plasma cortisol, in vitro interferon-gamma production, acute-phase proteins, feed intake, and growth. *J Anim Sci* 2006; 84:351–359.

34. Coetzee JF, Edwards LN, Mosher RA, et al. Effect of oral meloxicam on health and performance of beef steers relative to bulls castrated on arrival at the feedlot. *J Anim Sci* 2012; 90:1026–1039.

35. Roberts SL, Hughes HD, Sanchez NCB, et al. Effect of surgical castration with or without oral meloxicam on the acute inflammatory response in yearling beef bulls. *J Anim Sci* 2015; 93:4123–4131.

36. USDA. 2007. Dairy 2007, Part I: Reference of Dairy Cattle Health and Management Practices in the United States, 2007 USDA-APHIS-VS, CEAH. Fort Collins, CO #N480.1007.

37. USDA. 2008. Beef 2007–08, Part I: Reference of Beef Cow-calf Management Practices in the United States, 2007–08 USDA-APHIS-VS, CEAH. Fort Collins, CO #N512-10080.

38. Spurlock D, Stock M, Coetzee J. The impact of 3 strategies for incorporating polled genetics into a dairy cattle breeding program on the overall herd genetic merit. *J Dairy Sci* 2014; 97:5265–5274.

39. Vickers KJ, Niel L, Kiehlbauch LM, Weary DM. Calf response to caustic paste and hot-iron dehorning using sedation with and without local anesthetic. *J Dairy Sci* 2005; 88:1454–1459.

40. Winder CB, LeBlanc SJ, Haley DB, et al. Clinical trial of local anesthetic protocols for acute pain associated with caustic paste disbudding in dairy calves. *J Dairy Sci* 2017; 100:6429–6441.

41. Stafford KJ, Mellor DJ. Dehorning and disbudding distress and its alleviation in calves. *Vet J* 2005; 169:337–349.

42. Stock ML, Baldridge SL, Griffin D, et al. Bovine dehorning: Assessing pain and providing analgesic management. *Vet Clin North Am Food Anim Pract* 2013; 29:103–133.

43. Stewart M, Stookey JM, Stafford KJ, et al. Effects of local anesthetic and a nonsteroidal anti-inflammatory drug on pain responses of dairy calves to hot-iron dehorning. *J Dairy Sci* 2009; 92:1512–1519.

44. Coetzee JF, Nutsch AL, Barbur LA, et al. A survey of castration methods and associated livestock management practices performed by bovine veterinarians in the United States. *BMC Vet Res* 2010;6:12.

45. Moggy MA, Pajor EA, Thurston WE, et al. Management practices associated with pain in cattle on western Canadian cow-calf operations: A mixed methods study. *J Anim Sci* 2017;95:958–969.

46. Stilwell G, de Carvalho RC, Lima MS, et al. Effect of caustic paste disbudding, using local anesthesia with and without analgesia, on behaviour and cortisol of calves. *Appl Anim Behav Sci* 2009; 116:35–44.

47. Kleinhenz MD, Van Engen NK, Gorden PJ, et al. The pharmacokinetics of transdermal flunixin meglumine in Holstein calves. *J Vet Pharm Ther* 2016; 39:612–615.

48. McMeekan CM, Stafford KJ, Mellor DJ, et al. Effects of regional analgesia and/or a non-steroidal anti-inflammatory analgesic on the acute cortisol response to dehorning in calves. *Res Vet Sci* 1998; 64(2):147–150.

49. Sutherland MA, Mellow DJ, Stafford KJ, et al. Cortisol responses to dehorning of calves given a 5-h local anesthetic regimen plus phenylbutazone, ketoprofen, or adrenocorticotropic hormone prior to dehorning. *Res Vet Sci* 2002;73(2):115–123.

50. Milligan BN, Duffield T, Lissemore K. The utility of ketoprofen for alleviating pain following dehorning in young dairy calves. *Can Vet J* 2004; 45(2):140–143.

51. Duffield TF, Heinrich A, Millman ST, et al. Reduction in pain response by combined use of local lidocaine anesthesia and systemic ketoprofen in dairy calves dehorned by heat cauterization. *Can Vet J* 2010; 51:283–288.

52. Stilwell G, Lima MS, Carvalho RC, et al. Effects of hot-iron disbudding, using regional anaesthesia with and without carprofen, on cortisol and behaviour of calves. *Res Vet Sci* 2012; 92:338–341.

53. Stock ML, Barth LA, Van Engen NK, et al. Impact of carprofen administration on stress and nociception responses of calves to cautery dehorning. *J Anim Sci* 2016; 94:542–555.

54. Heinrich A, Duffield TF, Lissemore KD, et al. The impact of meloxicam on postsurgical stress associated with cautery dehorning. *J Anim Sci* 2009;92:540–547.

55. Coetzee JF, Mosher RA, KuKanich B, et al. Pharmacokinetics and effect of intravenous meloxicam in weaned Holstein calves following scoop dehorning without local anesthesia. *BMC Vet Res* 2012; 8:153.

56. Glynn HD, Coetzee JF, Edwards-Callaway LN, et al. The pharmacokinetics and effects of meloxicam, gabapentin, and flunixin in postweaning dairy calves following dehorning with local anesthesia. *J Vet Pharmacol Ther* 2013; 36:550–561.

57. Allen KA, Coetzee JF, Edwards-Callaway LN, et al. The effect of timing of oral meloxicam administration on physiological responses in calves after cautery dehorning with local anesthesia. *J Dairy Sci* 2013; 96:5194–5205.

58. Stock ML, Millman ST, Barth LA, et al. The effects of firocoxib on cautery disbudding pain and stress responses in preweaned dairy calves. *J Dairy Sci* 2015; 98:6058–6069.

59. Lay D, Friend T, Bowers C, et al. A comparative physiological and behavioral study of freeze and hot-iron branding using dairy cows. *J Anim Sci* 1992; 70:1121–1125.

60. Schwartzkopf-Genswein K, Stookey J, Passillé AD, Rushen J. Comparison of hot-iron and freeze branding on cortisol levels and pain sensitivity in beef cattle. *Can J Anim Sci* 1997; 77:369–374.

61. Tucker CB, Mintline EM, Banuelos J, et al. Effect of a cooling gel on pain sensitivity and healing of hot-iron cattle brands. *J Anim Sci* 2014;92:5666–5673.

62. Tucker CB, Mintline EM, Banuelos J, et al. Pain sensitivity and healing of hot-iron cattle brands. *J Anim Sci* 2014; 92:5674–5682.

63. Tucker CB, Fraser D, Weary DM. Tail docking dairy cattle: Effects on cow cleanliness and udder health. *J Dairy Sci* 2001; 84:84–87.

64. Schreiner DA, Ruegg PL. Effects of tail docking on milk quality and cow cleanliness. *J Dairy Sci* 2002; 85(10):2503–2511.

65. Lombard JE, Tucker CB, von Keyserlingk MAG, et al. Associations between cow hygiene, hock injuries, and free stall usage on US dairy farms. *J Dairy Sci* 2010; 93:4668–4676.

66. Weary DM, Schuppli CA, von Keyserlingk MAG. Tail docking dairy cattle: Responses from an online engagement. *J Anim Sci* 2011; 89:3831–3837.

67. Kroll LK, Grooms DL, Siegford JM, et al. Effects of tail docking on behavior of confined feedlot cattle. *J Anim Sci* 2014; 92:4701–4710.

68. Fajt VR, Wagner SA, Norby B. Analgesic drug administration and attitudes about analgesia in cattle among bovine practitioners in the United States. *J Am Vet Med Assoc* 2011; 238:755–767.

69. Newby NC, Pearl DL, LeBlanc SJ, et al. The effect of administering ketoprofen on the physiology and behavior of dairy cows following surgery to correct a left displaced abomasum. *J Dairy Sci* 2013;96:1511–1520.

70. Cattle and Calves Non-Predator Death Loss in the United States, 2010. USDA-APHIS National Animal Health Monitoring System. 2011.

71. Dairy 2007: Facility characteristics and cow comfort on U.S. dairy operations, 2007. USDA-APHIS National Animal Health Monitoring System. 2010.

72. Cook NB. Prevalence of lameness among dairy cattle in Wisconsin as a function of housing type and stall surface. *J Am Vet Med Assoc* 2003; 223:1324–1328.

73. USDA. 2010. Beef 2007–08, Part IV: Reference of Beef Cow-calf Management Practices in the United States, 2007–08. USDA:APHIS:VS, CEAH. Fort Collins, CO #523.0210.

74. Nicholson JDW, Nicholson KL, Frenzel LL, et al. Survey of transportation procedures, management practices, and health assessment related to quality, quantity, and value for market beef and dairy cows and bulls. *J Anim Sci* 2013; 91:5026–5036.

75. Wells SJ, Trent AM, Marsh WE, et al. Prevalence and severity of lameness in lactating dairy cows in a sample of Minnesota and Wisconsin herds. *J Am Vet Med Assoc* 1993; 202:78–82.

76. Whay HR, Main DCJ, Green LE, et al. Assessment of the welfare of dairy cattle using animal-based measurements: Direct observations and investigation of farm records. *Vet Rec* 2003; 153:197–202.

77. Whay HR, Waterman AE, Webster AJF et al. The influence of lesions type on the duration of hyperalgesia associated with hindlimb lamenss in dairy cattle. *Vet J* 1998; 156:23–29.

78. National Dairy FARM Program: Farmers Assuring Responsible Management, 2016. Available at: www.nationaldairyfarm.com/. Accessed September 28, 2017.

79. Validus Ventures, LLC, 2017. Available at: www.validusservices.com/. Accessed September 13, 2017.

80. NYS Department of Agriculture and Markets, New York State Cattle Health Assurance Program (NYSCHAP), 2002. Available at: http://nyschap.vet.cornell.edu/. Accessed September 28, 2017.

81. Anderson DE, Muir WM. Pain management in cattle. *Vet Clin North Am Food Anim Pract* 2005; 21:623–635.

82. Freedom of Information Summary: Flunixin Transdermal Solution. US FDA. Available at: https://animaldrugsatfda.fda.gov/adafda/app/search/public/document/downloadFoi/1944. Accessed September 29, 2017.

83. Chapinal N, de Passillé AM, Rushen J, et al. Effect of analgesia during hoof trimming on gait, weight distribution, and activity of dairy cattle. *J Dairy Sci* 2010; 93:3039–3046.

84. Chapinal N, de Passillé AM, Rushen J, et al. Automated methods for detecting lameness and measuring analgesia in dairy cattle. *J Dairy Sci* 2010; 93:2007–2013.

85. Rushen J, Pombourcq E, de Passillé AM. Validation of two measures of lameness in dairy cows. *App Anim Behav Sci* 2006; 106:173–177.

86. Schulz KL, Anderson DE, Coetzee JF, et al. Effect of flunixin meglumine on the amelioration of lameness in dairy steers with amphotericin D induced transient synovitis-arthritis. *Am J Vet Res* 2011; 72:1431–1438.

87. Whay HR, Webster AJF, Waterman-Pearson AE. Role of ketoprofen in the modulation of hyperalgesia associated with lameness in dairy cattle. *Vet Rec* 2005; 157:729–733.

88. Flower FC, Sedlbauer M, Carter E, et al. Analgesics improve the gait of lame dairy cattle. *J Dairy Sci* 2008; 91:3010–3014.

89. Flower FC, Weary DM. Effect of hoof pathologies on subjective assessments of dairy cow gait. *J Dairy Sci* 2006; 89:139–146.

90. Thomas HJ, Miguel-Pacheco GG, Bollard NJ, et al. Evaluation of treatments for claw horn lesions in dairy cows in a randomized controlled trial. *J Dairy Sci* 2015; 98(7):4477–4486.

91. Janssen S, Wunderlich C, Heppelmann M, et al. Short communication: Pilot study on hormonal, metabolic, and behavioral stress response to treatment of claw horn lesions in acutely lame dairy cows. *J Dairy Sci* 2016; 99:7481–7488.

92. Nagel D, Wieringa R, Ireland J, et al. The use of meloxicam oral suspension to treat musculoskeletal lameness in cattle. *Vet Med-Res Reports* 2016; 7:149–155.

93. Offinger J, Herdtweck S, Rizk A, et al. Postoperative analgesic efficacy of meloxicam in lame dairy cows undergoing resection of the distal interphalangeal joint. *J Dairy Sci* 2013; 96(2):866–876.

94. Coetzee JF, Mosher RA, Anderson DE, et al. Impact of oral meloxicam administered alone or in combination with gabapentin on experimentally induced lameness in beef calves. *J Anim Sci* 2014; 92:816–829.

95. Rushen J, Pombourcq E, de Passillé AM. Validation of two measures of lameness in dairy cows. *App Anim Behav Sci* 2006; 106:173–177.

CHAPTER 16

An Overview of the Segments of the Beef Cattle Industry and Animal Welfare Implications of Beef Industry Practices

Jason K. Ahola, John J. Wagner, and Terry Engle
Colorado State University

CONTENTS

SEGMENTS OF THE U.S. BEEF CATTLE INDUSTRY

The U.S. beef cattle industry is a very dynamic and robust industry. Local, regional, and international events can influence the U.S. beef cattle industry. Unlike most industries, the U.S. beef cattle industry is not controlled by one overall management program. Instead, it is composed of several independent operating segments that are loosely linked together in a supply chain fashion by live cattle and end products (Field, 2007). In general, the U.S. beef cattle industry is composed of seven segments: (1) seedstock producers, (2) commercial cow-calf producers, (3) yearling/stocker or backgrounding feedlot operators, (4) finishing feedlot operators, (5) packers, (6) retailers and foodservice distributors, and (7) consumers (Field, 2007).

Briefly, the seedstock producers are cow-calf producers that specialize in generating breeding stock with improved genetics to ultimately be utilized by commercial cow-calf producers. Commercial cow-calf producers purchase genetics (in the form of bulls, cows, embryos or semen) from seedstock producers to improve their herd genetics. In a commercial cow-calf operation, the goal is to have each cow produce one calf per year. The calf is raised by its mother on pasture until weaned (removed from mother). Calves are typically weaned at approximately 6–7 months of age (approximately 220–260 kg). After weaning, the majority of calves (male and female) are transported to a yearling/stocker operation or to a backgrounding feedlot operation, often after going through an ownership change. A proportion of females are retained on the commercial cow-calf operation as replacements for older, less-productive cows. Yearling/stocker operators typically graze weaned calves on pasture or crop residue (e.g., wheat pasture) for a prescribed amount of time to produce a heavier calf (350–400 kg) that is approximately 12 months old for entry into a finishing feedlot operation. Alternatively, weaned calves can enter a backgrounding feedlot operation after being weaned. In a backgrounding feedlot operation the cattle are housed in feedlot pens and fed a growing-type feedlot diet until they reach an approximate weight of approximately 350–400 kg. Depending on economic factors (feed price, crop residue availability, beef supply and demand, cattle price, etc.) weaned calves (particularly calves with heavier body weights) can bypass the yearling/stocker or backgrounding feedlot operations and be directly placed into a finishing feedlot where they are fed a growing diet for 30–60 days and then transitioned to a high grain-based finishing diet until harvest. At the end of either the yearling/stocker or backgrounding feedlot segments of beef cattle production, the cattle are moved into a finishing feedlot. During the finishing feedlot portion of beef production, cattle are typically fed a high-energy grain-based diet. Cattle will remain in the finishing feedlot approximately 130–220 days depending on their entry body weight and level of backfat at the time of harvest. When the cattle reach an approximate body weight of 600–780 kg, they are transported to an abattoir (packer segment of beef production) and slaughtered. The beef products are then sold to retailers and food service companies for purchase by consumers.

AN OVERVIEW OF ANIMAL WELFARE IMPLICATIONS OF BEEF INDUSTRY PRACTICES

An increasing number of consumers are making animal product purchasing decisions based on how animals were raised and cared for. This decision is, in most cases, based on labeling claims made on packaged products, point of purchase materials offered, and/or conversations with those selling the product. This issue is driven by consumers wanting to know more about how their food is raised and where it comes from. For the average consumer with no baseline knowledge, there appears to be a need for verification and validation that animals received appropriate care during their lifetimes and were treated as humanely as possible. As consumers want more information about their food products, and as beef product brands are working to differentiate themselves, the third-party verification of these credence attributes (i.e., those claims made about a product that can't be determined by simply looking at the product) continues to be in demand. Scrutiny over the use of traditional cattle industry practices (i.e., dehorning, castration, and branding) and whether pain mitigation is provided to cattle undergoing these procedures is increasing. Additional practices including abrupt removal at weaning and long-distance transportation are also being scrutinized, but at a lesser extent.

THE FIVE FREEDOMS

Much of the discussion about welfare of livestock over the years can be traced back to an early report (later known as the "Brambell Report") written in 1965 by a British governmental committee established as a result of widespread concern over the welfare of animals raised in intensive livestock production systems. This committee identified five "freedoms" that animals raised under intensive livestock production systems should have (Conklin, 2014). These five freedoms were identified as follows:

1. Freedom from hunger and thirst
2. Freedom from discomfort
3. Freedom from pain, injury, and disease
4. Freedom to behave normally
5. Freedom from fear and distress

It appears that over the last 50 years, the freedoms identified in this report have provided part of the foundation for welfare-based changes to livestock production systems, including components included in more recent animal welfare audits and questions related to traditional industry practices.

HISTORY OF ANIMAL WELFARE REGULATION

Before discussing changes occurring in the industry, a brief history of animal welfare regulation is in order. Currently, only regulations to oversee the handling of cattle during the slaughtering process are in place. Prior to the 1958 Humane Slaughter Act, no U.S. laws governed humane slaughter practices. The original law focused on ensuring that proper methods were used to render cattle insensible before shackling, hoisting, casting, or cutting. The Humane Methods of Slaughter Act of 1978 was passed as a follow-up to address additional cattle handling concerns associated with the slaughter process.

According to the USDA, the reason for the Act is to prevent needless animal suffering, improve meat quality, decrease financial losses, and ensure safe working conditions. Compliance with the Act is ensured in beef packing plants via a USDA Food Safety and Inspection Service (FSIS) veterinarian as well as FSIS inspectors on the kill floor. The veterinarian enforces humane slaughter methods throughout the plant by observing methods of slaughter, ensuring corrective action is taken, and reporting inhumane treatment of cattle.

Several animal handling and welfare regulations are associated with the Act, and include the following: (1) animal handling while unloading trucks, (2) appropriate and functional facilities to prevent animal injury, (3) animal handling in pens, alleyways, and chutes, (4) handling of disabled or non-ambulatory ("downer") animals, (5) access to water and potentially feed, and (6) stunning procedures.

Separate from the Humane Methods of Slaughter Act of 1978, the Animal Welfare Act has been in existence since 1966 (initially referred to as the Laboratory Animal Welfare Act of 1966). The Animal Welfare Act focuses on the use and treatment of animals in research laboratories, care and handling of pets, and prevention of cruel practices (including animal fighting). However, the Animal Welfare Act, which is administered by USDA Animal and Plant Health Inspection Service (APHIS), does not cover farm animals that are used for food and fiber. In place of federal laws addressing farm animal care, in recent years several states have passed legislation associated with farm animal welfare.

DEMAND BY RETAILERS, FOOD SERVICE, AND CONSUMERS

Separate from governmental regulation, in the 1990s the meat packing industry—led by the American Meat Institute (AMI)—initiated voluntary efforts to improve the handling of animals around the time of slaughter. In 1991, AMI published "Recommended Animal Handling Guidelines for Meat Packers," which was authored by Dr. Temple Grandin at Colorado State University. These guidelines focused on animal handling, including the identification of problems with animal handling during the slaughter process (e.g., moving animals, stunning, achieving insensibility on the bleed rail, and managing non-ambulatory animals) and possible solutions to these challenges.

In 1997, a follow-up document, "Good Management Practices (GMP) for Animal Handling and Stunning," was also authored by Dr. Grandin. This resource included information to help packing plants conduct self-audits of animal well-being. Ultimately, in 2004 AMI created one animal welfare document ("Animal Handling and Audit Guidelines for the Meat Industry"), which included AMI-approved forms for auditing beef plants.

Only since the mid-1990s has significant animal welfare research occurred, much of which has been led by Dr. Grandin. Her efforts began with a USDA-funded survey of stunning and handling practices in 24 federally inspected beef packing plants in 1996. This initial audit served as a baseline for animal welfare. Each plant was objectively scored for the percentage of cattle that:

- Had to be stunned more than once with the captive bolt stunner.
- Were sensible or partially sensible on the bleed rail.
- Fell down or slipped.
- Vocalized in the stunning chute area, stunning box, or restrainer.
- Were prodded with an electric prod.

A series of follow-up annual animal welfare audits began in the late 1990s, which were initiated by fast-food retailers (including McDonald's initially, followed by Wendy's and Burger King), and utilized Dr. Grandin and her approach to auditing packing plants. In those audits, a set of objective, measureable audit techniques were used to document incidence of the above traits, but primarily only around the time of slaughter. Targets identified for self- and third-party packing plant audits address the same categories that were included in the initial audit (i.e., stunning, bleed rail insensibility, falling/slipping, vocalization, prodding, etc.).

MARKET-DRIVEN ON-FARM ANIMAL WELFARE AUDITS

In recent years, one retailer of beef products has initiated a market-driven approach to addressing animal welfare in livestock. In 2010, Whole Foods Markets, a successful retail leader in natural, organic and locally produced foods, was the first grocery retailer to require that their beef, pork and poultry products would all be required to meet a set of standards. These standards consist of the 5-Step Animal Welfare Rating Standards program created by the Global Animal Partnership (GAP), a non-profit organization. Whole Foods launched the program in February 2011, and now requires a minimum of Step 1 compliance for all fresh chicken, pork and beef, and presently, there are 5-Step standards only for those three species.

The responsibility of GAP is to set animal care and well-being standards, and then to conduct training of independent certification companies to audit meat suppliers to the established standards. The program requires independent, third-party certification companies to evaluate locations (farms/ranches, backgrounders, feedlots) to the 5-Step standards, and the GAP certification process will determine the Step Level achieved. The certifier follows ISO Guide 65 guidelines, which requires auditors to be independent of the operation for which they are conducting the audit, employ qualified and trained auditors, and ensure consistency in the audits being conducted. Once the audit is conducted, the certifier determines compliance to the step within the standard and determines the location's approval for a 15-month period of time. The structure of the program is similar to other ISO Guide 65 models, such as certified organic or the USDA Processed Verified Program's (PVP) model for claims that could change with the animal over time, such as non-hormone treated cattle (NHTC), NeverEver3, etc.

In addition to GAP, there are other cattle welfare standard certification programs that are being used by food suppliers, such as Humane Farm Animal Care (HFAC), which is promoted with a "Certified Humane" label at food service and retail levels. The Animal Welfare Institute Standards promotes its program using the "Animal Welfare Approved" label. From an industry perspective, AMI has recommended its Animal Handling Guidelines and Audit Guide (as mentioned above), National Cattlemen's Beef Association (NCBA) has animal care and well-being guidelines as part of their Beef Quality Assurance (BQA) Program, and the Beef Marketing Group has a standard called the "Progressive Beef Standard."

It has been documented that USDA supports the concept of the beef industry creating and voluntarily implementing objective criteria for GMPs to ensure humane handling of cattle. For most beef cattle producers, this is preferred to the implementation of mandatory state or federal regulations. However, in the newly expanding market for verification of humane handling of cattle, there is a great need for science-based research related to the development of uniform consumer-driven standards for the proper management, care, handling, and transportation of animals. This includes specific research into the effects of typical on-farm and on-ranch practices (i.e., castration, dehorning, branding, etc.) on animal welfare.

The U.S. beef industry, compared to other more intensive animal production industries (poultry, swine, etc.), is generally viewed as having a much more solid track record in regard to the humane handling of cattle. This is due, in part, to the beef industry's voluntary grassroots BQA Program—created by cattle producers to assure consumers that the safe, high-quality, and wholesome beef they are purchasing is from cattle raised, cared for, and handled in a humane manner.

BEEF QUALITY ASSURANCE (BQA) AND CATTLE WELFARE

Nearly 20 years ago, in an effort to address recent consumer concerns about animal welfare in the beef industry, the National BQA program developed the "Producer Code for Cattle Care" in 1996 (NCBA, 2014a). The code is a set of ideals consistent with proper cattle care (Table 16.1).

The concepts included in the code were used by the National BQA Program in 2003 to create the "Cattle Industry's Guidelines for the Care and Handling of Cattle" best practices guide (NCBA, 2014b). This guide provides recommendations for cattle production, including self-evaluation checklists to improve production practices.

In 2014, the National BQA Program developed a set of "Supplemental Guidelines," which directly address animal welfare issues in the beef industry that are related to traditional industry practices. The guidelines address castration, dehorning (including disbudding), branding, tail docking in beef cattle, dairy calf management, and euthanasia. Development of the recommended guidelines was initiated by the national BQA program.

Table 16.1 Beef Quality Assurance (BQA) Producer Code for Cattle Care

Provide adequate food, water, and care to protect cattle health and well-being.
Provide disease prevention practices to protect herd health.
Provide facilities that allow safe and humane movement and/or restraint of livestock.
Use appropriate methods to euthanize sick or injured livestock.
Provide personnel with training to properly handle and care for cattle.
Minimize stress when transporting cattle.
Persons who willfully mistreat animals will not be tolerated.

Source: www.beefusa.org/uDocs/factsheet-bqaandanimalwelfare.pdf.

The BQA Program Advisory Committee was composed of veterinarians, animal scientists, cattle industry leaders, production managers and producers. The intent, as stated in the document, was for the guidelines to "focus on the animal and are aimed to satisfy scientifically valid and feasible approaches to meeting cattle health and welfare needs."

Castration: The guidelines recommend that castration be done prior to 3 months of age, prior to leaving the farm of origin, and by trained personnel, as well as to utilize methods that promote "well-being and comfort of cattle." While the guidelines do not indicate that analgesia or anesthesia have to be used, they encourage producers to seek veterinary guidance on this, particularly in older animals.

Dehorning: Relative to dehorning (including disbudding), the guidelines recommend the selection of polled cattle (cattle that do not grow horns) to avoid having to deal with dehorning, but also recognize that if dehorning is necessary, it should be done by trained personnel while horn development is at the horn bud stage to limit the amount of tissue trauma (which increases with horn development). As with castration, the suggestion is to discuss the use of anesthesia and/or analgesia with a veterinarian, particularly in older animals with advanced horn development.

Branding: Of importance to western cattle producers, branding is also addressed in the guidelines as it relates to permanent identification. Branding is a process of permanently identifying cattle to identify ownership. In the Western U.S., all cattle must be brand inspected (by the state of residence brand inspection division) when cattle ownership changes (i.e., they are sold) and/or they are transported to a different location beyond a certain distance (e.g., 75 miles). The recommendation is to brand (freeze or hot-iron) quickly and expertly with trained personnel and proper equipment. Further, cattle should never be jaw or face branded. Consistent with castration and dehorning, the guidelines suggest that use of pain mitigation can be discussed with a veterinarian.

The areas of tail docking in beef cattle and neonatal dairy calf management do not directly apply to the vast majority of beef cattle producers. However, the supplemental guidelines do address euthanasia protocols to limit animal welfare problems.

AMERICAN VETERINARY MEDICAL ASSOCIATION (AVMA) PERSPECTIVE

Since 2011, the AVMA has published several documents summarizing literature reviews of the welfare implications of castration, dehorning/disbudding, and hot-iron branding as they relate to cattle. In addition to summarizing the literature, AVMA clearly states their concerns with these industry practices and possible solutions to address them.

For instance, the AVMA indicates that "all methods of castration induce pain and physiological stress in animals of all ages," based on the available scientific literature (AVMA, 2014a). Furthermore, they suggest that pain and stress should be minimized, likely via both the use of anesthesia and analgesia. For branding, the AVMA recommendation is not as clear, in that the suggestion is that animal welfare should be considered when choosing a branding method, and that effort should be made to use methods that cause less pain and distress (AVMA, 2011). For dehorning/disbudding, the AVMA acknowledges there is pain associated with these practices and limiting pain/stress is important (AVMA, 2014b). In addition to suggesting the use of polled genetics to avoid pain/stress associated with dehorning/disbudding, the AVMA suggestion is to use pre-emptive analgesia when dehorning/disbudding is done.

However, lack of a unified and well-communicated commitment by the industry to mitigate pain induced by production practices has led to the public demand for production programs that document pain mitigation and appropriate animal husbandry. Until such time that the public has confidence in a unified animal welfare plan from beef producers, third-party auditors and humane association will continue to increase in number.

CONCLUSIONS

Humane treatment of animals has always been an ingrained social ethic among beef producers. Beef producers do understand the need for minimizing pain (from painful procedures) as it relates to production efficiency and animal well-being. However, the lack of a unified and well-communicated commitment by the industry to mitigate pain induced by production practices has led to public demand for such verification programs. Governmental regulation of on-farm animal welfare has been extremely limited; however, over the last 20 years retail and food service companies have demanded substantial animal welfare auditing in packing plants, much of which has been addressed via regular in-plant auditing. On-farm auditing has only occurred more recently via a market-driven approach by one retailer where consumers pay a premium for assurance of proper animal welfare via third-party auditing of cattle producers. Coupled with this, in the past few years large livestock industry organizations have identified some on-farm procedures that are associated with pain and/or stress. These organizations have taken different approaches as to recommendations for using pain mitigation when the procedures are conducted, due in part to situational variables including animal age, physiological development, and methods used. Objective scientific research related to pain and stress associated with these procedures is limited. Regardless, consumers will likely continue to pressure the beef industry to produce beef while limiting animal welfare issues associated with the 5 freedoms identified in Britain some 50 years ago.

ONLINE RESOURCES

Beef Quality Assurance (BQA) Supplemental Guidelines (www.bqa.org/Media/BQA/Docs/supplemental_guidelines_2014.pdf).

Global Animal Partnership (GAP) 5-Step Animal Welfare Rating Standards (www.globalanimalpartnership.com).

USDA Animal Welfare Information Center (http://awic.nal.usda.gov), includes animal welfare programs that are available.

REFERENCES

AVMA. 2011. Literature Review on the Welfare Implications of Hot-Iron Branding and Its Alternatives. American Veterinary Medical Association. Available at: www.avma.org/KB/Resources/LiteratureReviews/Documents/hot iron_branding_bgnd.pdf.

AVMA. 2014a. Literature Review on the Welfare Implications of Castration of Cattle. American Veterinary Medical Association. Available at: www.avma.org/KB/Resources/LiteratureReviews/Documents/castration-cattle-bgnd.pdf.

AVMA. 2014b. Literature Review on the Welfare Implications of the Dehorning and Disbudding of Cattle. American Veterinary Medical Association. Available at: www.avma.org/KB/Resources/LiteratureReviews/Documents/dehorning_cattle_bgnd.pdf.

Beef Production and Management Decisions. 5th Edition. 2007. Ed. Thomas G. Field. Upper Saddle River, N.J.: Pearson Prentice Hall: ISBN-13: 978–0131198388; ISBN-10: 0131198386.

Conklin, T. 2014. An Animal Welfare History Lesson on the Five Freedoms. Michigan State University Extension, East Lansing, MI. Available at: http://msue.anr.msu.edu/news/an_animal_welfare_history_lesson_on_the_five_freedoms.

NCBA. 2014a. Beef Quality Assurance Program and Animal Welfare. National Cattlemen's Beef Association. Available at: https://www.beefusa.org/uDocs/factsheetbqaandanimalwelfare.pdf.

NCBA. 2014b. Cattle Industry's Guidelines for the Care and Handling of Cattle. National Cattlemen's Beef Association. Available at: www.beefusa.org/uDocs/guidelinesforthecareandhandlingofcattle.pdf.

CHAPTER 17

Ranching with Integrity and Intentionality

Tom Field
University of Nebraska-Lincoln

CONTENTS

The objective of this chapter is not to prescribe how ranching should be conducted but rather to create a starting point for discussing the purpose, philosophy, and fulfillment that can be obtained from the intentional and thoughtful practice of stewardship, husbandry and stockmanship. The content of the chapter is largely driven by the experiences and observations of the author based on a lifetime of work on an extensive ranching enterprise in western Colorado plus some three decades of engagement with ranchers as a teacher, researcher, and consultant.

The ranching landscape is a subject of almost infinite complexity. But the essence of 21st Century ranching—and the cowboy, and the ranch economy, and the landscape of the ranch is complicated adaptation.

Paul Starrs, Geographer

Let the Cowboy Ride

OVERVIEW OF RANCHING

Ranches are characterized by a series of components—people, landscapes, livestock, wildlife, financial resources, and infrastructure—that comprise an integrated system. None of these elements can be viewed in isolation, creating value from a ranch depends on the ability to understand the multifaceted connections between them. However, the ranching enterprise fails if any of the aforementioned components is allowed to remain in an unhealthy state for a prolonged period of time. Ranching in extensive environments is complicated and unlike many downstream participants in the supply chain such as fast-food enterprises, grocers, or restaurants, management cannot be conducted based on a recipe model of duplication. A one size fits all approach driven by an operational manual treated as the last word would yield disastrous results for the people, livestock and ecosystem of a ranching community.

Creating a healthy ranch system (Figure 17.1) requires the application of creativity to processes such as strategic planning, operational management, design thinking, and development of organizational culture. Healthy ranches are created when managers make decisions that take into account long- and short-term time horizons, avoid creating undesirable consequences either directly or indirectly, and ultimately recognize the interrelationships of the system as they make strategic trade-offs in search of optimal solutions. Excellence in ranching also requires intuition and insight built on skilled observation of landscapes, livestock, and people.

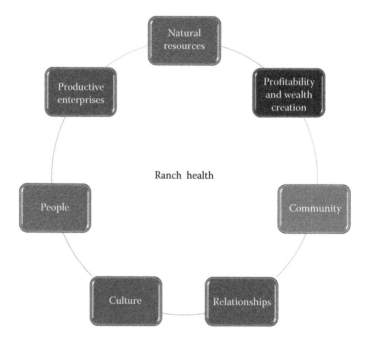

Figure 17.1 Components of a healthy ranch system.

COMPETING PRESSURES

Building and sustaining a healthy ranch also requires that decision makers effectively manage a set of broad trade-offs, competing interests, and multiple forces that pressure the ranch system.

Internal vs. External Pressures

Pressures from within and outside the ranch system must be recognized as they exert influence, either directly or indirectly, on the choices available to managers. Some of these pressures are within the circle of the manager's influence while many are beyond their direct influence. Internal pressures can be characterized as being related to the landscape and environment in which the ranch is located while external pressures are related to markets, consumer demand and governmental policies.

Short vs. Long-term Outcomes

All enterprises are driven by outcomes. For example, publically traded companies are driven by the quarterly report and their mission to return value to investors; thus, leaders of these organizations are under substantial pressure to take a shorter term view. Family-owned businesses are motivated to create wealth and to transfer it to the next generation—a longer term horizon. Ranch managers must make choices that effectively balance the need to attain performance in the short term—low mortality rates during calving season, annual profit or loss, or high pregnancy rates during breeding season—with longer term measures—creating productive grasslands, healthy riparian zones, productive herds, and strong organizational culture. Wyoming rancher Bob Budd captured the essence of this dilemma when he stated that the key to ranching is "to think at the pace of rocks and mountains while learning to act in our own lifetimes."

Proactive vs. Reactive Choices

Superior ranch management is founded on intentional choices made in accordance with the strategic vision of the business and in alignment with a set of deeply held core values. Proactive decisions are most easily made when conditions are relatively stable, resources are available, and goals are clearly defined. However, condition stability and resource availability are not always ideal in landscape based enterprises. Weather events, variability of precipitation volume and timing, disease occurrence, market disruptions, disruptive government policies, or shifting economic conditions may force managers into reacting to unexpected changes. Long-term success requires that managers be able to maintain a proactive perspective in the midst of chaos without ignoring the realities of the short term. As Ayn Rand once wrote, "you can ignore reality but you cannot ignore the consequences of ignoring reality."

Known vs. Unknown

While ranching is steeped in tradition and lore, the enterprise can ill afford to fall into the trap of "knowing." Franchised, recipe-style management is poorly suited to ranching; while core principles provide a foundation for decision-making the most successful ranches are characterized as adaptable, curious, continuous learning environments where asking the right questions is highly valued across all levels of the enterprise.

Extensive ranching is faced with the challenge of dealing with a series of unknown variables that lie beyond the influence of management. For example, the extent and timing of precipitation has a substantial impact on managerial flexibility—the availability of forage is a key variable that emerges in real time and is very difficult to predict. Weather events such as blizzards, temperature fluctuations that impact the growth of warm and cool season grasses, and market disruptions due to trade wars and the like are part of the "unknowns" to which managers must adapt. Chaotic conditions are not unique to agriculture and the importance of adaptive decision-making is clarified by H. L Hau (2004).

"We've discovered that those companies are great not because they were focused on cost or flexibility or speed but because they have the ability to manage transitions—changing market conditions, evolving technology, different requirements as the product moves through its life cycle. The companies that can adapt are the ones that will be here for the long term."

And so it is for ranching, learn then adapt then learn some more.

CONNECTING HEAD AND HEART

Daniel Pink (2012), in his examination of human behavior, suggests that motivation is the result of people's deep desire for purpose, mastery, and autonomy. His model is applicable to ranching on many levels—the most successful ranch managers are deeply tied to a sense of purpose—a belief that their lives and work have deep personal meaning and provide them an opportunity to practice their most cherished beliefs—work ethic, stewardship, dedication, and loyalty. Purpose drives the practitioners of extensive ranching to see beyond hardships and challenges and to invest their talent, creativity and sweat into the work of forging both enterprises and a way of life from the landscape.

The pursuit of mastery is at the forefront of successful ranches. Active, ongoing learning is essential and developing mastery drives a sense of continuous improvement coupled with a desire to attain craftsman level skills and abilities. The challenge and excitement of ranching is the opportunity to develop mastery in specific skill sets as well as in the realm of making strategic choices in the context of complicated systems. Developing exceptional skill in horsemanship, grassland management, cattle handling, equipment operation, and fence construction are examples of arenas requiring exceptional skill sets.

People who invest their lives in ranching embrace autonomy with the realization that living and working in remote areas demand a strong sense of self-reliance, accountability and problem-solving. The desire for autonomy is sufficiently strong that it drives the selection of a lifestyle where convenience is a secondary consideration, responsibility for performance is clearly assigned, and getting the job done is a cultural expectation. Ultimately, the outcome of embracing the power of purpose, the joy of mastery, and the freedom of autonomy is the creation of a ranch where human beings have proactively connected head and heart in search of meaning.

RANCH CULTURE

Considerable attention is typically focused on the "how" and "what" of ranching and certainly developing effective processes and tactics is central to the sustainability of the business. However, defining and living the "why" of ranching yields benefits that extend across all activities of the ranch team and ultimately provides a framework for balancing the diverse and competing components of extensive management systems.

At the heart of any sustainable business, ranching or otherwise, is a deeply seated set of core values—a small group of principles that guide decision-making in the organization and more importantly define the character of the enterprise. A clear set of values allow organizations to attract like-minded people and to provide a set of standards upon which to determine hiring, firing, performance reviews and rewards. These core values, although few in number, may be the most

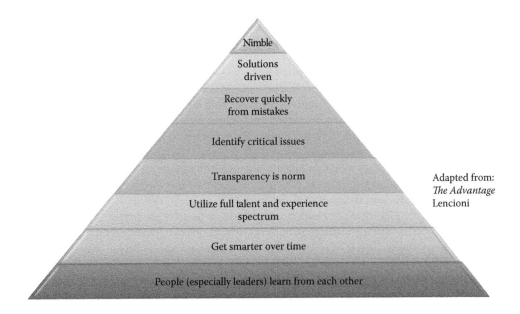

Nimble

Solutions
driven

Recover quickly
from mistakes

Identify critical issues

Transparency is norm

Utilize full talent and experience
spectrum

Get smarter over time

People (especially leaders) learn from each other

Adapted from:
The Advantage
Lencioni

Figure 17.2   Healthy organizations capable of driving innovation.
(*Source*: **Adapted from Lencioni**).

important decision for a ranch. Core values define the soul of a ranch and define its culture. When these values align around concepts such as integrity, stewardship, and husbandry then the stage is set for a healthy ranch.

Business writer Patrick Lencioni makes the case for the healthy organization in *The Advantage* (Figure 17.2). Lencioni's premise is straightforward—healthy organizations are capable of higher performance than "smart" ones. Ranching organizations are no different; success depends on the ability of people to learn from each other, to leverage each new experience and observation into a bank of wisdom, and to value the knowledge, skills and abilities of people across the organization. It is the third capability, to value knowledge beyond the confines of formal education that character-izes healthy ranches. As a case in point, the most common mistake made by wealthy investors who purchase ranches is to undervalue the collective knowledge of the people who have lived and worked there. The new owner seeks out a "pedigreed" manager, superimposes their biases and uninformed expectations on the decision-making process, listens only perfunctorily to local knowledge and then wonders why the ranch underperforms.

Transparency is vital within a ranching enterprise—information, mistakes, successes, and observations must be acknowledged and shared seamlessly. A working landscape is a complex entity and effective decisions require information to flow up and down the organization. Given the scope of the ranch landscape as well as the dynamic nature of the environment, an organizational culture that values information sharing is necessary for success. Hierarchy, fear of failure, and a lack of accountability undermines ranch performance. Transparency underpins the ability of the orga-nization to identify critical issues and to then allocate resources to their resolution. Furthermore, a culture that values learning and transparency is better suited to recovering from mistakes, develop-ing and implementing solutions, and retaining the advantage of nimbleness. Of course, none of the aforementioned outcomes is possible without a deep sense of trust among all members of the ranch team and the willingness of each to assume a high level of personal accountability.

VALUES

At the very center of ranching is a set of working values that provides direction, guides decision-making, and forms behavior. While each ranch may define their core values differently, two guiding principles are universal for success—husbandry and stewardship. Husbandry is the application of human creativity and labor to the care, cultivation, and propagation of crops and animals. Stewardship is the wise and responsible management of valuable resources that has been entrusted to one's care. It is difficult, if not impossible, to imagine a successful ranch that is devoid of commitment to husbandry and stewardship.

Husbandry values the symbiotic relationship between humans and livestock; the timeless contract that in exchange for the gifts of food, fiber, and by-product obligates the stockman to providing devoted care. Husbandry gets to the core of the work to be done; in the case of a ranch this generally means attention to the care, cultivation and propagation of forage (grass, forbs and browse) and livestock. Good husbandry requires sustained effort in planning and executing strategies and tactics that produce desired levels of performance from rangeland and livestock. William Danforth, the founder of Purina Mills, was driven by the belief that people are judged by the animals they keep. His vision frames the importance of husbandry and recognizes that superior management is required to create excellence in agricultural pursuits. However, husbandry without stewardship may produce desired outcomes in the short term but potentially with consequences that diminish the long-term success of the ranch.

Stewardship is the foundational ethic that drives long-term ranch performance. Stewardship is not a set of activities. Rather it is a philosophic underpinning that provides a spiritual sense of purpose to those who dedicate their lives to ranching. Stewardship is the "why" of ranching. Embracing stewardship calls people to a purpose greater than themselves and lays the foundation for a life that is deeply fulfilling and equally challenging.

Embracing the stewardship ethic explains why men and women choose to immerse themselves in a working landscape and to pursue a life that is so very visceral—where life and death; beauty and hardship; work and love; success and failure are ever present and mysteriously intertwined. The pursuit of unraveling the mystery, learning the pathways that connect the system, and building extraordinary resilience drive the practitioners of extensive ranching.

Stewardship must be applied to multiple resources on a working ranch (Figure 17.3) and while each resources is vital in its' own right; none of them can be considered individually without

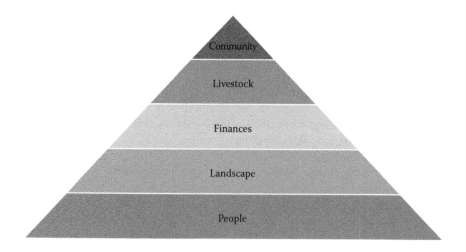

Figure 17.3 Focal points for stewardship.

consideration of the others. However, it is possible to produce a hierarchy that guides the process of building an integrated approach to stewardship.

The process begins with effective leadership committed to the development of people in alignment with the core values of the enterprise. To attain a high level of husbandry and steward-ship on a ranch requires that every individual working in the enterprise is committed to these principles. Without this fundamental alignment, there will always be actions taken by people that are counter to the guiding values and the rewards of stewardship will be very difficult to obtain. Once the team is assembled around the core values, the growth and development of each person must be addressed. Providing and investing in an environment where people can align around a great purpose, develop mastery, and have autonomy is the bedrock upon which the principles of husbandry and stewardship can be successfully applied. No ranch is sustainable without being people centered. Ultimately the care of all other resources is dependent on human attitudes, talent and behavior.

Stewardship of the landscape is the second priority as the natural ecosystem provides the nutri-ents and inputs that nourishes the enterprise over the long term. Healthy soils, watersheds, plant and wildlife communities are necessary for the extensive ranch to thrive. Grazing livestock depend on the well-being of the ecosystem and its ability to provide sustenance. Extensive management systems require that the choice of the type and kind of livestock is carefully matched to the capacity of the ecosystem. Expecting success from forcing the ecosystem to adapt to the needs of the animal is unrealistic.

Financial stewardship is the third tier of the pyramid. Ranching is a business enterprise and the thoughtful management of monetary resources is critical to all other elements of the business. While some might list financial resources as the first order of business, money is not the core reason for the ranch's existence. It is a tool and certainly a critical instrument but the pursuit of annual profitability does not drive the extensive ranch. Wealth is multifaceted on a ranch—certainly profit-ability is important but it must be balanced with other elements—staying in harmony with land-scape conditions and weather's variability, trading short-term gains for long-term wealth creation, and decision-making that is focused on accepting short-term sacrifice in exchange for long-term opportunity.

Applying the lessons of the good shepherd to the production of livestock is the next layer of stewardship. Some might suggest that the needs of the animal comes first but the reality is that quality livestock care is only possible when human, landscape and monetary resources are adequately addressed. The quality of husbandry applied to livestock is directly related to the care of the aforementioned resources and no ranch can thrive without commitment to the well-being of livestock.

The final focal point for the application of stewardship is the community. Ranching does not occur in a vacuum and great ranches have exceptional relationships with neighbors and the larger community. The value of these relationships cannot be overstated as the community and ranch share a common future fueled by shared labor, a willingness to share information about stray animals, predators, downed fences, and a host of other issues related to the ranch, and being vigilant in sup-porting other ranch families in times of joy and duress.

As important is the relationship with the greater community. Most extensively managed ranches interact with citizens who log, mine, recreate and utilize the natural resources of the community in a number of diverse ways. Managing in multiple use environments requires not only the ability to execute ranch strategies and tactics but to also maintain an effective approach to public relations with other users on the landscape.

Finally ranchers who invest stewardship into the community volunteer their time and treasure in youth development, local leadership, service to those in need, and a host of other philanthropic initiatives. The synergy with a community is critical to the well-being of the people working in the ranching sector. Healthy ranches need healthy communities to attract and retain talent.

STRATEGIC MANAGEMENT AND DECISION-MAKING

Decision-making processes in an extensive ranch environment can break down under the chaos of rapid changes such as weather events or market movements so it is important to build a framework that prevents emotional or irrational choices when conditions are less than ideal. Figure 17.4 provides a model where core values are at the center of the process followed by concentric rings representing sound strategic planning, access to appropriate expertise, and effective communication across the organization, and finally correct execution of the plan.

Developing a disciplined approach to decision-making allows the ranch to stay focused on the critical issues, maintain a systems perspective, collect useful metrics, and to react appropriately to changing conditions. In the realm of ranching, on a daily basis there are a multitude of decisions to be made that ultimately impact key goals. But these decisions cannot be made in isolation—they must be made in alignment with core values while recognizing the trade-offs that may be required as well as the unintended consequences on other components of the ranch (Figure 17.5). For example, the emergence of drought conditions may require that stocking rate is reduced through sale of at least some cattle to assure that rangeland health is protected and that the available feed will be sufficient to meet the needs of the appropriate number of animals. The positives of the decision include protecting natural resources and assuring that livestock will have adequate feed. However, there are trade-offs that must be made—reduced breeding herd inventories will negatively affect cash flow in the future and fewer females will be available as potential replacements thus potentially lowering the genetic progress of the cattle herd. Failure to make the trade-off by being immobilized under difficult conditions erodes the well-being of animals, landscapes and people.

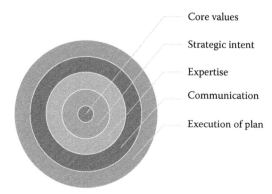

Core values

Strategic intent

Expertise

Communication

Execution of plan

Figure 17.4   Decision process for effective ranch management.

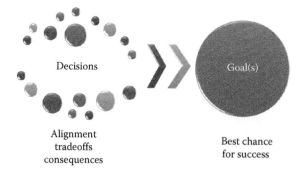

Decisions

Goal(s)

Alignment
tradeoffs
consequences

Best chance
for success

Figure 17.5   The role of trade-offs, consequences, and values alignment in decision-making.

The extensive ranch decision maker serves two primary masters—the natural environment that has a set of dynamic limitations imposed by geographic, seasonal, and weather conditions; plus a marketplace characterized by price volatility, shifting demands by supply chain participants, and diverse consumer wants and demands. To thrive over the long term, a management system must be developed by each individual ranch that fits the local ecosystem while aligning with a set of market targets that create profit potential. To ignore either of these masters—ecosystem or market—is to invite disaster.

STOCKMANSHIP

In as much as people shape a ranch, the ranch shapes the people. Perhaps the most formative interaction is that between livestock and people. The quality and spirit of these interactions shapes character and largely defines the person. The practice of stockmanship demands thoughtful decisions and proactive behavior.

Stockmanship is the specific set of practices designed to provide care to a set of livestock; it is an art form supported by science; and it is a way of thinking and problem-solving. The philosophy and practice of stockmanship is a life's work and is available to all who choose to work with livestock. However, it is an intentional choice made day after day under conditions both favorable and difficult. Like the artisan guilds of old, it is a community of practitioners who aspire to improve their level of mastery, to learn from one another, and to live by its code.

Stockmanship is not a new idea. In his memoir of ranch life in the Big Bend region of Texas, long time cowboy Roland Warnock wrote of his utter disgust at how Hollywood portrayed ranch life with horses and cattle always on the run with clouds of dust and lots of noise—"that's NOT the way we handled cattle, or the way they are handled today on the ranches. They're handled quiet." Just as Warnock was frustrated in his day, the images of ranching and cattle handling used by advertisers and entertainment media today are equally disconnected from reality. Purina Mills was founded in 1894 on a four pronged approach to stock raising—good breeding, good feeding, proper sanitation, and sound management. These four principals represent truth today just as they did at the time of the company's founding.

Stockmanship is an intentional and effective approach to interacting with livestock founded on the concept of continuous improvement. Stockmanship is more than a set of skills and a compilation of best management practices; it is a code for how to approach ranch life. There are numerous benefits that accrue from the practice of stockmanship but the primary is less stressful interactions between humans and livestock. Other desired outcomes are summarized below:

- Better safety for people and livestock by reducing stress
- Enhance on-ranch productivity of livestock
- Improve downstream performance of livestock throughout the supply chain
- Improve the quality of the products delivered to consumers
- Minimize damage to facilities and equipment
- Increase the satisfaction that comes from ranching

Stockmanship provides both qualitative and quantitative benefits that are fundamental to good ranching.

RANCHING WITH INTENT

Ranches that achieve excellence are led by people who are capable of complex problem-solving, strategic planning, and tactical delivery. These leaders also express a high level of emotional intelligence and personal integrity that drive the tripartite practice of stewardship, husbandry, and stockmanship. Specific to livestock management are five spires of decision-making—livestock handling, breeding and genetics, nutrition, animal health, plus education and training that are fundamental to animal husbandry and stockmanship. An integrated management plan that focuses on these five elements increases well-being of the enterprise as well as the humans and animals involved.

Livestock Handling

Attaining excellence in interacting with livestock is dependent on attitude, skill, and good facility design. Effective livestock handling is predicated on the desire of the handler to create a positive outcome by working with the natural behavior of animals to create desired movement of individuals or herds. An attitude of patience and empathy is foundational to good handling. Good handling techniques can be learned but they can only be implemented with good leadership. For example, if employees are unnecessarily stressed by unrealistic expectations and deadlines, poor working conditions, or a lack of resources; poor handling is likely to occur. If the handlers are under undue pressure they will likely transfer that stress to the livestock under their care. Thus, good livestock handling depends on quality care of the people who are expected to do the work.

Development of handling skills and facilities depends on an awareness of specific behavioral tendencies. For example, cattle...

- Don't handle complexity and have poor depth perception
- Prefer to stay and move with other cattle
- Are prey animals so keeping the handler within sight reduces stress
- Tend to move around the handler in a circular fashion
- Respond to pressure

Given the aforementioned behaviors, the great enemies of good handling are excessive stress and speed coupled with impatience. Whether moving a herd of cattle from one pasture to another, sorting cattle in an alleyway, or vaccinating a set of calves; handlers that recognize the value of slowing down, maintaining quiet and calm, and reducing stress are most likely going to achieve their objectives most efficiently.

Good handling technique can overcome facility limitations in many cases but when the infrastructure has been correctly designed then harmonious interactions between humans and livestock can be achieved. The work of experts such as Temple Grandin, Bud Williams, Anne Burkholder, and Curt Pate have demonstrated the power of good handling technique coupled with excellent facility design. While these thought leaders may have differences of opinion, their combined wisdom and example have changed ranching for the better. A ranch environment without a deep understanding of animal behavior manifested in the intentional design of handling systems and techniques to improve the well-being of both livestock and people will not be healthy. Stockmanship practiced correctly creates positive interactions between livestock and humans—cattle and their handlers learn to move in concert at a comfortable pace and without duress. Observing a master stockman at work provides an experience akin to attending a great ballet—movement is graceful, effortless, and intentional and in the moment time slows and a state of harmony is achieved.

But not all production practices are perfect and the decision to retain or discontinue them requires a thoughtful approach to the trade-offs involved. Branding is a practice used to create a permanent identification mark to help deter livestock theft and to facilitate sorting cattle from multiple ranches into correct ownership groups should herds become mixed. Branding is not a perfect practice as it induces stress onto an animal and permanently damages the hide—the most valuable by-product of beef production. Eventually, an innovative solution will provide an alternative but until then most extensively managed ranches will continue the practice.

Breeding and Genetics

Selection of breeding stock is a critical control point relative to several specific traits of importance that impact animal well-being. These include polledness (natural absence of horns), dystocia, traits affecting nutritional requirements, soundness, and disposition. Polled cattle (without horns) are desirable on a ranch for a variety of reasons—elimination of the need for mechanical dehorning, less bruising and injury of cattle during handling and transport, enhanced safety for people, and enhanced ease of handling cattle. The polled allele is dominant and thus horns can be eliminated from a herd by the use of homozygous polled herd sires over several generations. This decision is simple to implement and in addition to improving animal well-being, the market place discriminates against horned cattle as they sell for discounted prices compared to polled cattle. The industry has steadily decreased the number of horned cattle in favor of those naturally polled.

Dystocia or calving difficulty is particularly prevalent in young breeding females calving for the first time. However, through intentional and disciplined selection of sires to be mated to first-calf heifers, the incidence of dystocia can be minimized. While malpresentation of the calf during parturition is random and not under the influence of genetic control, birth weight and other factors contributing to ease of calving can be influenced through genetic selection. Ease of calving is important as it reduces stress on the calving females, her offspring, and caretakers. An additional advantage is that females that calve without undue stress or human intervention are more likely to experience better lifetime reproductive rates.

The nutritional requirements of livestock are influenced by a number of factors. Two of these are traits that are influenced through genetic selection are mature size and milk production. Animals that have high levels of milk production and large mature sizes require more energy to meet metabolic maintenance than their smaller and lower milk-producing contemporaries. Matching the appropriate level of mature size and milk production to the nutritional resources of a specific ranch is important to assure that livestock can have their nutritional needs met within the constraints of the ranch's resource capability. When livestock have metabolic needs that frequently exceed the ability of the grazing resource to meet those needs, supplemental feeds must be purchased to assure their continued health. Under extreme circumstances this situation forces managers into decisions that compromise either the economic sustainability of the ranch or the short-term well-being of their livestock. Appropriate matching of livestock type to the nutritional capacity of the ranch reduces the likelihood of being faced with the aforementioned trade-off.

Physical soundness and the absence of defects can be influenced through breeding stock selection. Livestock with correctly formed hooves, leg and joint structure, udders, and other functional traits are more likely to thrive under extensive ranch management. Animals that are functionally sound require less intervention from their handlers to treat injuries, lameness, and udder problems with the dual benefit of less stress for animals and lowered costs for the enterprise. Selection for livestock that are best suited to a particular ecosystem is a critical decision for managers. For example, cattle raised at high elevations (above 5,000 feet) may experience brisket disease resulting from excessive pulmonary arterial pressure. Fortunately, there is a protocol to measure differences between animals so that sire selection can be focused to choose those animals least likely to experience distress under high elevation conditions. Similarly, livestock production in those regions closer to the tropics requires selection for animals that are well suited to conditions where higher temperatures, humidity, and parasite loads are prevalent. In the cattle industry, the use of Brahman-influenced genetics has been an effective solution.

Disposition or level of docility is also under genetic influence. The practice of effective handling is always a requirement as even docile animals will become combative under conditions of undue stress caused by poor handling and management techniques. However, by selecting for more docile animals, the quality of interactions between livestock and humans can be enhanced. Effective stock handling techniques are heightened by the opportunity to engage with docile animals as opposed to those more likely to display aggressive or flighty, nervous behavior.

NUTRITION

Meeting nutritional requirements is a primary pivot point for a ranch manager. When feedstuffs are sufficient to meet the needs of livestock then a number of desirable outcomes are more likely—high reproductive rates, improved immune response, and better overall herd and individual animal performance. Thus nutrition is a central pillar upon which animal well-being is constructed.

A survey of 100 successful cow–calf managers and nearly 100 consultants, veterinarians, and other advisors working with profitable cow–calf enterprises found that the highest two management priorities for these professionals were herd nutrition and pasture/range management (Field, 2006). Strategies incorporated by ranch managers to enhance nutritional status of livestock include matching calving season with the seasonality of green forage production, extending the grazing season and enhancing the nutritional value of standing forage by incorporating rotational grazing systems, stockpiling forage for late fall, winter and early spring grazing.

Because weather conditions can be variable, effective nutritional management includes the ability to make appropriate adjustments based on precipitation, effective temperature, wind and other variables. Stockmanship is founded not only on superior planning and matching the production system to ranch conditions but also it is the practice of deep observation and intuition that allows managers the ability to change course and to take corrective action.

Animal Health

The goal of intentional ranching is to use preventative approaches to avoid disease while retaining the ability and capacity to deal with a disease event should it occur. Healthy livestock result from good nutrition, a strong immune system, and mitigation of stress. These factors can all be influenced by the application of sound management including appropriate mineral supplementation, a correctly designed vaccination program, and the application of low-stress livestock handling techniques.

Essential to providing high-quality care to livestock is establishing and maintaining a strong relationship with a veterinarian(s). The veterinary–client–patient relationship is the basis for nearly all of the key decisions specifically related to the prevention and treatment of disease. However, it is critical that the selection of a herd veterinarian includes criteria related to the themes of stewardship, husbandry, and stockmanship. Without alignment on the values embraced by the ranch, the relationship with the veterinarian will not yield the highest value.

Stress mitigation is one of the most important approaches available to enhance animal health and well-being. The elimination of stress from an extensive management system is unrealistic and impossible. Natural systems are filled with stress points—predation, seasonal changes in temperature, weather events, and pathogens. Stockmanship and intentional management can be incorporated to reduce stress in the system as well as to avoid situations where stress is compounded. For example, a vaccine is likely to stress the animal by introducing a substance designed to elicit an immune response. However, if stress from poor handling technique and bad facility design are added to the mix, then the accumulation may actually render the vaccine ineffective.

Identifying stress points in the enterprise and development of strategies to mitigate stress is an ongoing process where intentional management is in place. Implementing management protocols based on beef quality assurance principles developed by teams of ranchers, veterinarians, and subject matter experts is another effective strategy to reduce stress in the ranch system.

Education and Training

Mastery of stockmanship is a continuous process dependent on active learning. Habits and behaviors as well as ranch management protocols and tactics must be examined on an ongoing basis to assure that opportunities for improvement aren't missed. Providing ongoing educational opportunities and learning experiences are mission-critical for both ranch managers and their employees. Additionally, ranch leaders must foster an environment where there is organizational wide focus on seeking solutions, developing more effective processes, and growing the capabilities of people.

Ranches driven to help their people grow both professionally and personally are more likely to find innovative solutions, to more thoughtfully practice stewardship, husbandry, and stockmanship; and to let go of old practices and habits in favor of better approaches.

FINAL THOUGHTS

Healthy ranches do not happen by accident; rather they are the result of a commitment to nurturing landscapes, livestock, and people. People who choose to live out their talents and values in the realm of extensive ranching are motivated by a variety of factors but those who have found the most joy and fulfillment have drawn on all dimensions of their humanity—creative, intellectual, physical, and spiritual.

Extensive ranching provides a studio and laboratory for the practice of the triad of stewardship, husbandry, and stockmanship. In this environment, theory has little value unless it manifests itself in some useful manner; perfection is nearly impossible to attain given the chaotic and random nature of working with the vagaries of nature but its pursuit is worth a lifetime of effort; there is nowhere to hide from accountability, the visceral experience is valued far more than the virtual; and the possibility of failure and of wonder is always present.

The extensive ranch, regardless of geography, is not a genteel place but it is always a combination of raw beauty and challenge. To be successful in such an environment means letting go of convenience, instant gratification, and certainty. The ranch takes on a life of its own and usually becomes the sun around which everything else orbits—birthdays, holidays, homework, recreation, and family gatherings are organized to fit into the ranch schedule that has no respect for the clock. If newborn calves in the utility room, frozen bags of colostrum in the freezer next to the ice cream, and days that begin before the sun rises and conclude in the dark aren't welcome, then ranching should be avoided.

There are no perfect ranches nor perfect ranchers. For those who engage in the work of caring for livestock and landscapes, not all days have happy endings. Even the most experienced and skilled practitioners experience setbacks and make mistakes. Lightning strikes kill livestock and people; range fires, blizzards, and drought wreak havoc on the best formulated plans of well- managed ranches, and best practices don't always work. Nonetheless, there are few professions where people can immerse themselves in nature, take on the responsibility for the stewardship of landscape, livestock, and people; and enter into the ancient contract of the good shepherd—it is in this realm that a life worth living can be forged.

REFERENCES

Field, T. G. 2006. Priorities First. American Angus Association.

Lee, H. L. 2004. The Triple-A Supply Chain. *Harvard Business Review*. October, 102–112.

Lencioni, P. 2012. *The Advantage*. Josey-Bass.

Pink, D. H. 2011. *Drive: The Surprising Truth About What Motivates Us*. Riverhead Books.

Warnock, K. F. 1992. *Texas Cowboy: The Oral Memoirs of Roland A Warnock and His Life on the Texas Frontier*. Trans Pecos Productions.

Wickman, G. 2012. *Traction: Get a Grip On Your Business*. BenBella Books.

CHAPTER 18

Range Beef Cow and Calf Health and Welfare

Franklyn Garry
Colorado State University

CONTENTS

OVERVIEW OF BEEF CALF PRODUCTION

In order to understand the health and welfare challenges beef cattle face in the US beef cattle industry, it is important to first understand how beef production systems operate. The primary producers of commercial beef animals are called cow/calf producers. These are the foundation of the beef industry because they own and manage beef cows with the purpose of breeding them and raising their calves. Most commonly cow/calf producers keep the calves with their dams until weaning at approximately 7 months of age, and then sell these weaned calves to someone else who raises them to larger size.

Once calves are sold from the home ranch they will commonly be grown further by grazing grass or fields from which crops have been harvested. Managing calves during this period of growth is called "backgrounding" or "stocker feeding." From such an operation the calves will be sold to a feedlot operation for further growth, typically being fed increasing amounts of cereal grains or the byproducts from ethanol or human food production. Alternatively the calves may be sold directly from the home ranch to a feedlot. After the animal has achieved slaughter weight, either from grass feeding or grain feeding, it will go to slaughter. A small number of cow/calf producers maintain the calves throughout the entire production cycle, eventually taking them to slaughter at about 18 months of age and selling the beef products to consumers. More commonly the calves are sold at each step of this beef supply chain. Therefore, the beef industry is "segmented," meaning that each of these growing steps is done under separate ownership.

The cattle health challenges that producers deal with are very different in these segments of the industry. Stocker and feedlot cattle are very uniform in age, and they are young, growing animals. On a cow/calf operation, the beef cattle range in age from newborn to very old mature cows. They undergo all of the life changes of a reproducing population, including growing, breeding, gestating, birthing, and lactating. Therefore, the health challenges they face are very diverse.

Beef cow/calf production occurs throughout the continental US. East of the 100th Meridian there is sufficient rainfall to have relatively lush pastures and production of beef cattle may require as little as an acre per cow/calf pair. In the Western US, much of the landscape is more arid and larger land base is required to grow sufficient forage for a cow/calf pair, sometimes equaling 20–80 acres per pair. Therefore, there can be large differences in some aspects of production between the East and West. Although much of the information in this chapter is applicable to any cow/calf operation, the focus of the chapter is on beef production in the western range.

COW/CALF MANAGEMENT

An important aspect of cow/calf production is that the final product a producer markets is the calf. In other words, a commodity rather than a finished and differentiated product bought directly by a consumer. Calves are priced by their weight, with little difference in the value of each calf beyond its size. This is different than selling a finished product like a choice steak, or a marinated roast, where the seller can set different prices for their products based upon what it cost to produce them. The price a cow/calf producer receives for their calves is mainly determined by the commodity markets, and that price is beyond the producer's control.

This means that a producer may have invested a considerable amount of money in producing and raising the calf to weaning weight and then find that the money received for the calf is less than the cost to produce it. Many of the costs are substantial, such as owning land, buying feed, and maintaining buildings and equipment. Most producers strive to produce their animals for as little cost as possible, in order to optimize their chances of making a living.

Most cow/calf producers manage their cattle "extensively," rather than "intensively." Intensive management of livestock is typical of poultry, pork, dairy, and feedlot production. The term means that the feeding and housing of animals is very intensively overseen. In such systems, all feed elements are gathered and processed and fed to maximize animal production. Housing is designed to control the environment. This increases the cost of production, but the production gains that are achieved outweigh these costs and are more profitable.

By contrast most beef cows and calves are managed with limited control of what they eat and how they are shielded from the elements. Cows are left outside to face the elements, graze the forage that is available and find shelter in a more naturalistic way. This keeps costs lower, and relies on having animals that can fend for themselves. While this somewhat feral and naturalistic process can be appealing to some members of our society, it also carries many risks, as would be seen in a population of wild animals. Some of the health and welfare challenges that face beef cattle are direct results of this style of production. For example, it is highly unusual for intensively managed poultry, pigs, or dairy animals to experience blizzards or predation, but these are relatively common challenges for beef cattle.

The significance of these aspects of cattle management relative to cattle welfare is profound. This is not a highly profitable industry. Producers are involved in the industry for numerous reasons, including love of the land, enjoyment and great pride in their cattle, enjoyment of working with the cattle, and appreciation of this style of living. For all of these reasons producers typically work very hard to ensure the well-being of their animals. It is important to make money to remain in business, even though this is not a business where people can reasonably expect to get rich. The investment of time, energy, and money in the business is very large. In an industry where monetary return on investment is typically in the low single digits, and where revenue comes from selling animals, producers have additional compelling reasons beyond emotions and moral commitment to assure that every animal is as healthy and robust as possible.

BEEF COW PRODUCTION CYCLE

Cattle have a long generation time compared with poultry or pigs. The gestational period of a beef cow is approximately 280–285 days. After a cow conceives, she will carry the calf for about 9 and a half months. After delivering the calf she needs approximately 60 days until her reproductive tract has recovered from the pregnancy and she returns to fertile reproductive status. Therefore, a full cycle of reproduction is approximately a year.

Cows are capable of becoming pregnant at any time of year—they are not seasonal breeders like sheep and goats. If bulls are left commingled with the breeding herd throughout the year, then heifers can conceive as they reach puberty and cows will be bred whenever they come into heat (ovulate), and calves can be born at any time of year. There are some producers who allow such year-round access of cows to bulls, but the overwhelming majority of cow/calf producers in the Western US focus on managing a breeding season of their choice. Managing a breeding season establishes a "production cycle" that can improve herd productivity.

Having a breeding season allows a producer to control the birth time of calves, which in turn helps focus labor efforts and forage availability, and develops a calf crop of relatively similar age and size for marketing. Smaller scale producers and producers in the Southeastern US are more prone (approximately 50% of producers) to not have a defined breeding season, probably because of more year-round forage availability, or the desire to not manage the herd very much. Larger scale producers and especially producers in the Western states mostly manage for a single calving season (approximately 80% of producers), and most of these calve in the Spring, between February and May (approximately 65%–70% of producers).

A spring calving season provides cows green grass with more nutrient density during lactation, the time of most intense nutrient demand. Calving at other times of year means that cows need more supplemental feed, which is very costly, if they are to maintain good milk production for their calves, and still remain in good body condition so they can breed back again.

Having a successful defined calving season requires the producer to manage bulls and cows fairly closely. But this effort can be rewarded by having most of the calves born in a short period of time. This means that problems during the calving season can be more closely monitored, and it produces a good crop of calves with very similar ages and sizes, which usually leads to better prices for the calves when they are sold.

ANIMALS ON A COW/CALF RANCH

The cattle on a cow/calf operation represent both genders and a wide variety of ages and life circumstances. The cows range in age from 2-year-old primiparous heifers (first calf heifers) to very old dams. Typically around 15% of cows will be sold each year. Some may be sold to another ranch as breeding animals, especially in circumstances where a ranch has too little feed to maintain current herd size (e.g., during drought). More commonly cows are sold as slaughter animals (culled from the herd) because they are no longer productive.

The two most common reasons for culling cows are that they do not produce calves (fail to conceive, abort, or produce an unthrifty calf), or that they wear their teeth down as they get old and cannot chew food well and maintain good body condition. This says a lot about cow health in the cow/calf production system. The majority of cows are healthy and productive throughout their lives and only leave the herd when they become unproductive or old. It is common for many cows in a herd to be 10–15 years old before they are culled, which is very old for a cow.

To produce calves, bulls are also included in the herd. Most commonly a producer keeps about one mature bull for every 25 cows. When young bulls are introduced to the herd, they are expected to mate with about 17–18 cows. Most producers buy young bulls from a seedstock, or purebred breeder to maintain genetic diversity in the herd. Bulls are selected for certain traits desired by the cow/calf producer, such as growth efficiency, frame size, ability to perform in harsh conditions, foot and leg conformation, or ability to produce calves that birth easily. Seedstock production is similar to commercial cow/calf production except that the breeder maintains genetically pure stock that are carefully managed and monitored to demonstrate the special traits desired by the buyer. This chapter is focused on commercial cow/calf producers who commonly crossbreed to get hybrid vigor and sell animals into meat production channels.

The third animal population on a cow/calf ranch are the growing youngstock. At calving season, these will be newborn calves and will face the health challenges common to neonates of all species. These calves will nurse the dams for the majority of their early nutrition and graze on forage as they grow. They are most commonly weaned at about 7 months of age. Owners of Spring calving herds strategize to breed so that calves are born before new grass is growing and then grass is available for cows and calves during the nursing period. As the grass dies in the Fall, calves are weaned.

Some heifer calves are kept on the ranch to grow as "replacement heifers," selected as the most likely best cows of the future, to replace the mature cows that are culled. These heifers are grown and bred the next breeding season so that they calve during the regular calving season the year they turn 2 years old. The remainder of the heifer calves not kept as breeding animals are usually sold in the fall after weaning to enter the beef production system as stockers or feedlot animals.

The majority of all bull calves are castrated at an early age and marketed after weaning as animals to be grown and harvested for beef. Some select bull calves may be kept for breeding if the producer can manage them in a way that avoids breeding with closely related dams. Bull calves going into meat production systems are routinely castrated because steers are easier to manage as they do not go through puberty and develop adult male behaviors. This is desirable for the stocker and feedlot industries as well, avoiding fighting, injury, and metabolic and behavioral changes that adversely affect meat quality at slaughter.

OVERVIEW OF HEALTH PROBLEMS

Beef cattle on a cow/calf operation are exposed to a wide variety of health risks. Because of the extensive nature of management on a rangeland operation, the problems most commonly seen are different from populations managed intensively, such as feedlot cattle or dairy cattle. In an extensively managed system, the cattle are exposed to more environmental risks such as extreme weather.

In an intensively managed system, animal nutrition is very well controlled and diets can be adjusted to meet animal needs at an exquisite level. Range cattle, however, may have sufficient nutrients when forage is lush and highly available, but weather and soil conditions may predispose to imbalanced diets, plant toxicities and nutrient deficiencies. During the 6 months or so when range plants are not growing, or are unavailable, cattle must be supplemented with additional feed, but this can still result in dietary insufficiencies.

Cattle in intensive production are usually seen individually at least once a day, and sometimes more frequently. Appetite, attitude, general well-being and musculoskeletal soundness are continually monitored. An animal with a health problem is usually promptly identified, treated and monitored for recovery. In a range setting, cattle are seen from a distance and some disease conditions may be overlooked until they are well advanced problems. Cattle may only be individually scrutinized occasionally. Some sick animals will hide from sight, a normal response of sick wild animals trying to avoid being seen by predators. In such cases, disease may be severe and chronic before it is identified, and it is common for ranchers to discover that a calf has died rather than having the opportunity to treat it.

One of the reasons horses are ridden when managing a herd of beef cattle is that the cattle do not move away from another four-legged herbivore. Cattle and wild herbivores see and respond to people on foot and people on horseback differently. Cowboys not only use horses for convenience, but also because they can manage cattle more easily and in closer proximity on horseback, more easily identifying sick animals. Cattle can be "gentled" to accept humans in closer proximity, and some producers make efforts to do this in order to make handling during cattle processing easier, but in general beef cattle on range are not managed very often by people on foot.

All disease problems represent significant animal production issues, and some present significant welfare problems. Undeniably, any animal with an illness or injury is compromised to some degree. On the other hand, animals with different diseases do not suffer equally. There is a big difference between diseases that damage sensitive tissues and cause pain, compared with diseases where tissues may not function well, but where the problem is relatively imperceptible to the animal. For example, diseases that damage the musculoskeletal system not only impair the animal's ability to move and get feed and water but also cause overt animal suffering. By contrast, many diseases of the reproductive system affect sperm or ova production or the rate of fertilization, but do not affect the animal's overall well-being. Such problems limit the animal's productive capability, and typically lead to culling, but affected animals can be otherwise physically fit and overtly healthy. As another example, many infectious diseases are subclinical in affected animals, meaning that the animal carries the disease agent and can potentially infect other animals, but otherwise show no signs of disease or physical impairment.

The challenge that all cow/calf producers accept is to implement preventive and monitoring practices that strive to assure good animal health, welfare, and productivity. For subclinical diseases that affect animal performance, testing programs are instituted to identify the problem and control or eliminate it. When overt clinical disease occurs, producers need to identify and treat it, or, if untreatable, to euthanize the animal.

HEALTH PROBLEMS OF COWS AND BULLS

As is true for other populations, including humans, mature cattle are larger, more robust in the face of environmental challenges, and more resistant to disease problems than babies or young growing adolescents. On the other hand, they have some other liabilities that come with aging.

Relatively few beef cows experience problems that cause extreme suffering or risk of death. The overall death rate of cows in beef herds is commonly about 1%, quite close to the death rate in human populations. Extreme weather, such as blizzards or extreme cold can injure or kill adult cattle, but this level of weather challenge is unusual, and it is typically the younger animals that experience frostbite, death by hypothermia, or respiratory disease associated with extreme dust and wind. Nevertheless, of the 1% of cows that die on ranches, approximately 16% of these deaths are attributed to weather-related causes such as lightning, chilling and drowning.

As noted above, the main reasons for culling beef cows are reproductive failure or old age. The causes of reproductive failure include disease, but while reproductive diseases produce infertility, it is unusual for reproductive disease to cause animal pain or suffering. If worn teeth are seen as a disease, then it is the most common disease of old beef cows, leading to loss of body weight. Cow body condition, which equates to amount of muscle and body fat, is visually appraised when cows are checked for pregnancy at the end of the breeding season. Such thin cows are unproductive and are culled before the condition leads to suffering or risk of death.

The third leading cause of culling is economic constraints, such as poor market conditions or drought that limits profitability and obliges a producer to sell cows to reduce herd size. Another common reason to cull cows also has nothing to do with health and performance. Some cows are culled due to poor temperament—they may be too wild or break through fences, or are dangerous to work with.

Beyond these top four reasons, the other health reasons that cows are culled, euthanized, or die on the ranch occur at very low rates. One notable health challenge for both cows and bulls is lameness or injury. These problems can include damage to feet and legs in a rugged range environment, injuries to joints that can occur during mating, or injuries from fighting. An infectious foot disease known as footrot occurs when certain bacteria invade into the skin between the claws, causing severe pain, and swelling of the foot. This is an easily treated condition if it is caught early in the disease process, but as noted above, some cases escape detection until they are advanced and the bones and joints of the toes are also infected. Such untreatable conditions usually result in culling or euthanasia. Footrot is far more common in intensively managed cattle where feet are exposed to mud and water. It is relatively uncommon on beef operations. Only 3%–5% of culled cows leave the herd because of physical unsoundness.

Cows exposed to lush legume pasture can develop life-threatening bloat of the rumen. This is an emergency because bloat limits respiratory efforts and can kill cows by suffocation. Cows can also eat solid chunks of feed, such as turnips or potatoes and if not chewed properly they can obstruct the esophagus, preventing eructation of gas, producing ruminal bloat, and similarly presenting a life-threatening situation. These types of conditions are certainly problematic when they occur, but they are infrequent, and most ranch managers are familiar with how they occur and can prevent them. Of the 1% of cows that die on a ranch about 5% of these deaths are attributable to bloat or other digestive tract emergencies.

Problems such as infectious mastitis, infectious respiratory disease, and gastrointestinal infections do occur in beef cows and bulls, but again they are infrequent compared with cattle raised in intensive operations. About 3%–4% of all cattle deaths are attributable to these causes. These types of diseases are more common when animals are housed or gathered in close proximity. Such disease problems are associated with hygiene and ventilation issues, and these circumstances are unusual in range conditions. There are several infectious diseases that can be insidious in a herd because some animals carry the infectious agent subclinically, but then serve as reservoirs for infection of other animals. Such problems represent a significant threat to herd production, but it is uncommon for animals to suffer from the condition. Examples include Bovine Diarrhea Virus, Johne's disease, and Bovine Trichomoniasis. These problems can have severe adverse effects on reproduction or weight gain, or predispose to other disease conditions and affect the herd very negatively, but few animals suffer overt disease and suffering.

Cattle on a ranch are almost always exposed to the outdoor environment. This means that compared with more intensively raised animals they are also exposed to flying and biting insects. These are very difficult to control in an outdoor setting, although producers use a variety of strategies to minimize their impact. Much of that impact is the nuisance and discomfort of dealing with the insects. But some insect pests also transfer disease agents and can cause disease. Examples include the transfer of blood parasites such as Bovine Anaplasmosis, or other disease agents like Vesicular Stomatitis Virus. As discussed above, such infectious diseases are problematic, but only occasional animals are affected with problems that produce overt suffering.

In contrast, there are a couple notable ocular diseases that directly relate to the environment. One is Pinkeye (Infectious Bovine Keratoconjunctivitis). This is a severe problem caused by bacteria carried in dust or by flies that damages the cornea of the eye, causing extreme pain, and in severe cases resulting in blindness. This can be a major problem for the affected cattle and the herd. Fortunately, the disease can be treated effectively, but its highly contagious nature means it can cause major problems and it is difficult to control. Another ocular disease is Cancer Eye (Bovine

Ocular Squamous Cell Carcinoma). This occurs mostly in cattle with white skin around the eyes, and therefore tends to occur in some breeds more than others, but it is also promoted by extreme sun exposure. Caught early, the cancer can be treated and eliminated. In prolonged or severe cases, it can spread to other body regions and may warrant euthanasia. Eye problems account for about 1%–2% of cows that are culled.

The extensive management and feeding of beef cattle presents other challenges. Although these are not common, they can affect large numbers of animals in a herd, or even the entire herd. Such problems include plant poisonings and nutrient deficiencies. There are a wide variety of toxic plants. Some affect the cardiorespiratory system, such as larkspur poisoning. Others affect kidneys (oxalates), neurologic system (locoweed), liver (senecio species, kochia weed), or skin (kochia weed, buckwheat). Some plants that are commonly eaten and nutritious can become poisonous in certain environmental conditions such as drought or too much moisture, including nitrate poisoning, cyanide poisoning, bluegreen algae poisoning, and fungal endophytes on some grass species (ryegrass poisoning, fescue grass). These problems may lead to severe animal health issues and even kill some animals. Surveys estimate that about 1%–2% of all cattle deaths are attributable to poisoning.

Nutrient imbalances also can affect large numbers of animals. Common hazards for range cattle include copper deficiency, hypomagnesemic tetany, sulfate toxicity, selenium poisoning, selenium deficiency, and others. These are usually specific to particular types of soils or particular plants that grow on the soil. Cow/calf producers and their veterinarians or nutrition advisors try to prevent these problems by sampling the feed or sampling the animals and then feeding supplements to try and balance the ration. But it remains a liability of extensive animal management that the entire diet is difficult to control. Cows on range are obliged to eat the forage that is available. Sometimes deficiencies and excesses can be identified and avoided, but commonly the problem is already occurring before it can be identified. Of the 1% of cows that die on the ranch, about 2% of these deaths are attributed to such mineral/metabolic diseases.

Another hazard for beef cattle is the grazing of plants that have spines, seeds, or awns that injure the lining of the mouth. The initial injury is usually minor, but commonly the injury is then invaded by bacteria that cause inflammation of the oral cavity, or that spread through adjacent tissues and the lymphatic system to cause severe cellulitis or abscessation. Since the problem occurs around the mouth and jaws, it is commonly referred to as lumpy jaw. In some cases, the problem can be treated and resolves, but in severe cases the animal may be culled or die.

Perhaps the most important occupational hazard of beef cows, and particularly first-calving beef heifers is difficult calving, called dystocia. If the calf is too large for the size of the maternal pelvis, or if the calf is presented for birth in an abnormal position, or in the case of twin birth, the dam may have trouble delivering, or the delivery may be impossible. This presents a life-threatening problem for the calf, and will be discussed below, but also a major problem for the dam. One of the reasons that having a defined short calving season is preferred by most ranchers is that it facilitates gathering the dams for a discreet number of weeks to monitor delivery and assist when problems occur. The majority of difficult births can be successfully assisted if the difficulty is identified, and alternatively a caesarean section surgery can be performed. This requires more-or-less round the clock monitoring and calving season is the busiest time of year for ranch managers.

Dystocia is more common in 2-year-old dams, because they have never delivered before and are not fully grown to mature cow size. This problem can be a major source of animal injury including torn tissues and crippling neurologic or musculoskeletal injuries. Bulls can be selected to produce smaller calves at birth, called "calving ease sires," and many ranchers focus on this feature when they select breeding bulls for their heifer breeding. Dystocia is a fact of life in any population and cannot be completely eliminated. But over the last several decades, with better bull selection procedures, increased focus on heifer selection and growth, and better education about both prevention and resolution of the problem the problem now occurs in only 2%–3% of mature cows and approximately 10%–12% of 2-year-old heifers. Still, injuries occurring during calving account for 15%–20% of the 1% of cattle that die on the ranch.

HEALTH PROBLEMS OF CALVES AND GROWING YOUNGSTOCK

Babies and young growing animals of any species are more at risk of injury or disease than older animals, and that is true of beef animals as well. The highest risks of injury, problems with environmental exposure, and infectious disease episodes occur at and shortly after birth, and then decline with age. Baby calves are smaller and more fragile, and they have less immune protection than older counterparts.

In addition to the availability of growing grass, it is this susceptibility to disease of young calves that makes seasonal management of breeding and calving a more productive way to manage beef herds. By focusing on breeding for a limited and intense period of time, a producer can synchronize the timing of all of the management practices of the ranch. An intense breeding season means that calvings will occur in a similarly compressed time period. Since all of the cows are synchronized in time, the producer can focus on assuring good calf delivery and attention to newborn calf health during a short period of a couple months. This also makes it easier to feed the cow herd to meet their nutritional needs as these needs change with pregnancy, delivery and lactation status.

The term "extensive management" used frequently in the preceding text refers to extensive management of animal housing and environment, and to extensive management of cow forage supply. But it does not mean that management is random and helter-skelter. To the contrary, most western range cow/calf producers are very focused managers of breeding and reproduction, of calving time and newborn calf management, of weaning time and weaned calf management, of breeding bull selection and management. Having a defined calving season facilitates attention to these herd details because most animals are in a similar stage of production and life.

Many managers of very small cow/calf herds do not have a defined calving season and calve year-round. Most likely this is because the herd and its management are hobbies rather than the main occupation of the owner. In these situations, it is common that calf disease and death losses are higher than on larger, well-managed ranches. Similarly, many cow/calf operations in the Southeastern US do not have a defined calving season, probably because of the availability of grass throughout much of the year. Similarly, such management is associated with higher rates of calf disease and death loss.

Birth is a very traumatic and risky experience. During delivery all calves experience a period of limited oxygen supply, and then need to start breathing and performing the physiologic functions of any animal outside the protected environment of the uterus. These functions include breathing, delivering oxygen to all tissues, generating body heat, mobilizing energy reserves, standing, finding a teat to get milk, and nursing. This process does not always go smoothly. In the event of difficult or prolonged birth, the challenge of adapting to life outside the uterus can be even more difficult. During dystocia delivery calves may be injured, and following such difficult delivery they will have a higher oxygen deficit and greater trouble making these physiologic adaptations.

Mature cows have tremendous influence in assuring calf survival because they dry the calf off and help stimulate it to make it get up and find the udder. First-time mothers often have a more difficult time delivering the calf, they may be injured themselves during the process, and they lack the experience to know how best to help the newborn calf. For all of these reasons, having a capable human available to assist the delivery and then to help the dam take care of the calf can mean the difference between life and death for the newborn.

The highest risk of calf death is during and immediately after calving. On cow/calf operations of 200 cows or more, approximately 96% of calves born to heifer dams are born alive, while 4% die during delivery. Smaller operations, with 50 cows or fewer, have double this amount of calf loss at birth. For mature cows on larger ranches, the percentage of calves lost during delivery is less than 2%, and again, on smaller ranches more calves die, at approximately 3%. Most likely, the difference in success between such ranches is the amount of observation and assistance provided by ranch personnel.

Some calves suffering from difficult delivery will survive for a short while but are still suffering the consequences of the birthing trauma and may die within the first 24–48 hours. Again, human intervention can substantially impact survival rates. Ranch personnel can provide the calf with supplemental oxygen, warmth, protection from the elements, and assure that the calf is obtaining colostrum and then milk from the dam. Although healthy newborn calves are typically very robust and can survive in pretty harsh weather conditions, there is a limit to their survivability. Calves suffering from dystocia are far more susceptible to the adverse effects of bad weather. Of all calves that die within the first 3 weeks after birth, approximately 50% of deaths are attributable to birth-related problems or weather-related causes.

Frostbite can occur in newborn calves less than 3 weeks old during bouts of extreme cold. Frostbite will most commonly affect ears and tail because these extremities have very low blood flow when calves are chilled. If only the ears and tail are affected, calves will lose the tips of these body parts, but typically survive without further liability. Sometimes feet are also affected, and these calves have to be euthanized. In more extreme cold, some calves will die, and calves affected by dystocia are particularly at risk, as described above.

There are numerous management tactics used to decrease the likelihood and the impact of dystocia. Replacement heifers are managed and fed for optimal growth to assure they are appropriately sized and prepared for delivery of calves. Heifers are bred to calving ease bulls to minimize the likelihood of excessively sized calves at birth. Heifers near the time of delivery are monitored for impending delivery and assisted when needed. Newborn calves are monitored for vitality and managed more intensively if they appear compromised by the delivery. Most ranches have a calving facility where a limited number of compromised dams and calves can be temporarily protected from extreme weather. Breeding management that produces a discreet calving season makes all of these objectives easier to accomplish because all the cattle are in similar stages of gestation and calf delivery at approximately the same time so that personnel effort can be focused.

The first 3 weeks after birth are referred to as the calf neonatal period. The primary challenges after the first 48 hours are infectious conditions, and the most common are gastrointestinal infections with viruses or protozoal agents. These cause diarrhea and loss of body fluid and electrolytes, which in turn make the calf very sick and these problems can potentially be lethal. If they are identified early, which requires careful observation and oversight by ranch personnel, calves can be successfully treated. Bacterial infections of the gastrointestinal tract also occur and these can be more life-threatening because these enteric bacterial pathogens such as *Escherichia coli*, *Salmonella*, or *Clostridia* produce toxins and are capable of invading and damaging other organ systems. On most ranches, these infections occur at a lower frequency than in other cattle management systems. Ranch management can limit the occurrence and spread of enteric disease agents in calves by changing the location of calving dams or of cow/calf pairs during the calving season to limit the amount of contamination of the environment, and thus the exposure of calves to the infectious agents.

After the first 3 weeks, calves are still susceptible to a variety of infectious challenges, but during this period the most likely infections affect the respiratory tract. Susceptibility to respiratory infections continues to be the primary disease challenge through weaning and beyond, including growing stocker and feedlot calves.

Looking at the entire time period from birth through weaning, disease and death loss occur with decreasing frequency over time. On western ranches, 2%–3% of all calves will die during or at delivery. Of the calves born alive, approximately 4% of calves die between birth and weaning. A third of these losses of live-born calves occur during the first 24 hours, another third of the losses occur over the 3 weeks of the neonatal period, and the remaining third die between 3 weeks and weaning at 7 months. Therefore, the highest risk period for baby calves is during and shortly after birth, with decreasing challenges over the next 3 weeks. Maintaining a fairly short calving season with seasonal breeding helps the producer focus attention on the needs of the dam and baby calf during the high-risk period.

Other health challenges will occur during the neonatal period and prior to weaning, but they are far less common. Predation is a challenge on some operations, accounting for about 5% of all calf deaths, and lameness or injuries account for about 4% of deaths. All other causes represent very small percentages of death loss on most operations.

After the high-risk birthing and neonatal periods, cow/calf pairs are usually turned out onto the range. It is worth noting, as mentioned earlier, that when calves are out on range, and less frequently observed, some deaths occur that are unseen and undiagnosed. Such deaths are often discovered not by seeing a sick or dead calf, but by seeing a dam without a calf.

SUMMARY

The cow/calf industry in the US maintains mature cows and bulls who produce calves for sale to other segments of the beef industry. These operations have a wide variety of cattle in all different stages of life. The health and welfare challenges of these animals are diverse. The cattle on these ranch operations are extensively managed relative to housing and nutrition and are subject to environmental challenges and nutritional challenges, but overall they tend to be robust and healthy with low rates of disease and death loss. Most larger size commercial cow/calf operations intensively manage the reproductive program to establish a calving season, in order to more optimally maintain cow and calf health and productivity. This allows the producer to focus on specific disease challenges that occur at different stages of the animals' lives.

ADDITIONAL SOURCES

Specific numbers and percentages of animals affected and regional differences in cattle management are derived from the national surveys conducted by the USDA: Veterinary Services, National Animal Health Monitoring System, and the author believes they are the most representative numbers available about the management and health of the national beef cattle herd. The most recent survey was completed in 2008, and a new survey is being conducted in 2017–2018, but results are not yet available.

Beef 2007–08 Part I—Reference of Beef Cow-calf Management Practices in the United States, 2007–08. USDA–APHIS–VS, CEAH. Fort Collins, CO #N512–1008.

Beef 2007–08 Part II—Reference of Beef Cow-calf Management Practices in the United States, 2007–08. USDA–APHIS–VS, CEAH. Fort Collins, CO #N512.0209.

Beef 2007–08 Part III—Changes in the U.S. Beef Cow-calf Industry, 1993–2008. USDA–APHIS–VS, CEAH. Fort Collins, CO #518.0509.

Beef 2007–08 Part IV—Reference of Beef Cow-calf Management Practices in the United States, 2007–08 USDA–APHIS–VS, CEAH. Fort Collins, CO #523.0210.

Beef 2007–08 Part V—Reference of Beef Cow-calf Management Practices in the United States, 2007–08 USDA–APHIS–VS, CEAH. Fort Collins, CO #532.0410.

CHAPTER 19

Animal Care Issues in Beef Cattle Feedlots

John J. Wagner
Colorado State University

CONTENTS

INTRODUCTION

Cattle entered what is now the United States through four distinct paths starting in the early sixteenth century (Bowling, 1942). The Spanish brought cattle from the West Indies to the Atlantic or Gulf of Mexico coast of Florida. Spanish cattle also entered the Southwest, including present-day New Mexico and Texas, from Mexico. The third incursion of cattle came from French settlements in present-day Canada into states near the St. Lawrence River and Great Lakes region. The fourth entry was associated with the settlement of various colonies along the Atlantic coast by the Dutch, English, and Swedes. Interestingly, many of the cattle imported by the English were of Spanish origin and were purchased in the West Indies. Imported cattle were initially used as draft animals. Bowling (1942) concluded that the initial economic purpose for the importation of cattle by the Spanish was for hide production. Tongues and tallow were also important products; however, the production of beef remained a by-product of the hide industry. Cattle imported by the Dutch and Swedes were used primarily for dairy production.

The interest in beef production as a primary economic enterprise did not occur until the end of the eighteenth or the start of the nineteenth century in response to the production of surplus crops in the Ohio River Valley. In June of 1817, New York City received its first shipment of Ohio grain-fed steers (Matsushima and Farr, 1995). Early grain and cattle production systems were remarkably simple. Whitaker (1975) stated that in early-nineteenth-century Illinois and Iowa, corn required little effort to plant and even less to harvest, because cattle were often turned into the fields to harvest the crop themselves, and hogs often followed the cattle to salvage any grain that the cattle knocked down and wasted. Once fattened on surplus corn, livestock walked to markets located near the river systems because railroads were not established yet. The cattle feeding industry expanded as surplus crops were produced. Technological improvements, such as the invention of the John Deere steel plow in the 1830s, the increased availability of hybrid seed corn in the 1930s, and the development of deep-well irrigation in the 1940s, in addition to political decisions, such as the Homestead and

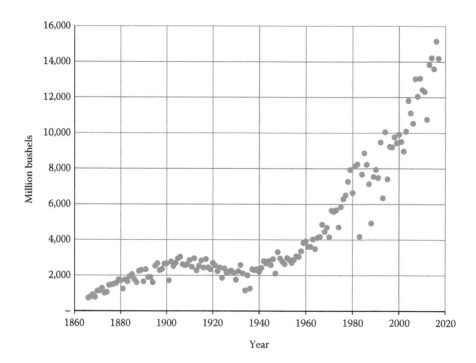

Figure 19.1 Corn production in the United States, 1866–2017. United States Department of Agriculture, Economic Research Service.

Morrill Acts of 1862, have resulted in large increases in corn production over the past 150 years (Wagner et al., 2014). Annual corn production in the United States has increased from an average of 2.65 billion bushels per year during World War II to over 15 billion bushels by 2016 (USDA Econ. Res. Ser., 2017, Figure 19.1). The post-World War II abundance of corn and other cereal grains fueled large surpluses of inexpensive feed which promoted the development of large-scale confined animal feeding operations.

Confined animal feeding operations, including beef cattle feedlots, have evolved into highly sophisticated, complex systems that are very dependent upon technology. Rollin (2004) called this evolution "a loss of husbandry to industry" and stated further that this change "threatened the traditional fair contract between humans and animals." The purpose of this chapter is to evaluate the feedlot industry regarding cattle well-being issues and identify areas where the "Fair Contract" described by Rollin has been either maintained or challenged.

THE FIVE FREEDOMS

As discussed in an earlier chapter, the well-being of an animal can be described by "The Five Freedoms." Conklin (2014) described these freedoms as follows:

1. Freedom from hunger and thirst: by ready access to fresh water and a diet to maintain full health and vigor.
2. Freedom from discomfort: by providing an appropriate environment including shelter and a comfortable resting area.
3. Freedom from pain, injury, or disease: by prevention or rapid diagnosis and treatment.

4. Freedom to express normal behavior: by providing sufficient space, proper facilities and company of the animal's own kind.
5. Freedom from fear and distress: by ensuring conditions and treatment which avoid mental suffering.

Hunger and thirst. Cattle feedlots are in the business of selling feed. This generally results in an abundance of fresh, highly nutritious feed being readily available for cattle during most hours of the day. Feed intake is highly correlated to water intake. Thus, cattle feedlots make a tremendous effort to provide clean, fresh water continuously to cattle. It appears as if the industry is largely successful at providing the freedom from hunger and thirst.

If one considers the entire statement "...a diet to maintain full health and vigor" there is considerable room for improvement concerning this freedom. Cattle are ruminant animals that evolved to consume forages. Current economic conditions, namely an abundance of inexpensive feed grains and a large demand for beef, has driven an increase in the feeding of high grain, low forage diets. Feeding high-grain, low-roughage diets, when taken to extreme, has led to increases in digestive upsets including feedlot bloat and ruminal acidosis in feedlot cattle. In addition, the feedlot industry is facing an epidemic of liver abscesses. Historically the industry used a pharmaceutical approach to control liver abscesses by providing the antibiotic tylosin in the diet at approximately 9.9 mg/kg dry matter. Concern over the potential development of antibiotic resistant strains of bacteria through routine feeding of antibiotics to animals has pressured the livestock industry to reduce or eliminate the feeding of feed grade antibiotics. This pressure is likely to increase in the future. Increasing the concentration of roughage in feedlot diets can reduce the incidence of digestive upsets and liver abscesses.

Feeding more roughage in the diet will likely reduce gain efficiency and increase production costs. Another approach could be to feed the same roughage concentration but increase roughage particle size in the diet. Increases in roughage particle size result in increased roughage effectiveness at reducing the rate of feed consumption and stimulating rumination and saliva production, thus resulting in reductions in rumen acidity. Currently, roughage particle size in many feedlots is limited by physical factors associated with the ability of feed to flow through feed mills. Incorporating roughages with larger particle size into feedlot diets could cause feed milling issues that likely will require changes to feed mill design.

Discomfort. The freedom from discomfort focusses on providing the animal with an appropriate environment including shelter and a comfortable place to lie down. In order to evaluate how successful feedlots are at providing this freedom, a discussion of how cattle in an extensive range situation fair under various environmental conditions is warranted. Cattle respond to heat or cold stress through various physiological or behavioral mechanisms. If cattle are hot, they seek shade, cooler ground surface temperatures, or exposure to the wind. If cattle are cold they bunch together, seek shelter from the wind, and prefer to lie down on warmer ground surface areas. Even under the best range conditions, cattle are less than 100% successful at completely alleviating discomfort.

In well-managed, properly designed feedlots, most cattle are comfortable most of the time. Cattle generally have the ability to select a dry place to lay down during most of the year. They are able to exhibit normal bunching behaviors when they are cold. Bunching reduces the surface area of the herd that is exposed to cold wind. This reduces body heat loss. During extreme weather, this bunching behavior can have negative consequences. Most cattle that die in blizzards suffocate rather than die of cold exposure. In many feedlots, particularly in northern climates, wind breaks have been constructed to provide cattle with shelter from the wind. Providing cattle with bedding during severe winter weather can also help maintain cattle comfort. Most large feedlots are reluctant to provide windbreaks and bedding to cattle. During summer months, wind breaks can have deleterious effects by blocking wind that could help keep cattle cool. With regard to bedding, the standard operating storm management procedure is to remove snow from feedlot pens as soon as possible following a storm. Feedlot managers fear that the placement of bedding in pens prior to and during a storm will slow the snow removal process from the pen once the storm subsides.

Many desert climate feedlots employ shade structures to help cattle remain cool. Most feedlots in other areas of the country are reluctant to build shade structures. Another procedure to help keep cattle cool is the use of sprinkler systems. Application of water to cattle increases evaporative cooling form the cattle body surface. However, continuous excessive water application can increase the relative humidity within the micro environment of cattle. A more desirable approach is to provide intermittent watering to the pen and animal surface area. Timing of the water application is also important. Nebraska research has shown that providing water to the pen surface early in the day, prior to the accumulation of a large heat load is more efficient than trying to cool the pen surface or cattle once they have become hot.

Most research in the use of shade, windbreaks, bedding, and sprinkler systems have focused on cattle performance attributes. In the short term, use of these interventions tends to improve animal performance. However, cattle are remarkable in their ability to compensate for times of stress-induced poorer performance by exhibiting tremendous rates of gain when the stressors are eliminated. Due to this compensatory performance, the use of environmental interventions has not been proven cost-effective over the long term. Whether an environmental intervention is cost-effective over the long term or not should not be the criterion to establish comfort and well-being of feedlot cattle. As demonstrated in other industries, maximum performance can be achieved using marginal living conditions such as gestation stalls and laying hen cages. Research using objective measures establishing the degree of cattle discomfort under various environmental stressors or interventions is needed. It is very likely that animals cannot be made 100% comfortable, 100% of the time. However, objective measures can be used to establish acceptable industry tolerances for environmental stress. A similar approach has been used by animal-handling specialists to improve cattle handling during processing and harvest.

Pain, injury, or disease. Advances in veterinary care and adoption of quiet cattle handling and beef quality assurance procedures have generally improved the health and well-being of most cattle in modern feedlots. Cattle in most feedlots are observed daily by trained animal health evaluators, generally removed from the pen in a timely fashion, and treated with the appropriate treatment protocol by trained feedlot veterinary technicians in a timely fashion. Frequently cattle treated for various maladies are housed in a hospital pen and allowed a chance to recover. In some feedlots, there is room for improvement. Sometimes the hospital area is crowded with too many cattle, not bedded, and less comfortable for the cattle than what their home pen would be. In addition, frequently one the normal feedlot step-up diets are used for the hospital diet. It may be advantageous to formulate a specific diet for the hospital program that provides additional nutrients and accommodates low feed intake by sick cattle.

An additional area of concern that warrants mention is the increase in "Production Diseases" experienced by cattle. Production diseases are either induced or exacerbated by certain types of management practices. Production diseases are often metabolic or nutritional in nature, and although unintended, production diseases are often an outcome of the deliberate methods used in livestock management and production. For example, cattle are ruminant animals that evolved to efficiently graze and utilize forages; however, current diets commonly fed to feedlot cattle contain low-roughage and high-feed grain concentrations. Providing high-grain diets to cattle can potentially create a biologically challenging nutritional environment that can result in subclinical or clinical rumen acidosis and lead to liver abscesses as described above.

Despite all of the advances seen in veterinary care and the millions of dollars spent on vaccine and antibiotic effectiveness and trace mineral supplementation research, the incidence of respiratory disease in feedlot cattle has increased over the past several years. Additional research is warranted to discover why this trend continues. Should bovine respiratory disease in feedlot cattle be considered a "Production Disease" that is strongly correlated to the use of high grain, lower roughage diets? Is the increase in respiratory disease the result of a loss in hybrid vigor in the beef cow herd as our efforts to increase marbling score and quality grade have concentrated Angus genetics throughout the commercial beef cow industry?

Expression of normal behavior. The feedlot industry's track record concerning this freedom is strong. Most feedlot pens provide sufficient space and plenty of pen mates to interact with. Reproductive behavior is controlled. This helps reduce pen injuries and calving problems due to unwanted pregnancies.

Fear and distress. Cattle in modern feedlots that have adopted low stress cattle-handling procedures are largely free from fear and distress. Unfortunately, there remain a few individuals who treat cattle harshly in the industry. These individuals need to be retrained or removed from positions of handling cattle. It is unlikely that all cattle will be free from fear and stress all of the time. Cattle-handling specialists have developed an audit system and industry standards to measure improvements in this area. Full adoption of these standards and audits by the industry is desirable.

CONCLUSION

Overall the feedlot industry does a decent job caring for cattle. There has been considerable progress made relative to the five freedoms. However, there are some areas for concern in the industry. A reduction in production diseases such as acidosis and liver abscesses through non-pharmaceutical means is needed. In addition, a better understanding of how cattle comfort is impacted by environmental stress and manipulations of the environment are warranted in order to design objective indicators of comfort to allow the industry to improve cattle housing.

REFERENCES

Bowling, G. A. 1942. The introduction of cattle into colonial North America. *J. Dairy Sci.* 25:129–154.

Conklin, T. 2014. An animal welfare history lesson on the Five Freedoms. Michigan State University Extension, East Lansing, MI, accessed September 29, 2017. http://msue.anr.msu.edu/news/an_animal_welfare_history_lesson_on_the_five_freedoms.

Matsushima, J. K., and W. D. Farr. 1995. *A journey back: a history of cattle feeding in Colorado and the United States.* Colorado Springs, CO: Cattlemen's Communications.

Rollin, B. E. 2004. Annual meeting keynote address: animal agriculture and emerging social ethics for animals. *J. Anim. Sci.* 82:955–964.

US Dep. Agric. Econ. Res. Serv. 2017. *Feed grains custom query.* Washington, DC: US Department of Agriculture National Economic Research Service, accessed September 28, 2017. https://data.ers.usda.gov/FEED-GRAINS-custom-query.aspx.

Wagner, J. J., S. L. Archibeque, and D. M. Feuz. 2014. The modern feedlot for finishing cattle. *Annu. Rev. Anim. Biosci.* 2:535–554. doi: 10.1146/annurev-animal-022513–114239.

Whitaker, J. W. 1975. *Feedlot empire: beef cattle feeding in Illinois and Iowa, 1840–1900.* Ames, IA: Iowa State University Press.

CHAPTER 20

Welfare Issues in Feedlot Cattle

Karen Schwartzkopf-Genswein
Lethbridge Research and Development Centre

Désirée Gellatly, Daniela Mélendez Suarez, and Sonia Marti
Agriculture and Agri-Food Canada

Eugene Janzen
University of Calgary

CONTENTS

    In order to limit the redundancies with other sections within this book the contents of this chapter will focus on those welfare issues that are relevant to feedlot cattle production and will include current findings from recently published and unpublished studies from a Canadian perspective. A brief overview of the North American feedlot production system will be given to provide a better understanding of the welfare issues related to different phases of the feedlot system and the specific management factors within each phase that contribute to poor welfare outcomes. Some topics such as nutritional needs, health and disease, pain control and painful procedures, and transportation and euthanasia will be covered briefly again in this chapter as they are all relevant to feedlot production and will be considered in the context of rearing cattle in an intensive setting.

THE FEEDLOT PRODUCTION SYSTEM

Cattle production in North America can be broken into two sectors: the cow–calf sector which represents an extensive production system and the feedlot sector representing an intensive production system. Intensive systems typically require more vigilant and regular management specific to animal housing and environment, feeding, and health and welfare related to animal confinement. The main goal of feedlot production is to grow (background) and or fatten (finish) cattle (as efficiently and economically as possible) until slaughter weight where they are sold to processors who sell beef into the retail market. For the purpose of this chapter, a feedlot will be defined as a feeding operation where cattle are fed for backgrounding and finishing in a confined area. Feedlots can range in size from as few as 50 and up to 150,000 head of cattle or greater at a single point in time (one time capacity). It should be noted that the welfare issues described in the following sections are relevant to all feedlot cattle regardless of the size of the operation.

Feedlot cattle (steers and heifers) are predominantly beef breeds; however, dairy steers can also be fed for slaughter as well as cows culled from the breeding herd, although this is less common. Heifers and steers are penned separately to avoid issues associated with sexual behaviors and feeding management. Bulls are typically not fed for slaughter due to management issues associated with aggression, wear on facilities and discounts at the time of marketing due to perceived issues in meat quality. A very small portion of feedlots specialize in feeding bulls. This has the welfare benefit of not having to castrate as well as the economic benefit of improved market weights similar to management practices in Europe. The welfare benefits of not castrating must be balanced against potential injuries associated with fighting, if the cattle are not managed properly, in small stable groups. Cattle typically enter the feedlot as weaned calves or as yearlings that have been grazed on pasture to a higher weight. Calves are at greater risk of welfare concerns due to their management and more fragile condition. Welfare issues associated with the different phases of feedlot production as well as management practices prior to entering the feedlot will be discussed.

WELFARE ISSUES RELATED TO MARKETING AND THE TRANSITION OF CATTLE BETWEEN THE RANCH AND FEEDLOT

One cannot discuss the potential welfare issues associated with feedlot production without a thorough understanding of the impact that the previous management and marketing of calves and yearlings has on their health and welfare once they have entered the feedlot. The following section will provide an overview of specific management practices that are known to increase stress in calves/yearlings and which have the potential to lead to negative welfare outcomes in the feedlot. The way in which the transition between the ranch and the feedlot is managed is a strong predictor of cattle health and welfare in the feedlot.

Pre-marketing Management

Processing on the Ranch

Processing refers to the handling of cattle for the purpose of conducting routine procedures necessary for optimal health, welfare, and management. The most common routine procedures include castration, dehorning, branding, ear tagging, parasite control, and vaccination. It should be noted that providing pain mitigation to calves that are castrated, dehorned, or branded does not compensate for improper or poor technique or lack of experience in conducting these procedures.

Processing occurs at specific points in time after birth with the most common times being less than 1 week of age and between 1 and 3 months of age, but this can vary widely by producer and geographic location. A study by Moggy et al. (2017a) indicated that 53%, 51%, and 52% of Canadian calves are castrated, dehorned, and branded at <1 week of age, respectively. The welfare impact of each of the routine management procedures indicated will be briefly discussed below.

Castration

The Canadian Beef Codes of practice (NFACC, 2013), US National Cattlemen's Beef Association (NCBA) beef quality assurance (BQA) program (NCBA, 2014), American (AVMA, 2014a), and the Canadian (CVMA, 2014) Veterinary Medical Associations recommend that bull calves be castrated as soon as practically possible after birth. This is because the trauma, and complications including increased risk of infection, blood loss, and death in some cases, is believed to increase as the testicles develop (Schwartzkopf-Genswein et al., 2012a). Recent studies on castration in beef cattle provide evidence confirming that the older the calf, the greater the number of acute (Meléndez et al., 2017a) and chronic (Marti et al., 2017a) pain/stress indicators of pain are observed. Regardless of age or method (surgical and nonsurgical), there is significant evidence that castration causes pain and stress in beef cattle (Coetzee, 2011; Meléndez et al., 2017a). It is because of this that the use of pain control drugs is now an industry requirement in Canada (not regulation in most provinces) for calves castrated at 6 months of age and older but is currently not required for calves younger than that. The NCBA guidelines for castration do not indicate that analgesia or anesthesia have to be used; however, they encourage producers to seek veterinary guidance on this, particularly in older animals (NCBA, 2014). This is in contrast to the World Organization for animal Health (OIE) guidelines which indicate that castration should be conducted with pain control regardless of animal age (OIE, 2010). There is also very little information on postoperative pain management which is necessary given that indicators of pain can be measured for several days (Meléndez et al., 2017a) and even weeks (Marti et al., 2017a) post-procedure. Only one published study has assessed healing aids following surgical castration. The study concluded that the time taken to heal lesions post-castration was not reduced in calves administered commercially available wound-healing agents compared to calves receiving no wound-healing agents (Marti et al., 2017b).

As it is unlikely that the North American feedlot industry will change to finishing bulls rather than steers (in part due to current infrastructure which can house a large number (up to 350) of cattle in a single pen), solutions for reducing or eliminating the stress of castration are still very relevant. An alternative method of castration without physically removing the testicles includes the use of an immuno-castration vaccine. The vaccine induces antibodies against GnRF (gonadotrophin releasing factor) which has been shown to successfully decrease testosterone concentrations, testicular development, and physical activity but not reduce weight gain compared to intact bulls (Janett et al., 2012). The vaccine is labeled for use in postpubertal bulls and to be effective must be administered approximately 30 days prior to entering the feedlot and again at about 7 months of age once bull calves have entered a feedlot (Janett et al., 2012). Another study demonstrated that an anti-GnRF vaccine was a viable animal welfare-friendly alternative to traditional band castration in beef cattle under North American feedlot practices (Marti et al., 2015). Although such a vaccine is registered for use in other countries, there are none currently registered for use in North America. Until alternative methods of castration are available, reliance on pain mitigation strategies for castration using anesthetics and analgesics registered for use in cattle must continue so a standardized method of pain management for surgical and nonsurgical castration can be determined (Schwartzkopf-Genswein et al., 2012a).

Disbudding and Dehorning

Disbudding is usually conducted soon after birth when the horn is not yet formed (still in bud stage) while dehorning is most commonly done after 2–3 months of age when the horn buds attach to the frontal bone (AVMA, 2014). It is well documented that both procedures cause significant stress and pain in cattle regardless of their age (Heinrich et al., 2009; Duffield et al., 2010; Stilwell et al., 2010). The negative effects of dehorning are so great that they almost always result in weight loss, especially in older calves (Goonewardene and Hand, 1991). The older the calf and the larger the horns, the greater the tissue trauma, possibility of complications, and increased welfare concern (Schwartzkopf-Genswein et al., 2012a). Consequently, disbudding is recommended to be done as early as possible after birth (AVMA, 2014; CVMA, 2016); however, if it is conducted after horn formation (between 2 and 3 months of age) pain control is required in Canada (NFACC, 2013). A welfare-friendly alternative to physical dehorning/disbudding is the use of genetically polled (having no horns) sires which produce polled calves eliminating the need to disbud or dehorn (CVMA, 2016). The 1991 National Beef Quality Audit (Lorenzen et al., 1993) and the 2009 Canadian Beef Quality audit (van Donkersgoed et al., 2001) found that 31.3% and 40% of cattle, respectively, were observed to have horns at the time of slaughter. However, a recent US study (Youngers et al., 2017) reported that the percentage of feedlot cattle with horns averaged 7.7% (ranging between 1% and 26%) indicating that more producers are dehorning on the ranch or that there is an increased use of polled genetics.

Branding

Branding is still a very common method of animal identification which is used as proof of animal ownership and its practice has deep historic and cultural roots within the North American beef industry. In addition, Canadian cattle crossing the US border currently require a brand for export purposes used to track reportable diseases. Branding is typically done at the same time as some other routine management procedures such as castration and dehorning or weaning. Branding remains relevant for producers as surprisingly, cattle rustling continues to be an issue in beef production, particularly on the ranch as cattle graze on large tracts of land that are not highly monitored. A recent interview with a Royal Canadian Mounted Police (RCMP) officer indicated the number of missing cattle rose by 5% between 2015 and 2016 in Alberta and by 40% between 2013 and 2014 in Saskatchewan which he attributed to the high value of cows and calves (between $1,500 and $2,000) (Graveland, 2016).

Both hot-iron and freeze branding are well known to cause pain and distress (Schwartzkopf-Genswein et al., 1997a, b, c; 1998; Watts and Stookey, 1999). Hot-iron branding has been shown to produce more acute pain than freeze branding (Schwartzkopf-Genswein and Stookey, 1997; Schwartzkopf-Genswein et al., 1997a, b, c; 1998), and an inflammatory response (as measured via infrared thermography) was observed for up to 6 days post freeze branding and up to 7 days post hot-iron but most likely persists much longer because the temperature at the hot-brand site was still significantly greater than the control site at that time (Schwartzkopf-Genswein and Stookey, 1997). Hot-iron branding is commonly used in feedlot cattle while freeze branding is used more in purebred cattle. There is currently a lack of research on pain mitigation for either branding method with the exception of one study assessing the anti-inflammatory effects of aloe vera gel applied onto the brand immediately after hot-iron branding. The study found no improvements in brand healing in cattle receiving the gel compared to those who did not (Tucker et al., 2014). There are other more welfare conscious methods of identification including ear tags, and or radio-frequency identification (RFID) tags. Unfortunately, these are not considered permanent methods since tags can be removed or inadvertently torn off which ultimately reduces their use as the sole method of identification, especially in range cattle. Another alternative is retinal identification, using the unchanging vascular patterns on the retina.

Weaning

Weaning in beef calves typically occurs between 6 and 8 months of age (United States Department of Agriculture [USDA], 2008a). The combined effect of the abrupt separation (physically and visually from the cow) and loss of milk is well known to be stressful for the calf (Haley et al., 2005; Hötzel et al., 2010). Alternative methods of weaning include fence-line weaning where auditory and visual contact is retained between the cow and calf or two-stage weaning where a nose-flap first stops the calf from suckling and is removed when the cow and calf are separated (Haley et al., 2005). Weaning is typically done with one or more of the procedures discussed earlier including castration, dehorning, branding, ear tagging, and vaccination. Newly weaned calves are at an increased risk for greater morbidity especially when other stressors such as transportation and commingling with unfamiliar calves are added (Wilson et al., 2017).

Preconditioning

It is well documented that routine management procedures (castration, dehorning, branding, and weaning), on their own, cause stress and or pain in calves to some degree. However, it is common for several or all of the procedures listed above to be conducted at the same time to reduce animal handling. It should be noted that handling itself is a known stressor for cattle (Woiwode et al., 2016), particularly if rough (Grandin, 1997). The welfare implications of handling cattle numerous times to conduct one routine management procedure must be weighed against the effects of conducting multiple routine procedures at one time. Research studies assessing this are currently lacking in the literature.

The most important and relevant point regarding ranch processing and the welfare of feedlot cattle is that the effects of multiple acute and/or chronic stressors are additive. A study conducted assessing the effects of castration and branding done at the same time compared to each procedure alone confirmed that calves receiving both procedures had greater physiological and behavioral indicators of pain than those that only received one of the procedures (Meléndez et al., 2017b unpublished data). This was also confirmed in another studies assessing the combined versus single effects of castration and dehorning (Sutherland et al., 2013). Ultimately, the greater the stress response the more substantial the immunosuppressive effect resulting in increased calf morbidity and mortality (Dhabhar et al., 1997; Duff and Galyean, 2007; Wilson et al., 2017).

Preconditioning is a management strategy used to reduce welfare issues associated with bovine respiratory disease (BRD) in highly stressed, weaned and transported beef calves (Radostits, 2000). Although there is no standard protocol for what constitutes preconditioning, the most common recommendations are that calves be weaned for a minimum of 35–45 days, be familiar with eating from a feed bunk, be castrated and dehorned at least 3 weeks prior to being transported off the ranch and be vaccinated (no earlier than 4 months of age) and provided parasite control (Radostits, 2000). If only some of these preconditioning criteria are done the term "conditioned" is used instead of preconditioned. The welfare benefits of preconditioning are well known and include improved rate of gain (Karren et al., 1987) as well as the reduced incidence of BRD in the first 28 days in the feedlot (Macartney et al., 2003). It is assumed that conditioned and preconditioned cattle would experience similar health and welfare benefits but this has not been well studied. One of the few studies assessing the effect of conditioning (or not) combined with long and short haul transport on calf welfare, reported that conditioning calves prior to transport allowed them to better tolerate the stressors of transport and handling (Schwartzkopf-Genswein et al., 2006). This was concluded based on the fact that conditioned calves had reduced stress hormone (cortisol) concentrations pre- and post-loading and spent greater proportions of their time feeding and less time standing and walking after transport compared to nonconditioned calves (Schwartzkopf-Genswein et al., 2006).

Unfortunately even though the health and welfare benefits of preconditioning/conditioning have been shown, its use by North American cow–calf producers has been inconsistent due to the associated cost benefits for the cow–calf and feedlot producer (CanFax, 2015). This is because premiums are not typically paid for preconditioned calves marketed through an auction as buyers for feedlots cannot confirm if they have been conditioned/preconditioned or not (Endres and Schwartzkopf-Genswein, 2018). This ultimately has negative welfare consequences for non-preconditioned calves purchased through an auction who have considerably greater morbidity and mortality rates (Wilson et al., 2017). As the consumer demand for welfare-friendly practices increase, it is likely that preconditioning will be a required practice.

Welfare Issues Related to Marketing

Auction vs Ranch Calves

Two main purchasing strategies for feedlots exist (1) acquire poor quality, non-preconditioned calves at a discounted price or (2) acquire high-quality, preconditioned calves at a premium price. The first strategy usually means that the buyer is willing to gamble on higher morbidity and mortality rates in the feedlot in place of paying a lower price per calf. The second strategy means that the buyer is willing to pay more for conditioned calves which typically have better welfare outcomes. Calves in the first scenario are usually purchased from an auction while calves in the second scenario are usually purchased directly from a ranch and are commonly referred to as "ranch direct" calves. It should be noted that both ranch direct and auction calves may or may not be preconditioned. The main welfare benefit for ranch direct calves is that they are not transported to an auction market or mixed with other unfamiliar calves where the stressors of additional handling (loading and unloading) and exposure to respiratory pathogens from other stressed calves is removed. The effect of preconditioning alone compared to being ranch direct is not known and more research in this area is needed. An alternative to live animal auctions is the use of video or online cattle sales which has increased in recent years as a strategy to improve welfare in the feedlot. This marketing strategy has the benefit of selling cattle through a central location while removing the negative welfare outcomes associated with increased loading and unloading, transporting, and comingling stressed calves.

Transportation

The marketing of calves, yearlings, backgrounding, and finishing cattle ultimately means they must be transported off the ranch or feedlot. Several recent reviews and studies on the relationship between transportation and stress in cattle have been published (Schwartzkopf-Genswein et al., 2012b; Tucker et al., 2015; Schwartzkopf-Genswein et al., 2016). The reviews indicate that numerous factors (alone or in combination) can influence cattle welfare outcomes during and after transportation. The combined effects of the management practices conducted prior to transport (described in the previous sections) in addition to those occurring during transport are believed to reduce an animal's ability to cope with their environment. Similar to routine management procedures, the greater the number of transport-related stressors occurring in a single journey, the greater the likelihood that welfare could be compromised. For example, González et al. (2012a) reported that cattle transported for longer durations at greater temperatures experienced shrinkage at more than 10% of body weight and had the greatest mortality rates at the time of off-loading.

Transport factors that can impact cattle welfare outcomes include: loading density, transport duration, trailer design and ventilation, driving and handling quality, road and environmental conditions, and fitness of the animals (Schwartzkopf-Genswein et al., 2012b; Goldhawk et al., 2015; Tucker et al., 2015). Negative welfare outcomes have been associated with several different transport conditions. For example, the effects of environmental extremes (exposure to ambient temperatures significantly higher or lower than the thermal neutral zone for cattle) can be exacerbated by inadequate or excessive ventilation, as well as overloading or the presence or absence of bedding. Excessive weight loss (>10% of body weight) associated with the mobilization of body reserves and dehydration (Tarrant et al., 1992; Warriss et al., 1995; Knowles et al., 1999) can result from factors such as increased energy expenditure associated with maintaining balance for extended periods of time, limited or no ability to lie down, removal of access to feed and water or exposure to temperature extremes. These factors are also related to negative outcomes such as increased incidence of lameness, becoming nonambulatory or death (González et al., 2012). Finally poor handling, driving quality, and or facilities can increase slipping and falling which can lead to serious injury and lameness (Grandin, 1997). The most relevant findings from a Canadian transportation benchmark study (González et al., 2012) indicated that more welfare issues were observed when transportation durations exceeded 30 hours; longer journeys at higher temperatures increased shrink and poor welfare outcomes; and cattle shipped at loading densities lower than 0.5 m^2 or greater than 1.5 m^2 were more likely to die, become nonambulatory, or lame on the truck. In addition, cull cows and calves were found to be more susceptible to transport stressors as they had the highest incidence of lameness, becoming nonambulatory and dying at the time of off-loading compared to feeder and fat cattle. Finally, poor welfare outcomes were reduced in loads of cattle transported by drivers with more than 5 years of experience hauling livestock. The authors of that study attributed this to the drivers' heightened understanding of the negative effects that poor driving technique (sharp cornering and stopping) has on cattle condition as well as their ability to manage risk (i.e., minimizing delays, altering transport times or routes during periods of inclement weather).

Currently, there are no agreed upon set of physiological or behavioral criteria (i.e., cutoffs) that can be used to access the impacts of specific journey durations or any other conditions of transport on cattle welfare. Although these measurements provide insight into an animal's ability to cope with transportation it would be difficult if not impossible to identify specific cutoffs for determining the point at which an animal's welfare is compromised. In addition, it is unlikely that industry stakeholders, veterinarians, or scientists would agree on where the cutoffs would lie. Consequently, welfare evaluations during and immediately post-transport will most likely remain outcome based using criteria such as the incidence of lameness, non-ambulation, or deaths.

Stress responses associated with long-distance transport have been shown to be reduced by the provision of a resting period where animals can lie down and are given access to feed and water (Cooke et al., 2013). The question still remains as to what the most appropriate rest stop length is for long haul (>15 hours) cattle. One of the only published studies on the effects of varying rest stop durations on cattle found that rest periods ≥10 hours did not prevent increased short- and long-term stress (cortisol), and did not improve average daily gain measured up to 25 days after transport (Marti et al., 2017c). The authors of this chapter caution that replication of this study is required before definitive conclusions about rest stop length on cattle welfare can be made due to the fact that it was not conducted using commercial transport trailers and only assessed a small number of cattle. There is currently a lack of science-based information regarding the relationship between rest stop duration and quality, and health and welfare outcomes in cattle once they have arrived at the feedlot.

It is important to note that the likelihood of an animal having a negative welfare outcome during transportation relies heavily upon their fitness for transport (age and health condition at the time of loading) in combination with how they were managed previously and the quality of handling and driving they were exposed to (Schwartzkopf-Genswein et al., 2012b; Tucker et al., 2015). The transport of compromised or unfit animals is a major welfare concern for obvious reasons. Producers must be educated to differentiate between unfit and compromised cattle. In Canada, an unfit animal should never be transported, while a compromised animal can be transported with special provisions (CFIA, 2013).

It is noteworthy that the majority of North American cattle are transported without experiencing severe welfare issues; however, the potential for the increased risk of suffering is increased during long haul (>30 hours) transportation (González et al., 2012). The interplay between pre-transport management, cattle type (age, breed, condition), and specific transport conditions (i.e., duration, loading density, weather conditions) on cattle welfare is not yet fully understood and continued research is required to determine which of these factors alone or in combination have the greatest welfare impacts.

WELFARE ISSUES ASSOCIATED WITH THE BACKGROUNDING FEEDLOT

Backgrounding refers to the first 90–100 days in the feedlot after calves or yearlings have arrived. During this period cattle are fed high-forage/low-grain rations with the goal of maximizing growth and minimizing fat deposition. Welfare issues specific to the backgrounding phase will be discussed below.

Arrival Processing

All calves and yearlings that enter the backgrounding feedlot undergo initial processing which may include vaccination and or revaccination, growth implants, identification with an ear tag or tags, performing castration or dehorning on those calves that were not done on the ranch; or in the case of castration, had not been done correctly and either one or both of the testicles are still present. It should be emphasized that castration and dehorning are more invasive at this time due to the removal of larger and more developed testes and horns increasing tissue trauma, risk of infection, blood loss and death potential (Schwartzkopf-Genswein et al., 2012a). Calves can be further handled during arrival processing when allotting them to home pens to achieve the desired number of head per pen or more uniform groups.

Although many feedlots choose not to place a new brand on cattle entering the feedlot, branding remains common in custom feedlots where cattle are fed for multiple owners and some cattle may require a brand for export purposes. Many lending institutions still require that cattle purchased with their money bear their brand. It is becoming more common for feedlot managers to refrain from removing horns when calves come into the feedlot due to increased labor (and now required pain control) and negative effects on growth performance.

The High-risk Calf and BRD

Although there are numerous health issues that can afflict cattle entering and being reared in a backgrounding feedlot, it is beyond the scope of this chapter to cover all of them in detail. Instead, this section will focus on the major health and welfare problem occurring in this phase of feedlot production which is BRD—a leading cause of morbidity (70%–80%) and mortality (40%–50%) (Edwards, 2010) with a cost of approximately 2 billion dollars annually (Powell, 2013). Clinical symptoms of BRD include fever (>104°F) depression, reduced body condition, nasal/oral/ocular discharge, coughing, and lack of rumen fill or lack of appetite (Urban-Chmiel and Groom, 2012). Comprehensive reviews related to this topic have been published (Duff and Galyean, 2007; Edwards, 2010; Wilson et al., 2017).

The main risk factors for BRD include: being young (≤4 months of age), not vaccinated, recently weaned, castrated or dehorned, transported long distances, exposed to sudden or extreme changes in weather and commingled with calves from multiple locations (i.e., auction versus ranch), exposure to a new diet and a novel environment including pens, waterers, and feed bunks (Wilson et al., 2017). Most of these risks can be attributed to lack of preconditioning management; however, the added stressors imposed at the time of entry which include initial processing, novel environment, and pen mates likely contribute to increasing stress and susceptibility to disease. The link between these management factors and reduced animal health and performance has been reported by many researchers (Ribble et al., 1995; Price et al., 2003; Chirase et al., 2004; Stanger et al., 2005).

The more risk factors associated with a calf, the greater the likelihood that it will succumb to BRD or to other communicable diseases associated with raising a large number of cattle in a confined area. Calves that have been exposed to all or most of the previously listed risk factors are appropriately referred to as "high-risk" calves and feedlot managers need to be proactive in the prevention, early detection and treatment of these kinds of calves to reduce calf morbidity and mortality. The greatest incidence of BRD is observed within the first 30 days after calves enter the feedlot (Duff and Galyean, 2007; Edwards, 2010; Wilson et al., 2017). High-risk calves typically become ill sooner after arrival than low-risk calves, which is associated with the timing of the stressors in relationship to the exposure to pathogens. Wilson et al. (2017) concluded that calves that have spent several days in the marketing chain may develop clinical BRD before or very soon after they have entered the feedlot while calves spending less time in the marketing chain may get ill 2–4 weeks after arrival; related to the length of time it takes for BRD to develop. Consequently, the initial vaccination protocol used on entry to the backgrounding feedlot is determined by the level of illness risk.

Welfare Issues Associated with Feedlot Heifers

The most significant welfare issue associated with feedlot heifers is the chance that they may be pregnant on entry to the feedlot. Feedlot managers have little control over this but can minimize its occurrence by purchasing heifer calves from reputable and conscientious sources. Pregnancies often go undetected in heifers due to their small size, making detection and appropriate management difficult. Unlike cow–calf operations, the feedlot is not well equipped to manage pregnant animals as they require special facilities such as a calving chute, a sheltered, clean and dry area away from other cattle and equipment such as a calf jack, chains, etc. Although it is important to have trained staff to assist calving heifers, typically few feedlot pen riders have this experience. Without the basic calving facilities and trained staff, the potential for negative welfare outcomes for both the heifer and the calf are great. Calf survival rates are extremely low because the calf is typically removed from the heifer shortly after birth with little or no chance to nurse and receive colostrum. Heifers may also suffer complications following birth such as retained placental or vaginal prolapse as well as increased death losses and health costs (Habermehl, 1993; Rademacher et al., 2015).

In cases where pregnancies are too advanced for spaying or abortifacients are not used at the time of arrival, heifers will calve in the feedlot. Calving heifers are at high risk of dystocia because their pelvis is usually small which decreases the likelihood of calving naturally (Miesner and Anderson, 2015). If excessive traction with a calf puller is required, this can result in a down heifer which may become nonambulatory. In this situation, cesarean section is recommended and must be conducted by an experienced veterinarian with the use of appropriate anesthetics and analgesics to ensure good welfare outcomes (Miesner and Anderson, 2015).

A common method of managing feedlot pregnancies is to administer abortifacients to heifers on arrival to the feedlot. Although these agents are effective in eliminating pregnancy, induced abortions can result in numerous complications including dystocia, retained fetal membranes, acute toxic metritis, and death (Barth, 1986). An alternative practice to eliminate feedlot pregnancies is spaying (ovariectomy) which could be done as part of a preconditioning program prior to entering the feedlot or at the time of entry providing the pregnancy is still early. This procedure may help to improve heifer welfare by eliminating pregnancy as well as estrous/riding behavior which may make heifers more susceptible to injuries. There are two main methods of spaying; dropped ovary and flank techniques. The dropped ovary technique is done by inserting a special tool (ovaritome) into the vagina which facilitates the severing of the ovaries while flank spaying is done by making an incision in the flank to gain access to the ovaries which are then removed using a scalpel (Horstman et al., 1982; Petherick et al., 2013). Both methods are surgical procedures, which have been shown to cause some stress and pain in heifers and cows, leading to the recommendation that flank and drop method spaying should not be performed without pain management (Horstman et al., 1982; Petherick et al., 2013). In addition, the drop method is recommended over flank spaying as it is considered to be less invasive and was shown to cause fewer behavioral and physiological indicators of pain (Horstman et al., 1982; Petherick et al., 2013). Research on pain mitigation strategies for spaying is currently lacking.

WELFARE ISSUES ASSOCIATED WITH THE FINISHING FEEDLOT

The finishing phase (typically the last 100 days after backgrounding) focuses on feeding high-grain/low-forage rations to backgrounded calves or yearlings until they reach a prescribed finish (fat cover) before marketing for slaughter. While backgrounding cattle typically have more welfare issues related to pre-transport management and BRD, finishing cattle have more welfare issues related to feeding and diet.

Nutrition

Cattle are ruminants by nature, meaning they evolved consuming plant material containing fiber required for the optimal functioning of the rumen necessary for good health and welfare. The main focus of the backgrounding phase is to optimize growth while reducing fat disposition and this is accomplished by feeding diets that are higher in fiber (forage) and lower in energy (grain). However, the main goal of the finishing phase is to fatten cattle as quickly and efficiently as possible. In North America, this has been achieved by feeding diets that are high in grain (between 60% and 95%) but low (40% to 5%) or in some cases even void of forage. The transition from the backgrounding to the finishing diet is a challenging time which is characterized by significant changes in the rumen microbiota which may become severely unbalanced, particularly if the transition is too abrupt (Tajima et al., 2001). A negative welfare consequence of this rapid diet change is the accumulation of acid (lactate and volatile fatty acids) which lowers the ruminal pH resulting in either acute (pH < 5) or subacute acidosis (pH < 5.8) (Schwartzkopf-Genswein et al., 2003; Penner et al., 2011). Both of these conditions can result in further health issues such as liver abscesses and laminitis (Galyean and Rivera, 2003, Nagaraja and Lechtenberg, 2007). Cattle with acute acidosis become ill and typically stop consuming feed. In more severe cases, acidosis causes thickening of the ruminal mucosa causing lesions (rumenitis) which reduce the animal's ability to absorb feed nutrients and may result in death (Owens et al., 1988). Although the effects on the rumen environment, feeding and weight gain are obvious, it is currently not known if acidosis is a painful condition in cattle.

Feedlot managers are willing to assume a certain death loss associated with nutritional disease in order to maximize weight gain for the majority of the cattle being fed. Morbidity and mortality rates related to acidosis are difficult to find in the literature, likely due to the difficulty in diagnosing the disease in live animals. Liver abscesses found at the time of slaughter have been used as a proxy measure for acidosis and have been reported to be as high as 56% (Fox et al., 2009). There are several risk factors for acidosis including the length of the diet transition period, how fermentable the diet is, the frequency, amount and rate at which the feed is consumed (Owens et al., 1988, Schwartzkopf-Genswein et al., 2003, Tucker et al., 2015) and, the individual's ability to manage the effects of high acid production (Goad et al., 1998).

In order to mitigate the negative consequences of abrupt changes from low to high-grain diets, feed managers have developed "step-up" protocols whose main purpose are to gradually increase the amount of grain in the diet over a 3–4 week period (Aschenbach et al., 2011, Penner et al., 2011). One study found that most cattle can be rapidly adapted to high-grain diets in just a few incremental diet steps. However, the study concluded that reducing acidosis in the most susceptible cattle ultimately means increasing the time that the transition occurs for the entire group (Bevans et al., 2005).

Other approaches to reducing acidosis include feeding diets containing sufficient fiber (Nagaraja and Lechtenberg, 2007), adding ionophores (agents that moderate feed consumption) (González et al., 2009; 2012a) or buffers to the feed (i.e., sodium bicarbonate, seaweed) (Enemark, 2008). Although all of the approaches discussed above are reported to control acidosis to some degree, none of them have completely eliminated it.

Even though there are many anecdotal reports that cattle on low-fiber diets have been observed consuming straw or wood chip bedding, dirt or wooden fence boards, there are no published studies documenting what level of fiber deprivation triggers these behaviors or how important is the motivation to consume fiber and ruminate. Overall few studies have focused on the welfare aspects of feeding high-concentrate diets and therefore more work needs to done in this area.

Growth Promoting Agents and Cattle Welfare

One of the main goals of the feedlot industry is to grow and fatten cattle as efficiently as possible. In North America, this has been achieved, in large part, by the use of growth promoting agents including growth implants (Synovex-S, progesterone and estradiol benzoate and Revalor-S, trenbolone acetate and estradiol) and β-agonist compounds (Zilpaterol, Optaflexx, or ractopamine). Growth implants have been used in the feedlot industry for more than 40 years (commercially available in 1975), while β-agonists have only been commercially available since 2004 (Radunz, 2010).

The majority of feedlot cattle are implanted in the ear at least once and often twice which is largely dependent upon the animal weight, sex, and product used (USDA, 2013). Implants contain natural and synthetic hormones that alter the hormone status of the animal which increases average daily gain (10%–16%) and improves feed efficiency (Jones et al., 2016) with reported use in 84% of feedlot cattle (USDA, 2011; Stewart, 2013). β-Agonists are nonsteroid agents administered in the feed in the last 28–35 days before slaughter and work by altering growth at the cellular level by redirecting energy to protein rather than fat synthesis, resulting in increased weight gain, ribeye area, and total red meat yield (Lean et al., 2014). Reported use of β-agonist in US feedlot cattle was 57% (USDA, 2011); however, the actual use today is likely much greater than indicated in the 2011 report.

The effects of implants on cattle welfare have been poorly studied although research focusing on their production effects suggested that implanted cattle may be more susceptible to heat stress, especially in summer months which is late in the feeding period when the cattle are also heavier (Gaughan et al., 2005). Their use has also been implicated in triggering the onset of the buller steer syndrome (Blackshaw et al., 1997) but this has not substantiated. Some implants have been reported to increase mounting and aggressive behavior, (Lesmeister and Ellington, 1977; Stackhouse-Lawson et al., 2015) which has the potential to increase animal injury. However, this was not a consistent observation as other studies reported that implants had no effect on behavior (Baker and Gonyou, 1986; Godfrey et al., 1992).

The most recent welfare concerns regarding growth promoters have centered on β-agonist. These concerns were prompted by increasing anecdotal reports that cattle being fed one of the most potent β-agonist on the market (Zilmax) were arriving to some slaughter facilities with hoof (sloughing) and lameness issues and were slow and difficult to move (Thomson et al., 2015). Although the exact cause could not be substantiated, some slaughter facilities and industry stake holders refused to accept cattle fed the product and, in 2013 it was removed from the market (Centner et al., 2014). Some research has reported negative welfare outcomes from feeding β-agonist such as significantly increased odds of death (Loneragan et al., 2014); increased lateral lying and aggressive behavior

(Stackhouse-Lawson et al., 2015); and tachypnea with an abdominal breathing, lameness, and reluctance to move, as well as greater serum lactate and creatine kinase concentrations (indicators of muscle damage and dehydration, respectively) (Thomson et al., 2015) compared to those not fed β-agonist. Thomson et al. (2015) concluded that the clinical signs and serum biochemical abnormalities observed in affected cattle were similar to those observed in pigs diagnosed with fatigued pig syndrome, and suggested that the term fatigued cattle syndrome be used to describe cattle with the same clinical symptoms. Thomson et al. (2015) further concluded that this syndrome is likely the result of a combination of risk factors including; use of β-agonist and dosage, genetics, high ambient temperatures, being recently handled and transported, and being at or close to finishing weight. Feedlot managers need to be aware of the negative welfare consequences of this combination of factors and manage cattle accordingly and more research is needed to better understand the relationships between these factors.

WELFARE ISSUES COMMON TO THE BACKGROUNDING AND FINISHING FEEDLOT

The welfare issues described in the next section are those issues common to both backgrounding and finishing cattle. These issues are largely related to environmental factors since the housing and facilities are the same for both types of cattle. Likewise, the prominent welfare issue of timely and properly conducted euthanasia is relevant to all cattle regardless of the sector or phase of production.

Environment

The environmental factors that cause the greatest welfare concern in feedlot cattle include excessive cold, heat, or mud. At this time there is no agreed upon definition of what constitutes excessive from a welfare point of view for any of the three listed factors. By definition welfare is affected when these conditions impact an animal's ability to cope (i.e., inability to maintain constant body temperature). As with most factors that can negatively affect cattle welfare, the combination of degree and duration of exposure needs to be taken into consideration. The majority of studies published on the impacts of environmental factors on beef cattle were designed to assess production rather than welfare so discussion is limited in some instances.

Outdoor feedlots exist in both cold and hot regions within North America which means at some point the cattle housed within them will be exposed to extreme ambient conditions. In Canada and the northern US, cold conditions are the main focus as summer temperatures (and humidity) are not typically high enough or last long enough to cause a welfare concern. The biggest concern in cold climates is the high windchill effect which increases energy demands in cattle (Ames and Insley, 1975; Ames, 1988). This is why windbreak fencing is common, particularly in western Canada. Indicators of cold stress in cattle include shivering (Gonyou et al., 1979) and loss of body weight associated with increased energy required for thermoregulation and reduced energy availability for growth and fat deposition (Ames, 1988). However, temperature-related weight loss is typically not an issue in properly managed feedlots because feed deliveries are increased to compensate for the extra energy demands during cold periods. Recommended strategies to reduce cold weather effects on cattle include increasing feed deliveries, and the provision of ample amounts of clean dry bedding.

Greater cattle welfare issues are associated with extreme heat rather than cold. Excessive heat load can increase body temperature, respiration and panting (Brown-Brandl et al., 2003) as well as shade-seeking behavior, standing (Widowski, 2001), and death (Nienaber and Hahn, 2007).

The most common and effective methods of reducing heat load and ultimately the chance of heat-related death in feedlot cattle include the provision of either shade, water, or a combination of both (Mitlöhner et al., 2002; Kendall et al., 2007; Sullivan et al., 2011). In order for producers to better manage cattle during hot weather, it would be beneficial to provide them with a clear definition when (at what temperature and humidity levels) cattle may be at risk of heat stress. Unfortunately, the determination of such a threshold is difficult if not impossible due to large variations in factors such as animal acclimation and individual differences in heat sensitivity as well as the measure of heat load used (ambient temperature, temperature–humidity index, heat-load index) leading to the conclusion that animal responses such as high respiration and open mouth panting would be more useful than a single value of heat load (Brown-Brandl et al., 2003; Gaughan et al., 2010). There is currently a lack of information regarding the amount of shade required per animal relative to pen density, the frequency, and duration of cattle handling as well as the type and conformation of shade used, the type of sprinkling and availability of water through space (Tucker et al., 2015).

Muddy conditions are known to increase infectious (footrot, digital dermatitis) and injury-related (due to slippery condition) lameness (Stokka et al., 2001) reduce weight gain (Morrison et al., 1970) and locomotion (Degen and Young, 1993) and increase energy demands (Dijkman and Lawrence, 1997) as mud acts as a medium for heat loss. Cattle housed in outdoor lots were found to have more dry mud and manure adhering to their hair coats (tag) than those housed indoors (Honeyman et al., 2010) which could potentially reduce the isolative value of the hair and increase the chance of cold stress. Some studies have reported increased indicators of stress such as reduced body temperature and increased white blood cell counts and elevated cortisol and thyroxine levels in cattle exposed to mud compared to those that were not (Tucker et al., 2007, Webster et al., 2008), while other studies have found no differences in measures of immune competence or stress levels (cortisol response) in cattle housed in dry versus muddy pens (Wilson et al., 2017).

Strategies recommended to reduce the negative effects of mud include regular pen cleaning, provision of bedding during wet periods, adequate feedlot drainage, and dirt or straw mounds within each pen that provide a dry area where cattle can lie down (NFACC, 2013). Currently little information is available regarding the appropriate amount of dry lying space needed for each animal. Adequate lying space is likely a challenge for cattle housed in pens containing 200 animals or more.

Lameness

Although it is unlikely that the incidence of lameness in feedlot cattle has increased over the past 30 years, its acknowledgment as a major health and welfare issue in feedlot cattle has only recently been recognized (Schwartzkopf-Genswein et al., 2012a; Tucker et al., 2015; Marti et al., 2016). Lameness is characterized by a change in normal gait resulting from pain or discomfort associated with hoof and leg injuries, or disease (Greenough, 1997). The main and obvious reason that lameness is a significant welfare concern is its association with pain. Another welfare concern is the reduction or complete curtailing of an animal's ability to access feed and water, seek shelter, or escape from predators or dangerous situations resulting from reduced or loss of lost mobility (nonambulatory).

Information regarding the prevalence of lameness in feedlot cattle as well as the identification of the most common types of lameness is minimal. An older survey assessing lameness in feedlots in the mid-western United States found that 16% of all feedlot health issues were attributed to lameness and lameness accounted for 70% of all non-fit sales cattle (Griffin et al., 1993). The 2001 USA Beef Quality Audit reported that 31.4% of the cattle observed during lairage (prior to slaughter) were lame (Roeber et al., 2001). A pilot study conducted in the pens of a small number of Alberta feedlots found lameness prevalence to be between 32.8% and 52.8% (Tessitore et al., 2011). A recent unpublished Canadian study on the prevalence and characterization of lameness in feedlot cattle confirmed that lameness is an important disease in Canadian feedlots accounting for 30% of all health problems: only second to respiratory disease (40% of all health problems) in terms of its welfare impact (Marti et al., 2016). This same study reported that a high percentage (15.25%) of those cattle identified and assessed for lameness for the first time were severely lame and experienced high relapse rates (between 8% and 13%).

The main infectious causes of lameness in feedlot cattle include foot rot, infectious arthritis (Stokka et al., 2001), and more recently digital dermatitis (Döpfer et al., 1997; Marti et al., 2016). Foot rot is the result of a bacterial (*Fusobacterium spp.*) infection resulting in swelling of the interdigital tissue of the toes, coronary bands, and heels (Stokka et al., 2001), and is found more commonly in finishing than backgrounding cattle (Tibbetts et al., 2006). Digital dermatitis is an infection associated with spirochete bacteria that causes a raw lesion on the back of the heel just above or at the coronary band (Döpfer et al., 1997) and is found more commonly in cattle >430 kg (Marti et al., 2016). As implied, infectious arthritis is the result of a joint infection associated with mycoplasma bacteria (Stokka et al., 2001). Marti et al. (2016) reported that that the incidence of foot rot and joint infection was 71.77% and 19.25%, respectively. The above-described conditions are highly contagious and can spread rapidly affecting a large percentage of animals within a pen which greatly increases their significance as a welfare concern.

The most common noninfectious causes of lameness include toe-tip necrosis, laminitis, and injuries. Toe-tip necrosis is characterized by necrosis of the distal part of the third phalanx (P3) bone in the foot caused by trauma that allows bacteria to enter the claw as a consequence of white line separation and is commonly seen in highly agitated receiving cattle (Jelinski et al., 2016). Laminitis refers to inflammation of the connective tissue (corium) located between the pedal bone and hoof horn and has strong associations with acidosis which causes bacteria and their toxins to enter the blood stream resulting in inflammation of the corium (Stokka et al., 2001). Finally, injuries refer to such conditions as sprains, fractures, and lacerations all of which can cause mild-to-severe lameness and have a reported incidence of 3.51% (Marti et al., 2016). All of the previously listed conditions are believed to be extremely painful in advanced stages as evidenced by severely modified gait scores as well as weight loss and in some cases mortality (Greenough, 1997; Tibbetts et al., 2006; Marti et al., 2016).

Wet pen conditions and constant moisture with manure contamination have been implicated in softening and allowing bacterial organisms to penetrate the skin between the claws (Bergsten, 1997). This was confirmed by Marti et al. (2016) who found increased incidence of both foot rot and digital dermatitis during the spring (41.3%) and fall (28.7%) when pen conditions were wet. Other risk factors for both infectious and noninfectious causes of lameness include rough pen surfaces associated with frozen manure, rocks, or other sharp objects: slippery or poorly maintained pens and handling facilities and frequent or rough handling (Grandin, 1997; Stokka et al., 2001; Green et al., 2012). Little is known about the relationship between the incidence of lameness and housing on concrete or slatted floors although one study suggested that the claw health of beef cattle housed on straw or deep litter was better than those house on slatted floors (Tessitore et al., 2009).

The most significant risk factors associated with increased lameness and lameness severity reported in a recent Canadian feedlot study were as follows: being a yearling, being sourced from an auction market rather than a ranch, having a greater number of days on feed, being housed with a greater number of cattle within a pen, having reduced bunk space, having reduced forage content in the diet, being exposed to a greater number of handling events, and being housed in pens with increased mud depth (>5 cm) (Marti et al., 2016). The same study found that although the majority of cattle were lame in the hind limbs (approximately 68% of all lame animals); the severity of lameness was approximately 40% greater in cattle that were lame in the fore limbs.

Common mitigation strategies based on the risk factors listed above include improving pen drainage, increasing pen cleaning, and or provision of bedding to reduce wet and slippery pen conditions (Stokka et al., 2001; Marti et al., 2016); regular pen and handling facility maintenance to reduce rough pen surfaces and eliminate sharp objects (Stokka et al., 2001); reducing handling frequency and eliminating rough handling (Grandin, 1997; Lensink et al., 2000a, b; Green et al., 2012); reducing pen density and increasing bunk space availability; purchasing ranch direct calves and maintaining adequate forage levels in the diet (Marti et al., 2016). Vaccines may be useful in controlling the infectious agents associated with feedlot lameness; however, at this time those registered for use in North America have been marginally successful (i.e., in the case of foot rot; Checkley et al., 2005). It should be noted that studies assessing the relationship between reduced lameness and the majority of mitigation strategies indicated above are lacking in the scientific literature. Improved diagnosis and targeted treatment protocols as well as continued research focused on the testing and identification of effective mitigation strategies (based on known risk factors) will help to reduce the incidence, severity, and relapse rates of lameness in feedlot cattle, ultimately improving welfare.

Buller Syndrome

The feedlot environment imposes artificial social groupings both in composition and size. Feedlot pens typically hold between 100 and 300 head of cattle at a time. Under natural circumstances cattle form stable groups of up to 50 animals which include a mixture cows, bulls, and heifer and bull calves (Sowell et al., 1999). Consequently, it is not surprising that confinement rearing large groups of cattle may exacerbate some social problems.

Although it is not desirable to mix cattle between pens, it is common practice to sort finishing cattle into new groups (pens) based on weight (weight-breaks) and amount of fat cover, several times prior to marketing, with the ultimate goal of selling more uniform groups of cattle for slaughter. Several studies have shown that mixing unfamiliar cattle increases aggression and related injuries and may also limit individual animal access to resources such as feed and water (Tennessen et al., 1985; Mench et al., 1990).

The buller steer syndrome is a major welfare issue in some feedlots with an annual incidence between 2% and 4% (Irwin et al., 1979). It is characterized by excessive and sustained mounting of one particular steer (buller) by one or several other steers (riders) housed within the same feedlot pen. If appropriate interventions are not put in place the buller steer can suffer fatigue, trauma to its back and hind quarters, broken limbs and in some cases death (Blackshaw et al., 1997). One study reported that morbidity and mortality were 2.5 and 3.2 times greater in bullers than other cattle (Taylor et al., 1997). To the authors' knowledge, there are no reports of this behavioral problem in cattle housed on range.

The reported causes of buller syndrome are numerous indicating that its manifestation is complex and may involve multiple risk factors. These factors include: animal weight, group size and space per animal, mixing, social dominance, transportation, handling, weather, excessively muddy or dusty pens, improper or late castration, feeding management (Blackshaw et al., 1997), and the use of growth promoting implants and how they are managed (improper technique, extra dosing, and re-implanting) (Voyles et al., 2004; Bryant et al., 2008).

Few studies have been published on this topic—likely because of its erratic nature and the difficulty in inducing it for study purposes which has hindered the development of successful mitigation strategies. One of the few studies conducted on the relationship between pen space the incidence of bullers found that for every 9.3 m^2 increase in pen size bulling was reduced by 0.05% (Irwin et al., 1979). Some recommended management practices to reduce the occurrence of buller syndrome include: limiting the number of steers within a pen to 200 or fewer, minimizing the number of different groups of cattle used to form a pen, and implanting at the time of entry into the feedlot rather than later in the feeding period (Blackshaw et al., 1997). Some producers have successfully managed the problem by building "hides" which are overhead structures that prevent mounting and allow the buller to escape from the riders. Another commonly used strategy is the removal of the buller from its home pen to a smaller recovery pen for some days or weeks before introducing it into a new pen. It should be noted that none of the strategies described above are known to eliminate the problem.

On-Farm Euthanasia

Of all the welfare issues covered in this chapter, timely and properly conducted euthanasia would represent the single greatest improvement in feedlot cattle welfare. By definition, euthanasia is the ending of life in a way that minimizes or eliminates pain and distress (AVMA, 2013). Although the definition is simple, the decision to follow through with the procedure and the criteria used to make it are not. Consequently, animals may suffer unnecessarily if the criteria to euthanize are not clear for staff that are placed in charge of carrying out this task. The general reasons for euthanasia are well known and agreed upon by industry stakeholders and include poor health, disease, and injury (Woods and Shearer, 2015). It should be noted that marketing a "poor doing" animal before it reaches slaughter weight is not considered euthanasia but rather is known as "railing" or selling an animal (usually at a discount) to eliminate future welfare problems. Salvage value is the main reason producers will transport compromised and unfit cattle to slaughter even though it is inhumane and in some cases illegal (Endres and Schwartzkopf-Genswein, 2018). Many countries have government regulations or industry guidelines as to what constitutes fitness for transport. Some examples of these would include when calving is imminent, broken leg, severe cancer eye, severe lameness, and being newborn (CFIA, 2013).

Feedlots typically have designated pens (separate from the main feeding pens) where animals can be moved to help manage poor health conditions. These pens are usually smaller, provide more shelter, and are close to handling equipment to facilitate medical assessments and treatments. Cattle that are acutely ill and are expected to recover quickly are usually treated and placed back in their home pen while others needing follow-up treatments are placed in pens known as "hospital" pens where they can be easily handled as needed to complete their treatment regimen and where they are allowed to convalesce. Cattle that have not improved in condition following their stay in the hospital, have relapsed, or require considerable time for recovery are placed in "chronic" pens where, typically the prognosis for recovery is poor. Cattle that neither improve nor worsen in condition are placed in "rail" pens. Feedlot animals that are at greatest risk of suffering due to delayed euthanasia are those animals that are chronically ill.

While the general reasons for euthanasia are straightforward, the specific reasons used to decide an animal's fate are where many gray areas exist, predominantly related to the severity of the condition. Institutions such as the North American Meat Institute NAMI (2008), AVMA (2013) and CFIA (2013) have complied lists of conditions to guide producers, veterinarians, auction and abattoir staff as well as truckers and inspectors when euthanasia is required. The most common conditions include: emaciation, nonambulatory, severe cancer eye, chronic disease with poor prognosis (i.e., chronic respiratory disease or severe lameness), fractures of the leg, hip or spine that are irreparable, trauma or other conditions causing excruciating pain, transmissible disease (i.e., rabies, foot, and mouth), birth is imminent, and in heavy lactation. As it is difficult and impractical to have a reference list of all possible conditions and scenarios where euthanasia is required, producers must often use their own discretion or seek veterinary advice. Even when the condition is clearly defined, decisions to euthanize are usually based on additional criteria such as the severity of the condition as well as the likelihood of recovery. Both of these criteria are highly subjective and can be influenced by a producer's knowledge, past experience, sensitivity, emotional connection to the situation as well as gender, culture, religion, or age (Wood and Shearer, 2015). Wood and Shearer (2015) provide a comprehensive list of indicators of pain and distress such as inappetence, vocalization, etc., that can aid producers in assessing the severity of an animal's condition. It should be noted that indicators of pain should be used in conjunction with the duration they are exhibited to accurately assess an animal. Defining the cutoff between what is acceptable (or unacceptable) management of a compromised animal is highly important. As suggested by Grandin and Johnson (2009), the criteria for euthanasia must be strict enough to prevent suffering but not so strict that an animal capable of recovery is not allowed to be nursed back to health. Most producers do not like putting an animal down and describe it as an unpleasant experience which may unintentionally prolong animal suffering (Moggy et al., 2017b). It should never be acceptable to "let nature take its course" when death is inevitable or delaying euthanasia for reasons of convenience such as waiting for scheduled evaluations of cattle housed in chronic or sick pens whose conditions can decline rapidly (Wood and Shearer, 2015).

The other main factor that can have significant impacts on cattle welfare related to euthanasia is the method (and associated equipment) used. Numerous publications have described the techniques used to ensure loss of consciousness, cardiac arrest, and ultimately loss of brain function (AVMA, 2013; Human Slaughter Association (HAS), 2014; OIE, 2014). Although barbiturate and anesthetic methods are effective for euthanizing cattle their use is limited on-farm as they are controlled substances for use by veterinarians only and animals euthanized with these drugs may not enter the food chain. Gunshot (delivered via shot gun, rifle or handgun) is the preferred method of euthanizing feedlot cattle (Fulwider et al., 2008) and does not require veterinary presence which aids in ensuring timely euthanasia. To ensure effective humane euthanasia producers need to be educated in the proper positioning of firearm relative to the skull as well as the most appropriate type of ammunition and the size of the firearm (Grandin, 1994; Woods et al., 2010). One of the greatest breaches to welfare associated with euthanasia is lack of confirmation of death and therefore it is considered unacceptable to walk away from a euthanized animal without first testing for signs of death which include lack of corneal reflex, rhythmic breathing, and heartbeat (Woods and Shearer, 2015).

REFERENCES

Ames, D. (1988) Adjusting rations for climate. *Veterinary Clinics of North America: Food Animal Practice* 4:543–550.

Ames, D. R. and Insley, L. W. (1975) Wind-chill effect for sheep and cattle. *Journal of Animal Science* 40:161–165.

Aschenbach, J. R., Penner, G. B., Stumpff, F. and Gabel, G. (2011) Role of fermentation acid absorption in the regulation of ruminal pH. *Journal of Animal Science* 89:1092–1107.

AVMA (2013) *AVMA guidelines for the euthanasia of animals: 2013 edition*, Schaumberg, IL, American Veterinary Medical Association.

AVMA (2014a) *Welfare implications of castration of cattle*, Schaumberg, IL, American Veterinary Medical Association.

AVMA (2014b) *Welfare implications of dehorning and disbudding of cattle*, Schaumberg, IL, American Veterinary Medical Association.

Baker, A. M. and Gonyou, H. W. (1986) Effects of zeranol implantation and late castration on sexual, agonistic and handling behavior in male feedlot cattle. *Journal of Animal Science* 62:1224–1232.

Barth, A. D. (1986) Induced abortion in cattle. In: Morrow, D. A. (ed.) *Current therapy in theriogenology*, 2, 205–209. Philadelphia, PA: W.B. Saunders.

Bergsten, C. (1997) Infectious diseases of the digits. In: Greenough, P. and Weaver, A. D. (eds.) *Lameness in Cattle,* 3rd edition, 23–43, Philadelphia, PA: W.B. Saunders Co.

Bevans, D. W., Beauchemin, K. A., Schwartzkopf-Genswein, K. S., McKinnon, J. J. and McAllister, T. A. (2005) Effect of rapid or gradual grain adaptation on subacute acidosis and feed intake by feedlot cattle. *Journal of Animal Science* 83:1116–1132.

Blackshaw, J. K., Blackshaw, A. W. and McGlone, J. J. (1997) Buller steer syndrome review. *Applied Animal Behaviour Science* 54:97–108.

Brown-Brandl, T. M., Neinaber, J. A., Eigenberg, R. A., Hahn, G. L. and Freetly, H. (2003)Thermoregulatory responses of feeder cattle. *Journal of Thermal Biology* 28:149–157.

Brown-Brandl, T. M., Eigenberg, R. A. and J.A. Nienaber. 2006. Heat stress risk factors of feedlot heifers. *Livestock Science* 105:57–68.

Bryant, T. C., Rabe, F. M., Bowers, D. N., Hutches, D. B. and Jaragin, J. J. (2008) Effects of growth-promoting implant containing tylosin tartrate on performance, buller incidence and carcass characteristics of feedlot steers. *Bovine Practitioner* 42:128–131.

Canfax (2015) Economic considerations on preconditioning calves. Available at: www.canfax.ca/Samples/ Preconditioning%20Sept%202015.pdf.

Centner, T. J., Alvey, J. C. and Stelzleni, A. M. (2014) Beta agonists in livestock feed: Status, health concerns, and international trade. *Journal of Animal Science* 92:4234–4240.

CFIA (2013) Canadian food inspection agency. Compromised animals policy. Available at: www.inspection. gc.ca/animals/terrestrial-animals/humane-transport/compromised-animals-policy/eng/1360016317589 /1360016435110.

CFIA (2013) Livestock transport requirements in Canada. Available at: www.inspection.gc.ca/animals/ terrestrial-animals/humane-transport/transport-requirements/eng/1363748532198/1363748620219.

Checkley, S. L., Janzen, E. D., Campbell, J. R. and McKinnon, J. J. (2005) Efficacy of vaccination against *Fusobacterium necrophorum* infection for control of liver abscesses and footrot in feedlot cattle in western Canada. *Canadian Veterinary Journal* 46:1002–1007.

Chirase, N. K., Greene, L. W., Purdy, C. W., Loan, R. W., Auyermann, B. W., Parker, D. B.,Walborg, E. F., Stevenson, D. E., Xu, Y., Klaunig, J. E. (2004) Effect of transport stress on respiratory disease, serum antioxidant status, and serum concentrations of lipid peroxidation biomarkers in beef cattle. *American Journal of Veterinary Research* 65:860–864.

Coetzee, J. F. (2011) A review of pain assessment techniques and pharmacological approaches to pain relief after bovine castration: Practical implications for cattle production within the United States. *Applied Animal Behaviour Science* 135:192–213.

Cooke, R. F., Guarnieri Filho, T. A., Cappellozza, B. I and Boohnert, D. W. (2013) Rest stops during road transport: Impacts on performance and acute-phase protein responses of feeder cattle. *Journal of Animal Science* 91:5448–5454.

CVMA (2014) Castration of cattle, sheep and goats-position statement. Available at: www.canadianveterinarians. net/documents/castration-cattle-sheep-goats.

CVMA (2016) Disbudding and dehorning of cattle-position statement. Available at: www.canadianveterinarians. net/documents/disbudding-and-dehorning-of-cattle.

Degen, A. A. and Young, B. A. (1993) Rate of heat production and rectal temperature of steers exposed to stimulated mud and rain conditions. *Canadian Journal of Animal Science* 73:207–210.

Dhabhar, F. S. and McEwen, B. S. (1997) Acute stress enhances while chronic stress suppresses immune function in vivo: A potential role for leukocyte trafficking. *Brain Behaviour and Immunology* 11:286–306.

Dijkman, J. T. and Lawrence, P. R. (1997) The energy expenditure of cattle and buffaloes walking and working in different soil conditions. *Journal of Agricultural Science* 128:95–103.

Döpfer, D., Koopmans, A., Meijer, F. A., Szakáll, I., Schukken, Y. H., Klee, W., Bosma, R. B. Cornelisse, J. L., van Asten, A. J., ter Huurne, A. A. (1997) Histological and bacteriological evaluation of digital dermatitis in cattle, with special reference to spirochaetes and Campylobacter faecalis. *Veterinary Record* 140:620–623.

Duff, G. C. and Galyean, M. L. (2007) Recent advances in management of highly stressed, newly received feedlot cattle. *Journal of Animal Science* 85:823–840.

Duffield, T. F., Heinrich, A., Millman, S. T., DeHaan, A., James, S. and Lissemore, K. (2010) Reduction in pain response by combined use of local lidocaine anesthesia and systemic ketoprofen in dairy calves dehorned by heat cauterization. *Canadian Veterinary Journal* 51:283–288.

Edwards, T. A. (2010) Control methods for bovine respiratory disease for feedlot cattle. *Veterinary Clinics of North America: Food Animal Practice* 26:273–284.

Endres, M. and Schwartzkopf-Genswein, K. S. (2018) Overview of cattle production systems. In: Tucker, C. (Ed.) *Advances in Cattle Welfare*, Cambridge, MA: Woodhead Publishing.

Enemark, J. 2008. The monitoring, prevention and treatment of sub-acute ruminal acidosis (SARA): A review. *Veterinary Journal* 176 32–43.

Fox, J. T., Thomson, D. U., Lindberg, N. N. and Barling, K. (2009) A comparison of two vaccines to reduce liver abscesses in natural-fed beef cattle. *Bovine Practitioner* 43:168–174.

Fulwider, W. K., Grandin, T., Rollin B. E., Engle, T. E. and Dalsted, N. L. (2008) Survey of management practices on one hundred and thirteen north central and northeastern United States Daries. *Journal of Dairy Science* 91:1687–1692.

Galyean, M. L. and Rivera, J. D. (2003) Nutritionally related disorders affecting feedlot cattle. *Canadian Journal of Animal Science* 83:13–20.

Gaughan, J. B., Kreikemeier, W. M. and Mader, T. L. (2005) Hormonal growth-promotant effects on grain-fed cattle maintained under different environments. *International Journal of Biometeorology* 49:396–402.

Gaughan, J. B., Mader, T. L, Holt, S. M., Sullivan, M. M. and Hahn, G. L. (2010) Assessing the heat tolerance of 17 cattle genotypes. *International Journal of Biometeorology* 54:617–627.

Goad, D. W., Goad, C. L. and Nagaraja, T. G. (1998) Ruminal microbial and fermentative changes associated with experimentally induces subacute acidosis in steers. *Journal of Animal Science* 76:234–241.

Godfrey, R. W., Lunstra, D. D. and Schanbacher, B. D. (1992) Effect of implanting bull calves with testosterone propionate, dihydrotestosterone propionate or oestradiol-17β prepubertally on the pituitary–testicular axis and on postpubertal social and sexual behaviour. *Journal of Reproduction and Fertility* 94:57–69.

Goldhawk, C., Janzen, E., Gonzalez, L. A., Kastelic, J. P., Kehler, C., Ominski, K., Pajor, E. and Schwartzkopf-Genswein, K. S. (2015) Trailer temperature and humidity during winter transport of cattle in Canada and evaluation of indicators used to assess the welfare of cull beef cows before and after transport. *Journal of Animal Science* 92:1542–1554.

Gonyou, H. W., Christopherson, R. J. and Young, B. A. (1979) Effects of cold temperature and winter conditions on some aspects of behaviour of feedlot cattle. *Applied Animal Ethology* 5:113–124.

González, L. A., Correa, L. B., Ferret, A., Manteca, X., Ruíz-de-la-Torre, J. L. and Calsamiglia S. (2009) Intake, water consumption, ruminal fermentation, and stress response of beef heifers fed after different lengths of delays in the daily feed delivery time. *Journal of Animal Science* 87:2709–2718.

González, L. A., Schwartzkopf-Genswein, K. S., Bryan, M., Silasi, R. and Brown, F. (2012a) Relationship between transport conditions and welfare outcomes during commercial long haul transport of cattle in North America. *Journal of Animal Science* 90:3640–3651.

González, L. A., Manteca, X., Calsamiglia, S., Schwartzkopf-Genswein, K. S. and Ferret, A. (2012b). Ruminal acidosis in feedlot cattle: Interplay between feed ingredients, rumen function and feeding behavior (a review). *Animal Feed Science and Technology* 172:66–79.

Goonewardene, L. A. and Hand, R. K. (1991) Studies on dehorning steers in Alberta feedlots. *Canadian Journal of Animal Science* 71:1249–1252.

Grandin, T., (1994). Euthanasia and slaughter of livestock. *Journal American Veterinary Medical Association* 204:1354–1360.

Grandin, T. (1997) Assessment of stress during handling and transport. *Journal of Animal Science* 75:249–257.

Grandin, T. and Johnson, C. (2009) *Animals make us human*. Boston, MA, Houghton Mifflin Harcourt.

Graveland, B. (2016) Cattle rustling alive and well in the Canadian West. *The Canadian Press*. Available at: www.cbc.ca/news/canada/calgary/cattle-rustling-alberta-saskatchewan-still-a-thing-1.3422386.

Green, T. M., Thomson, D. U. Wildeman, B. W., Guichon, P. T. and Reinhardt, C. D. (2012) Time of onset, location, and duration of lameness in beef cattle in a commercial feed yard. Kansas State University. Available at: https://krex.k-state.edu/dspace/handle/2097/13556.

Greenough, P. R. (1997) *Lameness in cattle*. 3rd Edition. New York, USA: W.B. Saunders Company.

Griffin, D., Perino, L. and Hudson, D. (1993) Feedlot lameness. Animal diseases. Neb Guide, Institute of Agriculture and Natural resources, Cooperative Extension University of Nebraska-Lincoln. Available at: www.ianr.unl.edu/pubs/animaldisease/g1159.htm.

Habermehl, N. (1993) Heifer ovariectomy using the Willis spay instrument: Technique, morbidity and mortality. *Canadian Veterinary Journal* 34:664–667.

Haley D. B., Bailey D. W. and Stookey J. M. (2005) The effects of weaning beef calves in two stages on their behavior and growth rate. *Journal of Animal Science* 83:2205–2214.

HAS (Humane Slaughter Association) (2014) Humane killing of livestock using firearms. Wheathampstead, UK. Available at: www.hsa.org.uk/humane-killing-of-livestock-using-firearms-introduction/introduction-2.

Heinrich, A., Duffield, T. F., Lissemore, K. D., Squires, E. J. and Millman, S. T. (2009) The impact of meloxicam on postsurgical stress associated with cautery dehorning. *Journal of Dairy Science* 92:540–547.

Honeyman, M. S., Busby, W. D., Lonergan, S. M., Johnson, A. K., Maxwell, D. L., Harmon, J. D. and Shouse, S. C. (2010) Performance and carcass characteristics of finishing beef cattle managed in a bedded hoop-barn system. *Journal of Animal Science* 88:2797–2801.

Horstman, L. A., Callahan, C. J., Morter, R. L., Amstutz, H. E. (1982) Ovariectomy as a means of abortion and control of estrus in feedlot heifers. *Theriogenology* 17:273–292.

Hötzel, M. J., Ungerfeld, R. and Quintans, G. (2010) Behavioural responses of 6-month old beef calves prevented from suckling: influence of dam's milk yield. *Animal Production Science* 50:909–915.

Irwin, M. R., Melendy, D. R., Amoss, M. S. and Hutcheson, D. P. (1979) Roles of predisposing factors and gonadal hormones in the buller syndrome of feedlot steers. *Journal of the American Veterinary Medical Association* 174:367–370.

Janett, F., Gerig, T., Tschuor, A. C., Amatayakul-Chantler, S., Walker, J., Howard, R., Bollwein, H. and Thun, R. (2012) Vaccination against gonadotropin-releasing factor (GnRF) with Bopriva significantly decreases testicular development, serum testosterone levels and physical activity in pubertal bulls. *Theriogenology* 78:182–188.

Jelinski, M., Fenton, K., Perrett, T., Paetsch, C. (2016) Epidemiology of toe tip necrosis syndrome (TTNS) of North American feedlot cattle. *Canadian Veterinary Journal* 57:829–834.

Jones, H. B., Rivera, J. D., Vann, R. C. and Ward, S. H. (2016) Effects of growth promoting implant strategies on performance of pre- and postweaned beef calves. *Professional Animal Science* 32:74–81.

Karren, D. B., Basarab, J. A., Church, T. L. (1987) The growth and economic performance of preconditioned calves and their dams on the farms and of calves in the feedlot. *Canadian Journal of Animal Science* 67:327–336.

Kendall, P. E., Verkerk, G. A., Webster, J. R. and Tucker, C. B. (2007) Sprinklers and shade cool cows and reduce insect-avoidance behavior in pasture-based dairy systems. *Journal of Dairy Science* 90:3671–3680.

Knowles, T. G. (1999) A review of road transport of cattle. *Veterinary Record* 144:197–201.

Lean, I. J., Thompson, J. M. and Dunshea, F. R. (2014) A meta-analysis of zilpaterol and ractopamine effects on fedlot performance, carcass traits and shear strength of meat in cattle. *PLoS One* 9:5904.

Lensink, J., Boivin, X., Pradel, P., LeNeindre, P. and Veissier, I. (2000a) Reducing veal calves' reactivity to people by providing additional human contact. *Journal of Animal Science* 78:1213–1218.

Lensink, J., Fernandez, X., Boivin, X., Pradel, P., LeNeindre, P. and Veissier, I. (2000b) The impact of gentle contacts on ease of handling, welfare, and growth of calves and on quality of veal meat. *Journal of Animal Science* 78:1219–1226.

Lesmeister, J. L. and Ellington, E. F. (1977) Effect of steroid implants on sexual behavior of beef calves. *Hormones and Behavior* 9:276–280.

Loneragan, G. H., Thomson, D. U. and Scott, H. M. (2014) Increased mortality in groups of cattle administered the b-adrenergic agonists ractopamine hydrochloride and zilpaterol hydrochloride. *PLoS One* 9:1177.

Lorenzen, C. L., Hale, D. S., Griffin, D. B., Savell, J. W., Belk, K. E., Frederick, T. L., Miller, M. F., Montgomery, T. H. and Smith, G. C. (1993) National beef quality audit: Survey of producer-related defects and carcass quality and quantity attributes. *Journal of Animal Science* 71:1495–1502.

Macartney, J. E., Bateman, K. G., Ribble, C. S. (2003) Health performance of feeder calves sold at conventional auctions versus special auctions of vaccinated or conditioned calves in Ontario. *Journal of the American Veterinary Medical Association* 223:677–683.

Marti, S., Devant, M., Amataykul-Chantler, S., Jackson, J. A., Lopez, E., Janzen E. D., and Schwartzkopf-Genswein, K. S. (2015) Effect of anti-gonadotropin-releasing factor vaccine and band castration on indicators of welfare in beef cattle. *Journal of Animal Science* 93:1581–1591.

Marti, S., Janzen, E., Jelinski, M., Dorin, C. Shearer, J. and Schwartzkopf-Genswein, K. S. (2016) Occurrence, characterization and risk factors associated with lameness within Alberta feedlots. Final report to Alberta Livestock and Meat Agency project# 2013R035R.

Marti, S., Meléndez, D. M., Pajor, E. A. Moya, D., Heuston, C. E. M., Gellatly, D., Janzen, E. D. and Schwartzkopf-Genswein, K. S. (2017a) Effect of band and knife castration of beef calves on welfare indicators of pain at three relevant industry ages: II. Chronic pain. *Journal of Animal Science* 95:4367–4380.

Marti, S., Schwartzkopf-Genswein, K. S., Janzen, E. D. Meléndez, D. M. Gellatly, D., and Pajor, E. A. (2017b) Use of topical healing agents on scrotal wounds after surgical castration in weaned beef calves. *Canadian Veterinary Journal* 58:1081–1085.

Marti, S., Wilde, R. E., Moya, D., Heuston, C. E., Brown F., and Schwartzkopf-Genswein, K. S. (2017c) Effect of rest stop duration during long distance transport on welfare indicators of recently weaned beef calves. *Journal of Animal Science* 95:1–9.

Meléndez, D. M., Marti, S., Pajor, E. A., Moya, D., Heuston, C. E. M., Gellatly, D., Janzen, E. D. and Schwartzkopf-Genswein, K. S. (2017a) Effect of band and knife castration of beef calves on welfare indicators of pain at three relevant industry ages: I. Acute pain *Journal of Animal Science* 95:4352–4366.

Meléndez, D. M., Marti, S., Pajor, E. A., Moya, D., Gellatly, D., Janzen, E. D. and Schwartzkopf-Genswein, K. S. (2017b) Effect of subcutaneous Meloxicam on indicators of acute pain and distress after castration and branding in 2 month old beef calves (unpublished).

Mench, J. A., Swanson, J. C. and Stricklin, W. R. (1990). Social stress and dominance among group members after mixing beef cattle. *Canadian Journal of Animal Science* 70:345–354.

Miesner, M. D. and Anderson, D. E. (2015) Surgical management of common disorders of feedlot calves. *Veterinary Clinics of North America: Food Animal Practice* 31:407–424.

Mitlöhner, F. M., Galyean, M. L. and McGlone, J. J. (2002) Shade effects on performance, carcass traits, physiology, and behavior of heat-stressed feedlot heifers. *Journal of Animal Science* 80:2043–2050.

Moggy, M. A., Pajor, E. A., Thurston, W. E., Parker, S., Greter, A. M., Schwartzkopf-Genswein, K. S., Campbell, J. R. and Windeyer, M. C. (2017a) Management practices associated with pain in cattle on western Canadian cow–calf operations: A mixed methods study. *Journal of Animal Science* 95:958–969.

Moggy, M. A., Pajor, E. A., Thurston, W. E., Parker, S., Greter, A. M., Schwartzkopf-Genswein, K. S., Campbell, J. R. and Windeyer, M. C. (2017b) Management practices associated with stress in cattle on western Canadian cow–calf operations: A mixed methods study. *Journal of Animal Science* 95:1836–1844.

Morrison, S. R., Givens, R. L., Garrett, W. N., and Bond, T. E. (1970) Effect of mud-wind-rain on beef cattle performance in feed lot. *California Agriculture* 24:6–7.

Nagaraja, T. G. and Lechtenberg, K. F. (2007) Acidosis in feedlot cattle. *Veterinary Clinics Food Animal Practice* 23:333–350.

NAMI (2008) Cattle transport guidelines for meat packers, feedlots, and ranches. Available at: www.grandin.com/meat.institute.menu.html.

National Farm Animal Care Council (NFACC) (2013) Code of practice for the care and handling of beef cattle. Available at: www.nfacc.ca/pdfs/codes/beef_code_of_practice.pdf.

NCBA. 2014. Beef quality assurance program and animal welfare. National cattlemen's beef association. Available at: www.beefusa.org/uDocs/factsheetbqaandanimalwelfare.pdf.

Nienaber, J. A. and Hahn, G. L. (2007) Livestock production system management responses to thermal challenges. *International Journal of Biometeorology* 52:149–157.

OIE (2014) Chapter 7.5 Slaughter of Animals. In: *Terrestrial animal health code*. World Organization for Animal Health, Paris.

Owens, F. N., Secrist, D. S., Hill, W. J. and Gill, D. R. (1998) Acidosis in cattle: A review. *Journal of Animal Science* 76:275–286.

Penner, G. B., Steele, M. A., Aschenbach, J. R. and McBride B. W. (2011) Molecular adaptation of ruminal epithelia to highly fermentable diets. *Journal of Animal Science* 89:1108–1119.

Petherick, J. C., McCosker, K., Mayer, D. G. Letchford, P. and McGowan, M. (2013) Evaluation of the impacts of spaying by either the dropped ovary technique or ovariectomy via flank laparotomy on the welfare of *Bos indicus* beef heifers and cows. *Journal of Animal Science* 91:382–394.

Powell, J. (2013) Livestock health series: Bovine respiratory disease. Division of Agriculture Fact Sheet FSA3082. University of Arkansas. Fayetteville, AR. Available at: www.uaex.edu/publications/pdf/FSA-3082.pdf.

Price, E. O., Harris, J. E., Borgwardt, R. E., Sween, M. L. and Connor, J. M. (2003) Fenceline contact of beef calves with their dams at weaning reduces the negative effects of separation on behavior and growth rate. *Journal of Animal Science* 81:116–121.

Rademacher, R. D., Warr, B. N., and Booker, C. W. (2015) Management of pregnant heifers in the feedlot. *Veterinary Clinics of North America: Food Animal Practice* 31:209–228.

Radostits, O. (2000) Control and prevention of diseases in feedlot cattle. In: Marx, T. (ed.), *Alberta feedlot management guide*. Edmonton, Alta, Alberta Agriculture Food and Rural Development.

Radunz, A. E. (2010) Use of beta agonists as a growth promoting feed additive for finishing beef cattle State Beef Extension Specialist University of Wisconsin-Madison Fact sheet. Available at: https://fyi.uwex.edu/wbic/files/2010/11/Beta-Agonists-Factsheet.pdf.

Ribble, C. S., Meek, A. H., Shewen, P. E., Guichon, P. T. and Jim, G. K. (1995) Effect of pretransit mixing on fatal fibrinous pneumonia in calves. *Journal of the American Veterinary Medical Association* 207:616–619.

Roeber, D. L., Mies, P. D., Smith, C. D., Belk, K. E., Field, T. G., Tatum, J. D., Scanga, J. A. and Smith, G. C. (2001) National market cow and bull beef quality audit-1999: A survey of producer-related defects in market cows and bulls. *Journal of Animal Science* 79:658–665.

Schwartzkopf-Genswein, K. S. and Stookey, J. M. (1997) The use of infrared thermography to assess inflammation associated with hot-iron and freeze branding in cattle. *Canadian Journal of Animal Science* 77:577–583.

Schwartzkopf-Genswein, K. S., Stookey, J. M., de Passillé, A. M. and Rushen, J. (1997a) Comparison of hot-iron and freeze branding on cortisol levels and pain sensitivity in beef cattle. *Canadian Journal of Animal Science* 77:369–374.

Schwartzkopf-Genswein, K. S., Stookey, J. M. and Welford, R. (1997b) Behavior of cattle during hot-iron and freeze branding and the effects on subsequent handling ease. *Journal of Animal Science* 75:2064–2072.

Schwartzkopf-Genswein, K. S., Stookey, J. M., Janzen, E. D. and McKinnon, J. (1997c) Effects of branding on weight gain, antibiotic treatment rates and subsequent handling ease in feedlot cattle. *Canadian Journal of Animal Science* 77:361–367.

Schwartzkopf-Genswein, K. S., Stookey, J. M., Crowe, T. G. and Genswein, B. M. A. (1998) Comparison of image analysis, exertion force, and behavior measurements for use in the assessment of beef cattle responses to hot-iron and freeze branding. *Journal of Animal Science* 76:972–979.

Schwartzkopf-Genswein, K. S., Beauchemin, K. A., Gibb, D. J., Crews, D. H. Jr., Hickman, D. D., Streeter, M. and McAllister, T. A. (2003) Effect of bunk management on feeding behavior, ruminal acidosis and performance of feedlot cattle: A review. *Journal of Animal Science* 81(2):E149–E158.

Schwartzkopf-Genswein, K. S., Booth, M. E., McAllister, T. A., Mears, G. J., Schaefer, A. L., Cook, N. J., and Church, J. S. (2006) Effects of pre-haul management and transport distance on beef cattle performance and welfare. *Applied Animal Behaviour Science* 108:12–30.

Schwartzkopf-Genswein, K., Stookey, J., Berg, J. Campbell, J. Haley, D., Pajor, E. and I. McKillop, I. (2012a) Code of practice for the care and handling of beef cattle: Review of scientific research on priority issues. National Farm Animal Care Council. Available at: www.nfacc.ca/resources/codes-of-practice/beef-cattle/Beef_Cattle_Review_of_Priority_Welfare_Issues_Nov_2012.pdf.

Schwartzkopf-Genswein, K. S., Faucitano, L., Dadgar, S., Shand, P., González. L. A. and Crowe, T. G. (2012b) Road transport of cattle, swine and poultry in North America and its impact on animal welfare, carcass and meat quality: A review. *Meat Science* 92:227–243.

Schwartzkopf-Genswein, K. S., Ahola, J., Edwards-Callaway, L., Hale, D., and Paterson, J. (2016) Symposia paper: Transportation issues impacting cattle well-being and considerations for the future. *Professional Animal Scientist* 32:707–716.

Sowell, B. F., Mosley, J. C. and Bowman, J. G. P. (1999) Social behavior of grazing beef cattle: Implications for management. *Proceedings American Society of Animal Science*, pp 1–6.

Stackhouse-Lawson, K. R., Tucker, C. B., Calvo-Lorenzo, M. S. and Mitlöehner, F. M. (2015) Effects of growth-promoting technology on feedlot cattle behavior in the 21 days before slaughter. *Applied Animal Behaviour Science* 162:1–8.

Stanger, K. J., Ketheesan, N., Parker, A. J., Coleman, C. J., Lazzaroni, S. M. and Fitzpatrick, L. A. (2005) The effect of transportation on the immune status of bos indicus steers. *Journal of Animal Science* 83:2632–2636.

Stewart, L. (2013) Implanting beef cattle, University of Georgia Cooperative Extension.

Stilwell, G., Carvalho, R. C., Carolino, N., Lima, M. S. and Broom, D. M. (2010) Effect of hot-iron disbudding on behavior and plasma cortisol of calves sedated with xylazine. *Research in Veterinary Science* 88:188–193.

Stokka, G. L., Lechtenberg, K., Edwards, T., MacGregor, S., Voss, K., Griffin, D., Grotelueschen, D. M., Smith, R. A. and Perino, L. J. (2001) Lameness in feedlot cattle. *Veterinary Clinics of North America: Food Animal Practice* 17:189–207.

Sullivan, M. L., Cawdell-Smith, A. J., Mader, T. L. and Gaughan, J. B. (2011) Effect of shade area on performance and welfare of short-fed feedlot cattle. *Journal of Animal Science* 89:2911–2925.

Sutherland, M., Ballou, M., Davis, B., and Brooks, T. (2013) Effect of castration and dehorning singularly or combined on the behavior and physiology of holstein calves. *Journal of Animal Science* 91: 935–942.

Tajima, K., Aminov, R. I., Nagamine, T., Matsui, H., Nakamura, M. and Benno, Y. (2001) Diet-dependent shifts in the bacterial population of the rumen revealed with real-time PCR. *Applied and Environmental Microbiology* 67:2766–2774.

Tarrant, P. V., Kenny, F. J., Harrington, D. and Murphy, M. (1992) Long distance transportation of steers to slaughter: Effect of stocking density on physiology, behaviour and carcass quality. *Livestock Production Science* 30:223–238.

Taylor, L. F., Booker, C. W., Jim, G. K. and Guichon, P. T. (1997) Sickness, mortality and the buller steer syndrome in a western Canadian feedlot. *Australian Veterinary Journal* 75:732–736.

Tennessen, T., Price, M. A. and Berg, R. T. (1985) The social interactions of young bulls and steers after re-grouping. *Applied Animal Behavior Science* 14:37–47.

Tessitore, E., Brscic, M., Boukha, A., Prevedello, P. and Cozzi, G. (2009) Effects of pen floor and class of live weight on behavioural and clinical parameters of beef cattle. *Italian Journal of Animal Science Volume* 8(2):658–660.

Tessitore, E., Schwartzkopf-Genswein, K. S., Cozzi, G., Pajor, E., Goldhawk, C. Brown, F., Janzen, E., Klassen, P. and Dueck, C. (2011) Prevalence of lameness in 3 commercial feedlots in Southern Alberta during summer months. Proceedings of the Canadian Society of Animal Science Halifax NS May 4–5 p 75.

Thomson, D. U., Loneragan, G. H., Henningson, J. N., Ensley, S. and Bawa, B. (2015) Description of a novel fatigue syndrome of finished feedlot cattle following transportation. *Journal of the American Veterinary Medical Association* 247:66–72.

Tibbetts, G. K., Devin, P. A. S., Griffin, D., Keen, J. E. and Rupp, G. P. (2006) Effects of a single foot rot incident on weight performance of feedlot steers. *The Professional Animal Scientist* 22:450–453.

Tucker, C. B., Rogers, A. R., Verkerk, G. A., Kendall, P. E., Webster, J. R. and Matthews, L. R. (2007) Effects of shelter and body condition on the behaviour and physiology of dairy cattle in winter. *Applied Animal Behaviour Science* 105:1–13.

Tucker, C. B., Mintline, E. M., Banuelos, J., Walker, K. A., Hoar, B., Drake, D. and Weary, D. M. (2014) Effect of a cooling gel on pain sensitivity and healing of hot-iron cattle brands. *Journal of Animal Science* 92: 5666–5673.

Tucker, C. B., Coetzee, J. F., Stookey, J., Thomson, D. U., Grandin, T. and Schwartzkopf-Genswein, K. S. (2015) Beef cattle welfare in the USA: Identification of priorities for future research. *Animal Health Research Review* 6:107–124.

Urban-Chmiel, R. and Grooms, D. L. (2012) Prevention and control of bovine respiratory disease. *Journal of Livestock Science* 3:27–36.

USDA (2008a) Part I: Reference of Beef Cow-calf Management Practices in the United States, 2007–08. National Animal Health Monitoring System. Fort Collins CO: USDA. Available at: www.aphis.usda. gov/animal_health/nahms/beefcowcalf/downloads/beef0708/Beef0708_dr_PartI_rev.pdf.

USDA 2011. Feedlot 2011, Part I: Management practices on US feedlots with a capacity of 1,000 or more head, Fort Collins, CO, USDA-APHIS National Health Monitoring System.

USDA (2013) The use of growth-promoting implants in U.S. feedlots. Available at: www.aphis.usda.gov/animal_health/nahms/feedlot/downloads/feedlot2011/Feed11_is_Implant.pdf.

Van Donkersgoed, J., Jewison, G. Bygrove, J. Gillis, K. Malchow, D. and McLeod, G. (2001) Canadian beef quality audit 1998–99. *Canadian Veterinary Journal* 42:121–126.

Voyles, B. L., Brown, M. S., Swingle, R. S. and Karr, K. J. (2004) Case study: Effects of implant programs on buller incidence, feedlot performance, and carcass characteristics of yearling steers. *Professional Animal Scientist* 20:344–352.

Warriss, P. D., Brown, S. N., Knowles, T. G., Kestin, S. C., Edwards, J. E., Dolan, S. K. and Phillips, A. J. (1995) Effects of cattle transport by road for up to 15 h. *Veterinary Record* 136:319–323.

Watts, J.M., and Stookey, J.M., (1999) Effects of restraint and branding on rates and acoustic parameters of vocalization in beef cattle. *Applied Animal Behaviour Science* 62, 125–135.

Webster, J. R., Stewart, M., Rogers, A. R. and Verkerk, G. A. (2008) Assessment of welfare from physiological and behavioural responses of New Zealand dairy cows exposed to cold and wet conditions. *Animal Welfare* 17:19–26.

Widowski, T. M. (2001) Shade-seeking behavior of rotationally-grazed cows and calves in a moderate climate. In: Stowell, R., Bucklin, R. and Bottcher, R., eds. *Livestock Environment VI: Proceedings of the 6th International Symposium*, Louisville, Kentucky, USA. ASAE, 632–639.

Wilson S.C., Fell L.R., Colditz I.G. & Collins D.P. (2002) An examination of some physiological variables for assessing the welfare of beef cattle in feedlots. *Animal Welfare* 11:305–316.

Wilson, B. K., Richards, C. J., Step, D. L. and Krehbiel, C. R. (2017) Beef species symposium: Best management practices for newly weaned calves for improved health and well-being. *Journal of Animal Science* 95:2170–2182.

Woiwode, R., Grandin, T., Kirch, B. and Paterson, J. (2016) Effects of initial handling practices on behavior and average daily gain of fed steers. *International Journal of Livestock Production* 7:12–18.

Woods, J., Shearer, J.K., and Hill, J. (2010) Recommended on-farm euthanasia practices. In: Grandin, T. (Ed.), Improving Animal Welfare: A Practical Approach. CABI Publishing, Wallingford Oxfordshire, UK, pp. 186–213.

Woods, J. and Shearer, J. K. (2015) Recommended On-Farm Euthanasia Practices, 194–221. In: Grandin, T. (ed.), *Improving Animal Welfare: A Practical Approach*, 2nd Edition.

World Organization for Animal Health (OIE) (2010) Glossary. *Terrestrial Animal Health Code 1 (section7.6).* Available at: www.oie.int/index.php?id=169&L=0&htmfile=chapitre_1.7.6.htm.

Youngers, M. E., Thomson, D. U., Schwandt, E. F., Simroth, J. C., Bartle, S. J., Siemens, M. G. and Reinhardt, C. D. (2017) Case study: Prevalence of horns and bruising in feedlot cattle at slaughter. *The Professional Animal Scientist* 33:135–139.

CHAPTER 21

Sustainability, the Environment, and Animal Welfare

Shawn L. Archibeque
Colorado State University

CONTENTS

SUSTAINABILITY AND ENVIRONMENTAL PROTECTION

There has been a great deal of interest in the concept of sustainable food production for many different reasons. The predominant force driving this debate is the rise of the global human population first above 7 billion in 2011 and a projected increase to greater than 9 billion by 2050 (United Nations, 2016). To support this exponential growth, the Food and Agriculture Organization of the United Nations (FAO, 2013) has predicted that demand for fiber, food, and fuel will increase by 60% of current production quantities. Most of this growth is predicted to occur primarily in the less-developed regions of the world. The production of these increased needs will have to be conducted in a "sustainable" manner to avoid the collapse of a given population. Unfortunately, sustainability is what has been characterized as a "wicked problem." As defined by Churchman (1967), a "wicked problem" is problem where all of the underlying problems, limitations, and subjectivity associated with such problems that have no "correct" answer and where one can only do better or worse. With this understanding, and knowing that sustainability will include the preservation of environmental, economic, and social resources, rather than argue for the logic model that properly describes sustainability, for the purposes of this manuscript, we will assume that sustainability will include the ability to optimize nutrient use and minimize waste, while understanding that such a definition is limited. There is an extensive evaluation of the role of animal food versus human food production and the role in sustainability (CAST, 2013) that addresses these issues at great depth.

Animal domestication has been widely practiced around the globe for millennia (Zeuner, 1963; Clutton-Brock, 1999; Price, 2002). As evidenced in the social ethics arguments of Rollin (1992), the domestication of species by humans is dependent on the social contract of animal husbandry where "we take care of the animals and the animals take care of us." Therefore, it is not surprising that throughout history there are numerous examples of how humans have relied on this inherent contract to optimize resource utilization throughout the centuries. In fact, this very concept has allowed for the optimization of food usage by having animals consume foods that humans either did not want to consume or were deemed unfit for human consumption. A prime example of this concept is illustrated by the Great Irish Famine of the mid-1800s. While the famine is widely known as the "potato famine," it is the tangential impacts that the *Phytophthora infestans* caused that may have had as great an impact as the actual loss of the staple crop of potatoes. While there were likely many political issues that also led to the famine (O'Grada, 1995), it is interesting to note that while many food exports increased during the famine, the staples remaining for the working classes were greatly diminished. This included the production of pork and other livestock such as sheep and cattle that were also reliant on the potato as a component of their fodder (Bourke, 1968). Therefore, as there was a loss of the excess potato crop, not only did the Irish lose their staple food, but also much of their associated livestock production as well. This example along with innumerable others throughout history illustrates the dependence of humans upon a successful interaction with live-stock production systems and how management (or mismanagement) may play a role in the ultimate preservation and livelihood of this relationship.

With the global backdrop of growing concern over the rapid increase in the human population, there has been a growing social demand for food production services that help to address the relationship between humans, agricultural production systems and then natural environment. While not necessarily representative of the phenomenon in total, the growth of the Whole Foods Market serves as an example of this growth in concern over how food is produced in general, with a specific focus on environmental impacts. Whole Foods Market (2017) began as a small local market in Austin, TX, in 1980 which encompassed a total of $980\,m^2$. This has grown to become the largest single grocery chain serving this niche market in 2017, with an estimated 91,000 employees with stores in the US, Canada, and the UK. Many of the products and marketing strategies have been focused on products that are labeled as "Organic" or some other labeling program with a focus on defining particular practices that are used in the production of those foods and other products. Many of these labeling programs limit or forbid products or practices that are widely used in conventional livestock production. These may include the use of growth promotants, housing standards for live-stock, and other measures that are perceived to be harmful to either the environment or the welfare of the animals.

Conversely, there is a growing consensus among the scientific community that there are indeed some very real and objective measures that would indicate that much of the perceived "sustainability" of these niche programs may not hold up to deeper scrutiny (Capper et al., 2009, 2011; FAO, 2013). Many of these comparisons between conventional and niche livestock production systems focus on the advantages offered by conventional practices that optimize the use of approved technologies and management techniques for the production of livestock. However, many of these comparisons do not consider the additional cost borne by the animals for the sake of these efficiencies. For example, it had been a common practice in the swine industry to house gestating sows in crates that only allow the animal to stand up (sometimes not even fully) or lay down. While this crating method would undoubtedly make it easier for individual management of animals in a large scale production system, it does not take into account the pain or suffering that may be incurred by the animal in this highly efficient system. Not surprisingly many people find these compromises for production efficiency to be unacceptable and there have been growing efforts to improve many of the more egregious animal housing conditions such as veal crates, battery cages for hens and gestation crates for swine. Unfortunately, there is limited published data regarding life cycle assessments that truly determines if these practices actually do reduce resource use and efficiency of production when incorporating all of the costs, including population turnover of animals kept in these types of conditions or the infrastructure resources of steel, concrete, etc., required to build these types of structures. All of these considerations and more would have to be considered to allow for a "fair" comparison of the myriad methods that can and are used to raise livestock.

Needless to say, the reasons mentioned above and many more make it particularly difficult to describe what practices truly reduce environmental impact, increase sustainability and ultimately what the impacts these practices ultimately have upon the welfare of the animals produced for human consumption. While it is my personal opinion that we are trying to reach a target that we cannot see or describe adequately, there are some practices that have been fairly well accepted as being true for optimizing resource use and general preservation of the natural resources available to us for producing livestock based products. Most of these practices would be focused on optimizing the efficiency of nutrient use, thus minimizing loss of these nutrients into the environment where they may become concentrated and cause undesirable environmental effects. It is critical to note that the emphasis here is on "optimizing," not maximizing efficiency. It is possible to sometimes increase the efficiency, but as in most biological systems, there are some strains that may limit longevity and thus lead to a greater inefficiency over the long term. For example, if a producer could double the milk production from a cow using the same or similar amounts of resources, that would obviously be beneficial. However, if doubling the milk production of a cow required the producer to replace that cow after only one or two lactations with another cow that would require at least 2 years of development to become a suitable replacement, it becomes readily apparent that it is the best interest of sustainable use of resources to spread the resources used in the cows development (i.e., before they can give milk) over as many lactations possible. Therefore, in the current example it may be most beneficial to sacrifice the doubling of milk production in the current lactation so that the cow may be kept around for many more lactations.

One of the niche production areas that has received a large amount of attention for the potential to provide a form of sustainable food production is the Organic sector. On the cover of the Organic Trade Organization webpage (2018), it is stated that the program will "protect the environment." In addition, in their most recent survey of U.S. Families Organic Attitudes and Behaviors study (OTO, 2018), "concerns about the effects of pesticides, hormones and antibiotics" were listed as the top reasons for persons to support organic production. It is also a prominent topic that there is a strong animal welfare component associated with the production of Organic livestock. It is highlighted by OTO (2018) that "(The OTO) are standing up on behalf of the entire organic sector to protect organic integrity, advance animal welfare, and demand the government keep up with the industry and the consumer in setting organic standards." In a lawsuit against the USDA, the OTO also states that "The viability of the organic market rests on consumer trust in the USDA Organic seal, and trust that the organic seal represents a meaningful differentiation from other agricultural practices." To further examine this issue, one needs to consider what it is that differentiates the production of livestock under Organic standards from that of conventional livestock production standards. The standards are readily available via the website of the Code of Federal Regulations (eCFR, 2018). For the most part, the regulations are focused on aspects of animal care that include housing requirements and restriction of the use of many pharmaceutical agents that may be used in commercial livestock operations.

First we will consider the issue of housing, wherein the regulations clearly dictate a required amount of indoor and outdoor space required for differing species. These spatial requirements may also include "access" to areas with greater space or that may be outside for livestock that are primarily grown indoors or in pens as opposed to animals that predominantly spend time in larger enclosures such as pastures. The standard also indicates that "Year-round access for all animals to the outdoors, shade, shelter, exercise areas, fresh air, clean water for drinking, and direct sunlight, suitable to the species, its stage of life, the climate, and the environment." There are numerous caveats to this requirement in the standard to allow the producer to limit or suspend access as determined to avoid inclement weather, provide particular care for stage of life, etc. This is in contrast to many forms of conventional livestock production that will typically optimize the use of all space available, and in some species may lead to overcrowding and subsequent deterioration of animal welfare that accompanies not having appropriate space for normal growth and behaviors. These space issues seem to be most prevalent in the poultry and swine industries, which have the greatest extent of business vertical integration and emphasis on improving efficiency of production, thus limiting environmental impacts. While having too many animals in too small a space may compromise the health of individual animals and would not be tolerable to most individual's morality, providing additional space will not necessarily alleviate these issues in and of itself. In fact, if animals are given appropriate amount of space to avoid overcrowding and allow for natural behaviors, it is arguable that it may be in their best interest to remain indoors for the majority of the production cycle. This would allow for the reduction of certain vectors that may introduce disease, predation or potential loss or injury of animals which may escape from more extensive, less intensive confinements. By keeping animals in enclosures for longer periods of the year (if they are appropriately built to provide adequate space) we would allow animals to be produced in a wider variety of climates and locales. With the wider distribution of production, animals would not have to be transported as far to reach intended markets, which may also reduce impacts on animal welfare that inevitably arise when they are transported. Also, just because space is provided does not mean that the animals will use that additional space if there are factors that may inhibit them from using the space.

While there are many examples of how this may manifest, a few examples may be that animals are fed high-quality feed in an environmentally controlled enclosure with access to the outdoors that has a less appetizing array of feed and environment for the animal. This would preclude it from typically using the outdoors that it has access too. Additionally, if the livestock perceive a threat in the outdoor environment (i.e., raptors roosting near the outdoor areas of a poultry facility), they would tend to remain within the enclosure, which may lead to overcrowding and a subsequent decrease in animal welfare. While the intent of this regulation may hold true, the issue of space is not as simple as the standard attempts to make it and in fact does not guarantee an improved aspect of animal welfare, but it may provide the perception of such to the consuming public. As with most issues regarding the production of livestock under any labeling standard, while the regulations may be well intended, it is ultimately up to the husbandry of the livestock producer to provide proper animal welfare and animal welfare may thrive or fail under any livestock production system, Organic, conventional or otherwise.

In addition to concerns over the housing of livestock production, the Organic standard clearly limits the use of many pharmaceutical agents and states that "the producer of an organic livestock operation must not:

1. Sell, label, or represent as organic any animal or edible product derived from any animal treated with antibiotics.
2. Administer any animal drug, other than vaccinations, in the absence of illness;
3. Administer hormones for growth promotion;
4. Administer synthetic parasiticides on a routine basis;
5. Administer synthetic parasiticides to slaughter stock;
6. Administer animal drugs in violation of the Federal Food, Drug, and Cosmetic Act; or
7. Withhold medical treatment from a sick animal in an effort to preserve its organic status. All appropriate medications must be used to restore an animal to health when methods acceptable to organic production fail. Livestock treated with a prohibited substance must be clearly identified and shall not be sold, labeled, or represented as organically produced. (CFR, 2018)."

While all of these practices have been clearly shown to have dramatic impacts on the treatment of disease in livestock, they are banned from being used in Organic production for perceived benefits that are then marketed to consumer desires. However, this raises a particularly poignant concern over the treatment of animals that are known to be in a compromised state. In particular, the final point of the standard would indicate that other agents and methods of disease treatment could and would be used to preserve the organic status of an animal while a known treatment that actually works would not be used except as a last option. This means that an animal raised under the Organic standard would have a vastly increased time spent in a diseased state which would obviously be a detriment to the animal's welfare. Also, it has long been recognized that diseased animals are less productive as they are using biological resources to alleviate the disease (i.e., immune response) that would reduce efficiency and thus increase environmental burden. Also, these bans would increase the likelihood of parasite infestations, particularly in warmer climates that would further reduce animal productivity and welfare.

There are numerous practices that have environmental implications and in the interest of brevity, this chapter will largely focus on one of the most poorly understood products used that has substantial impacts on the animals and their potential environmental impact. These are the hormonal-based growth promotants. In light of the previous discussion regarding growth implants being banned from use in certain labeling programs, it is worth investigating this particular issue more thoroughly.

HORMONAL GROWTH PROMOTANTS

Throughout history, hormones and their metabolites that are naturally (endogenously) produced by animal and human populations have been reaching the environment. However, the quantity and concentration of hormones and their metabolites within a localized area that are excreted into the environment are increasing as populations grow. Furthermore, as livestock production becomes more concentrated there has been a recent increase in the interest surrounding the hormonal disrupting activity of compounds from both natural and anthropogenic sources. Several experiments have reported adverse impacts of steroid hormones in the environment (Tyler et al., 1998; de Voogt et al., 2003; Jobling and Tyler, 2003; Kolpin et al., 2002; Kidd et al., 2007). However, to date there have been few controlled experiments specifically addressing this issue.

While there is an increased awareness of environmental impacts associated with potential endocrine disruptor residues from livestock operations, the literature regarding this area of study is fraught with multiple deficiencies in the published knowledge about the subject. With such an increase in the demand for such knowledge, the Environmental Protection Agency (EPA) has funded several studies to address the source and fate of these potential endocrine disruptors. While there are numerous field level studies that address the issues, many are limited by lack of proper replication of experimental units, analytical difficulties, and inadequate controls of influential environmental factors. However, this area of research is rapidly growing and the knowledge of the role of potential endocrine disruptors released from livestock operations into the environment is increasing.

The U.S. Food and Drug Administration (FDA) first approved the use of hormone implants containing estradiol benzoate/progesterone in 1956 for increasing growth, feed efficiency, and carcass leanness of cattle. Subsequent implants containing testosterone, trenbolone acetate, zeranol, and a myriad of combinations of these hormones were later developed and approved for use in cattle by FDA. Currently, there are five hormones/xenobiotics (progesterone, testosterone, estradiol-17-β, zeranol, and trenbolone acetate) that have been approved for implants in cattle in the U.S. (Center for Veterinary Medicine, 1986, 1996, 1998, 1999, 2001, 2002, 2005a, 2005b). While there are additional growth promotants which may have unintended endocrine disruptive effects, the focus of this paper will be on endogenously produced sex hormones and hormones/ hormone analogs from subcutaneous growth implants.

HUMAN METABOLISM OF HORMONES/TOXICOLOGICAL IMPLICATIONS

Estradiol, progesterone, and testosterone are naturally occurring hormones that are present in humans, although the amounts vary with age, sex, diet, exercise, and with pregnancy and stage of the menstrual cycle in females. Additionally, humans may ingest steroid hormones as hormone replacement therapies, for contraception, or via contaminated drinking or food supplies. The systemic presence of these hormones may be as unbound compounds or attached to hormone binding proteins. Currently, the endocrine disrupting compound to receive the greatest amount of attention is estradiol-17-β. Estradiol-17-β is present in relatively high concentrations in newborn males and females, but concentrations drop rapidly after parturition. In adult males, estradiol (primarily in the bound form) concentrations are typically in the range of 20–40 pg/ml serum (Rubens and Vermeulen, 1983; Raben et al., 1992). In adult females, serum estradiol concentrations will vary between 40 and 400 pg/ml over the course of 28 day (Rubens and Vermeulen, 1983).

Children and the fetus in utero are considered at greater risk from exposure to hormones because their normal physiological hormone concentrations are much lower than adults. For example, a study using a radioimmunoassay indicated that estradiol concentrations in prepubertal boys and girls were 2.6 and 4.5 pg/ml, respectively (Potau et al., 1999). However, other researchers reported serum estradiol concentrations of 0.08 pg/ml in boys and 0.6 pg/ml in girls using a recombinant yeast assay (Klein et al., 1994). However, it should be noted that inter- and intra-assay variation of these very low estradiol concentrations was reported to be 50%–60% (Klein et al., 1994).

It should also be noted that oral intake of estrogens generally results in very poor bioavailability due to extensive metabolism after absorption from the gut. For example, a micronized estradiol dose was found to be only 0.1%–12% bioavailable and hormone replacement therapy with a daily dose of 0.625 mg of conjugated estrogens produces serum estradiol concentrations of approximately 40 pg/ml in postmenopausal women (O'Connell, 1995).

Although hormones are essential for various normal physiological processes in the body, excessive amounts may have detrimental effects. The most controversial and best documented is the effects of estradiol. Estradiol stimulates cell division in hormonally sensitive tissues, which may increase the possibility for accumulation of random errors during DNA replication. This increase in cell proliferation has also been documented to stimulate the growth of mutant cells (Henderson and Feigelson, 2000). While the literature in this area is too extensive to adequately cite in this manuscript, two epidemiological studies that may be used as an introduction are Ross et al. (2000) and Schairer et al. (2000).

SEX STEROIDS IN FARM ANIMALS (METABOLISM, EXCRETION, AND ENVIRONMENTAL IMPLICATIONS)

As previously mentioned, regulation of behavior, morphogenesis, and functional differentiation of the reproductive system are governed by sex steroids throughout the life cycle. The major endogenous sources of estrogens are the granulose cells of the ovarian follicles and placental tissues in female and the testis in males. Androgens are primarily derived from the Leydig cells of the male testes, but are also produced by the female ovary and the adrenal cortex. The gestagens (including progesterone) are synthesized by the corpus lutea, the placenta, and the adrenal cortex. After conjugation, hydroxylation, reduction, oxidation, or, to a minor extent, without metabolism, steroids are excreted via bile and feces, or in urine, primarily as water-soluble glucuronides or sulfates (Lange et al. (2002). According to Erb et al. (1977), metabolism and removal of steroids by the mammary gland is of minor importance.

Lange et al. (2002) provided a very thorough summarization of naturally excreted sex-steroid metabolites from a variety of livestock, including cattle, swine, and poultry. In this review, it was estimated that total daily excretion of estrogens from calves was 45 µg, 299 µg from cycling cows, and 540 µg from bulls. Total sex hormone excretions as summarized by Lange et al. (2002).

Pharmaceuticals often have similar physico-chemical behavior to endogenously synthesized steroids, for example, are lipophilic, in order to be able to pass through membranes and are persistent in order to achieve an effect. These properties may lead to bioaccumulation and provoke effects in the aquatic and terrestrial ecosystems (Halling-Sorensen et al., 1998). Estradiol is one of the main female sex hormones and the structural backbone for the engineering of some synthetic estrogens such as 17α-ethynyl estradiol utilized in human hormone treatments including birth control pills (Kuster et al., 2004). In addition, three potent synthetic chemicals with estrogenic (zeranol), gestagenic (melengestrol acetate), and androgenic (trenbolone acetate) action are widely used for growth promotion in cattle; however, these sex hormones and their metabolites do not occur naturally in animals and humans (Andersson and Skakkebaek, 1999). Table 21.1 presents a summation of the concentrations of hormones and hormone analogs in several popular implants. Additionally, USDA (2000) summarized that across all operations melengestrol acetate was fed to 78.8% of all female cattle within feedlots. While Randel et al. (1973) indicated that feeding melengestrol acetate to dairy heifers decreased estradiol excretion, there are very limited data to quantify this decrease in beef cattle. The data for this exercise used the data from USDA (2000) rather than more recent publications due to the greater degree of detail presented in the 2000 report relevant to this exercise.

To the best of our knowledge, the most thorough survey of implant usage and type is the Baseline Reference of Feedlot Management Practices, 1999 (USDA, 2000). Although practices within the industry have likely changed over the past 18 years, these data will provide a justifiable reference that will be somewhat representative of current industry practices. It is important to note that unlikely to be wholly accurate, the following summations are provided to allow a relative estimate of the quantity and type of implants in use.

To help provide a relative quantitation of hormones that are provided to U.S. feedlot cattle, a series of calculations was performed. Initially, the theoretical number of steers placed in feedlots within the weight categories of less than 700 lbs or greater than 700 lbs live weight at placement that received one, two, or three implants was derived by multiplying USDA (2017) summation of cattle within each of these weight classes by the proportion of cattle within all operation to have had 0, 1, 2, or 3+ implants. A similar operation was performed using placement data from USDA (2017) and the proportions of cattle to receive either anabolic or estrogen type implants, within categorization of implant number to generate a theoretical number of cattle to receive either an anabolic or estrogen type implant. The same procedure was also conducted with cattle implanted 2, and 3+ times, although these populations were, respectively, doubled or tripled to reflect the increased number of implants used (data not shown). For this calculation, it was assumed that the number of cattle receiving four implants was likely negligible. Then, for each subcategory of implantation number, an average concentration of growth promotant hormones within major androgenic and estrogenic implants (Table 21.1) was multiplied by the number of implanted cattle calculated in the previous step to determine the quantity of each growth promotant hormone provided within cattle given one, two, or three implants. The summation of these calculations is provided in Table 21.2.

Thus, estradiol and zeranol from implants provide only 1.54% of the estrogenic compounds that could potentially be excreted from the entire U.S. cattle herd, while progesterone from implants is 0.31% of the gestagens that would be produced by the entire U.S. cattle herd. However, testosterone and trenbolone acetate from implants could contribute 66.42% of the total androgenic hormones produced by the U.S. cattle herd. As indicated before, there are many obvious errors with this method; it assumes that all represented implants are used equally, it does not account for the almost complete cessation of sex-steroid hormone production during the extensive amount of time that dairy cattle spend in lactation, and it assumes that 100% of implanted hormonal

Table 21.1   Quantity of Hormones in Selected Growth Promoting Implants with Approval for Use in Feedlot Heifers and/or Steers

Implant	Estradiol (mg)[a]	Progesterone (mg)	Testosterone (mg)	Trenbolone Acetate (mg)	Zeranol (mg)	Anabolic Effect (days)
Synovex-C	7.2	100	0	0	0	120
Compudose	25.7	0	0	0	0	168
Encore	43.9	0	0	0	0	336
Synovex-S	14.4	200	0	0	0	120
Component-ES	14.4	200	0	0	0	120
Synovex-H	14.4	0	200	0	0	120
Component-EH	14.4	0	200	0	0	120
Ralgro	0	0	0	0	36	70
Ralgro magnum	0	0	0	0	72	70
Finaplix-H	0	0	0	200	0	105
Component TH	0	0	0	200	0	105
Revalor-IS	16	0	0	80	0	120
Component TE-IS	16	0	0	80	0	120
Synovex T80	16	0	0	80	0	120
Revalor-IH	8	0	0	80	0	120
Component TE-IH	8	0	0	80	0	120
Synovex-choice	10	0	0	100	0	120
Revalor-S	24	0	0	120	0	120
Component TES	24	0	0	120	0	120
Synovex T120	24	0	0	120	0	120
Revalor-H	14	0	0	140	0	120
Component TEH	14	0	0	140	0	120
Revalor-200	20	0	0	200	0	120
Synovex-plus	20	0	0	200	0	120
Average of implants with trenbolone acetate	14	0	0	129	0	—
Average of implants without trenbolone acetate	15	56	44	0	12	

[a] Represents quantity of actual estradiol (i.e., not estradiol benzoate).

Table 21.2   Theoretical Input of All Hormones into Steers and Heifers that Are Implanted with Growth Promotants Once, Twice, or Three Times[a] within U.S. Feedlot Operations using USDA NASS (2006) and USDA (2000) Data and Average Hormone Concentrations in Implants

	Hormone Type and Weight at Placement				
Number of Times Implanted	Estradiol (kg)[a]	Progesterone (kg)	Testosterone (kg)	Trenbolone Acetate (kg)	Zeranol (kg)
1	147	249	199	727	54
2	353	467	373	2,065	101
3	32	48	38	178	10
Total	533	764	611	2,970	165

[a] Assumes that the number of cattle implanted four or more times is negligible.
[b] An androgenic implant (trenbolone acetate containing product) alone or in combination with other growth promotants.
[c] An estrogenic implant containing estrogen, estrogen-like progesterone, testosterone, or a combination of these growth promotants.

material is excreted from the animal unchanged, etc. This "worst case" scenario intentionally magnifies the potential influence of natural versus anthropogenic sources of potential endocrine disruptors that are released from the U.S. cattle herd, and is not to be used as a "true" estimate of hormone release from livestock operations. By using this approach, it becomes much more apparent where greater research is needed to remediate potential deleterious environmental effects from implanted animals. However, the concentrated animal-feeding operations (CAFOs) still need to consider the problems associated with the high concentrations of natural hormones produced at these installations.

Environmental contaminants that adversely affect reproduction and development through alterations in endocrine functions in humans and wildlife have been identified as an issue of global concern (World Health Organization, 2002). Fish clearly have been affected by endocrine disrupting substances in the environment (Tyler et al., 1998; Thorpe et al., 2001; Jobling and Tyler, 2003; Woodling et al., 2006). For example, masculinization of fish exposed to discharges from pulp and paper mill effluents and runoff from beef feedlots has been reported and been associated with in vitro androgenic activity of water samples from affected sites (Larsson et al., 2000; Jenkins et al., 2003; Orlando et al., 2004; Larsson et al., 2006). Different lines of evidence suggest that steroidal chemicals could contribute to androgenic activity of feedlot discharges. Jensen et al. (2006) and Kidd et al. (2007) demonstrated the collapse of a fish population in an isolated lake system treated with a synthetic estradiol.

Trenbolone acetate is being released from ear implants to the blood where it is hydrolyzed to 17β-trenbolone and 17α-trenbolone whereafter they are being excreted through feces and urine (Schiffer et al., 2001). 17β-trenbolone and 17α-trenbolone are potent androgen receptor agonists in mammals and cause decreased fecundity (egg production) in fish (Wilson et al., 2004; Jensen et al., 2006), and both of the metabolites have long half-lives of about 260 days in liquid manure and can, therefore, lead to a potential ecological risk if discharged from feedlots (Schiffer et al., 2001; Jensen et al., 2006). However, it should be noted that the majority of these studies have occurred in an uncontrolled setting with the potential for multiple inputs of these potential endocrine disruptors. However, while there is some literature on the environmental influence of trenbolone acetate, it is clear from the data presented in this paper that while trenbolone acetate may not be considered by most to be as potent an endocrine disruptor as the estradiols, the proportion of androgenic hormones released from U.S. cattle herd that is associated with trenbolone acetate from implants could be vastly impacted by altering their use by feedlots. However, the "worst case scenario" presented in this manuscript also does not take into account the anecdotal information that would indicate that the U.S. cattle industry is trending toward less-aggressive implants (i.e., less trenbolone acetate), primarily to relieve certain behavioral and carcass issues.

Several studies provide evidence that testosterone and estrogens are strongly bound to soils (Hanselman et al., 2003; Lee et al., 2003; Das et al., 2004; Casey et al., 2005), and thus likely to accumulate in soils and in river sediments (Kuster et al., 2004). Hormone sorption has been correlated to soil particle size and organic matter content, although testosterone is not as strongly bound to soil as 17β-estradiol (Lange et al., 2002; Casey et al., 2004). It has been shown that testosterone is degraded in agricultural soils in conditions simulating a temperate growing season (Lorenzen et al., 2000). Shore and Shemesh (2003) measured testosterone in ground water but stated that estrogen remains in the topsoil due to its strong sorption potential.

Despite the quantity of literature indicating the strong sorption of hormones to soils, the presence of hormones in runoff and leachate from agricultural lands has also been well documented (Kjar et al., 2007; Kolodziej and Sedlak, 2007). Trenbolone acetate has been detected in beef feedlot runoff (Durhan et al., 2006; Jensen et al., 2006), testosterone and 17β-estradiol have been detected in runoff from poultry litter surface-applied to pastures (Finlay-Moore et al., 2000; Nichols et al., 1997), and estradiol has been found in ponds impacted by runoff from beef cattle on pastures (Cole et al., 1979). Contamination of water resources may occur through runoff and/or leaching of hormones in manure applied to agricultural fields and grazing cattle (Kuster et al., 2004; Kjar et al., 2007; Kolodziej and Sedlak, 2007). However, little is known about transport paths and mobility of hormones from livestock facilities and land application sites to water. However, under the 2005 final rule by the EPA, and current Nutrient Management Plan regulations, it is unlikely that feedlot runoff would be capable of directly reaching water resources, though the risk of leaching cannot be eliminated. However, the chance greatly increases as manure from these operations is applied to crop and pasture lands.

Currently, the degradation pathways of hormones are not clearly defined. The fate of estrogen conjugates is not well known, but it is often assumed that common fecal microorganisms such as *Escherichia coli* are capable of hydrolyzing them via glucuronidase and sulfatase enzymes to unconjugated forms (Belfroid, 1999); however, it is questionable if this assumption is valid for estrogen sulfates since they are often observed in sewage treatment works (Ternes et al., 1999a; Ternes et al., 1999b). Limited research has evaluated the stability of conjugated estrogens in manure (Hanselman et al., 2003). Degradation studies of unconjugated estrogens in soil, water, and manure have been conducted for several years, and the literature was recently reviewed by Hanselman et al. (2003). Recently, Jones et al. (2007) demonstrated photolysis of 17β-estradiol, testosterone, and progesterone by light in the UVA range (305–410 nm) in a phosphate-buffered media at pH 5.5. Additionally, it was indicated that both progesterone and testosterone were directly photolyzed, while 17β-estradiol was indirectly photolyzed in the presence of organic matter. The effectiveness of a lagoon-constructed wetland treatment system for producing an effluent with a low hormonal activity was recently investigated. Shappell et al. (2007) found that the nutrient removals were typical for treatment wetlands: TKN 59%–75% and orthophosphate 0%–18%. Wetlands decreased estrogenic activity by 83%–93% in the swine wastewater and estrone was found to be the most persistent estrogenic compound. Constructed wetlands produced effluents with estrogenic activity below the lowest equivalent E_2 (17β-estradiol) concentration known to have an effect on fish (10 ng/l) (Shappell et al., 2007).

HORMONES AND NUTRIENT USE

Producers are using hormonal growth promotants to stimulate improved performance in cattle. Typically, this will include an improved lean tissue deposition with a concomitant decrease in fat tissue deposition. This alteration in tissue deposition is also typically accompanied by improved gain:feed by 10%–15% while also increasing intake (Rumsey et al., 1981; Rumsey, 1982; Rumsey, 1985; Rumsey and Hammond, 1990; Rumsey et al., 1999). Additionally, by shifting growth to more lean tissue, which is rich in nitrogenous compounds, there is also a decrease in urinary N (Cecava and Hancock, 1994). Most of these changes in nutrient excretion from animals treated with growth promoting hormones are post-absorptive in nature and the ultimate decrease in urinary N excretion may range from 1.67 to 10.43 g N/ animal/ day (Lobley et al., 1985; Rumsey and Hammond, 1990; Cecava and Hancock, 1994; Lawrence and Ibarburu, 2006). Since most of the urinary N is present as urea, it may be readily degraded to NH_3. In a closed chamber system, 14%–15% of total manure N and 2%–37% of urinary N volatilized over a 7-day period (Cole et al., 2005; Archibeque et al., 2007). Due to the reactive chemical nature of NH_3, this represents a nutrient loss from an operation in a form that can no longer be managed in open air systems and may then subsequently react and cause changes in whatever ecosystem it may later deposit. To help put this environmental benefit into perspective, if we consider the mean reduction of urinary N to be 6.05 g/animal/day and that there are approximately 30 million cattle that will be fed in feedlots each year in the U.S., by implanting 95% of these animals, then approximately 172 tons of N would not be released into the environment each year, making this a truly substantial impact.

TAKE-HOME MESSAGE

As with many subjects, there are many factors to consider when discussing environmental inter-actions and animal welfare associated with livestock production. It is apparent that, in the short term, the implementation of new management techniques and new technologies improve animal productivity and can reduce environmental impacts of animal production. However, reduced long-term productivity, the lack of social acceptability, and the unknown long-term impacts of the use of new technologies to improve animal production efficiency may limit their use in the promotion of sustainable food production. Therefore, producers will need to balance the welfare of the animal along with the improvements in environmental impacts (both intended and potentially unintended) to determine the "optimum" management strategies.

REFERENCES

Andersson, A. M., and N. E. Skakkebaek. 1999. Exposure to exogenous estrogens in food: Possible impact on human development and health. *European Journal of Endocrinology* 140: 477–485.

Archibeque, S. L., H. C. Freetly, N. A. Cole, and C. L. Ferrell. 2007. The influence of oscillating dietary pro-tein concentrations on finishing cattle. II. Nutrient retention and ammonia emissions. *Journal of Animal Science* 85:1496–1503.

Belfroid, A. C. 1999. Analysis and occurrence of estrogenic hormones and their glucuronides in surface water and waste water in the Netherlands. *Science of the Total Environment* 225: 101–108.

Bourke, P. M. A. 1968. The use of the potato crop in pre-famine Ireland. *Journal of the Statistical and Social Inquiry Society of Ireland.* 12:72–96.

Capper, J. L., R. A. Cady, and D. E. Bauman. 2009. The environmental impact of dairy production:1944 com-pared with 2007. *Journal of Animal Science* 87:2160–2167.

Capper, J. L. 2011. The environmental impact of beef production in the United States: 1977 compared with 2007. *Journal of Animal Science* 89:4249–4261.

Casey, F. X. M., H. Hakk, J. Simunek, and G. L. Larsen. 2004. Fate and transport of testosterone in agricul-tural soils. *Environmental Science & Technology* 38: 790–798.

Casey, F. X. M., J. Simunek, J. Lee, G. L. Larsen, and H. Hakk. 2005. Sorption, mobility, and transformation of estrogenic hormones in natural soil. *Journal of Environmental Quality* 34: 1372–1379.

Cecava, M. J., and D. L. Hancock. 1994. Effects of anabolic steroids on nitrogen metabolism and growth of steers fed corn silage and corn-based diets supplemented with urea or combinations of soybean meal and feather meal. *Journal of Animal Science* 72: 515–522.

Center for Veterinary Medicine, Food and Drug Administration. Summary of NADA 138–612: Finaplix® (trenbolone acetate). 1986. www.fda.gov/cvm/FOI/736.htm.

Center for Veterinary Medicine, Food and Drug Administration. Summary of NADA 140–897: Revalor® G (trenbolone acetate and estradiol). 1996. www.fda.gov/cvm/FOI/1382.htm.

Center for Veterinary Medicine, Food and Drug Administration. Summary of NADA 009–576: Synovex® C and Synovex® S (estradiol benzoate and progesterone). 1998. www.fda.gov/cvm/FOI/498.htm.

Center for Veterinary Medicine, Food and Drug Administration. Summary of NADA 011–427: Synovex-H® (estradiol benzoate and progesterone). 1999. www.fda.gov/cvm/FOI/504.htm.

Center for Veterinary Medicine, Food and Drug Administration. Summary of NADA 140–992: Revalor® H (trenbolone acetate and estradiol). 2001. www.fda.gov/cvm/FOI/140-992.pdf.

Center for Veterinary Medicine, Food and Drug Administration. Summary of NADA 141–043: Synovex® Plus (trenbolone acetate and estradiol). 2002. www.fda.gov/cvm/FOI/141-043s100302.pdf.

Center for Veterinary Medicine, Food and Drug Administration. Summary of NADA 038–233: RALGRO® (zeranol). 2005a. www.fda.gov/cvm/FOI/038-233s030405.pdf.

Center for Veterinary Medicine, Food and Drug Administration. Summary of NADA 110–233: Component® ES and Component® EC (progesterone and estradiol benzoate). 2005b. www.fda.gov/cvm/FOI/110-315.pdf.

Churchman, C. W. 1967. Wicked problems. *Management Science* 14: B141–B142.

Clutton-Brock, J. 1999. *A natural history of domesticated mammals. Natural History Museum.* Cambridge University Press. London, England.

Code of Federal Regulations. 2018. Available at: www.ecfr.gov/cgi-bin/text-idx?c=ecfr&SID=9874504b6f102 5eb0e6b67cadf9d3b40&rgn=div6&view=text&node=7:3.1.1.9.32.7&idno=7#se7.3.205_1603. Accessed 1/2/2018.

Cole, N. A., R. N. Clark, R. W. Todd, C. R. Richardson, A. Gueye, L. W. Greene, and K. McBride. 2005. Influence of dietary crude protein concentration and source on potential ammonia emissions from beef cattle manure. *Journal of Animal Science* 83:722–731.

Cole, N. A., J. B. McLaren, and M. R. Irwin. 1979. Influence of pretransit feeding regimen and posttransit b-vitamin supplementation on stressed feeder steers. *Journal of Animal Science* 49: 310–317.

Council for Agricultural Science and Technology (CAST). 2013. Animal feed vs human food: Challenges and opportunities in sustaining animal agriculture toward 2050. Issue Paper 53. CAST, Ames Iowa.

Das, B. S., L. S. Lee, P. S. C. Rao, and R. P. Hultgren. 2004. Sorption and degradation of steroid hormones in soils during transport: Column studies and model evaluation. *Environmental Science & Technology* 38: 1460–1470.

De Voogt, P., B. Halling-Sorensen, B. van Hattum, P. T. Holland, F. Ingerslev, A. Johnson, M. Jurgens, A. Katayama, W. Klein, N. Kurihara, J. C. Leblanc, K. D. Racke, T. Sanderson, M. Shemesh, L. S. Shore, E. Vaclavik, M. van den Berg, and P. Verger. 2003. Environmental fate and metabolism: Issues and recommendations. *Pure and Applied Chemistry* 75: 1949–1953.

Durhan, E. J. et al. 2006. Identification of metabolites of trenbolone acetate in androgenic runoff from a beef feedlot. *Environmental Health Perspectives* 114: 65–68.

Erb R. E., B. P. Chew, H. F. Keller and P. V. Malven. 1977. Effect of hormonal treatments prior to lactation on hormones in blood plasma, milk, and urine during early lactation. *Journal of Dairy Science* 60: 557–565.

Finlay-Moore, O., P. G. Hartel, and M. L. Cabrera. 2000. 17 Beta-estradiol and testosterone in soil and runoff from grasslands amended with broiler litter. *Journal of Environmental Quality* 29: 1604–1611.

Food and Agriculture Organization of the United Nations (FAO). 2013. *World agriculture towards 2030/2050: The 2012 revision.* FAO, Rome.

Halling-Sørensen, B., S. N. Nielsen, P. F. Lanzky, F. Ingerslev, H. C. H. Lützhøft, and S. E. Jørgensen. 1998. Occurrence, fate and effects of pharmaceutical substances in the environment – A review. *Chemosphere* 36: 357–393.

Hanselman, T. A., D. A. Graetz, and A. C. Wilkie. 2003. Manure-borne estrogens as potential environmental contaminants: A review. *Environmental Science & Technology* 37: 5471–5478.

Henderson, B. E., and H. S. Feigelson. 2000. Hormonal carcinogenesis. *Carcinogenesis* 21: 427–433.

Jenkins, R. L., E. M. Wilson, R. A. Angus, W. M. Howell, and M. Kirk. 2003. Androstenedione and progesterone in the sediment of a river receiving paper mill effluent. *Toxicological Sciences* 73: 53–59.

Jensen, K. M., E. A. Makynen, M. D. Kahl, and G. T. Ankley. 2006. Effects of the feedlot contaminant 17 alpha-trenbolone on reproductive endocrinology of the fathead minnow. *Environmental Science & Technology* 40: 3112–3117.

Jobling, S. and C. R. Tyler. 2003. Endocrine disruption in wild freshwater fish. *Pure and Applied Chemistry* 75: 2219–2234.

Jones, J. M., T. Borch, R. B. Young, J. G. Davis, and C. R. Simpson. 2007. Photolysis of testosterone, progesterone, and 17β-estradiol by UVA light. In *Emerging contaminants of concern in the environment: Issues, investigations, and solutions.* Drewes, J. E., Battaglin, W. A., Kolpin, D. W., Eds. *American Water Resources Association:* Middleburg, VA.

Kidd, K. A., P. J. Blanchfield, K. H. Mills, V. P. Palace, R. E. Evans, J. M. Lazorchak, and R. W. Flick. 2007. Collapse of a fish population after exposure to a synthetic estrogen. *Proceedings of the National Academy of Sciences* 104: 8897–8901.

Kjar, J., P. Olsen, K. Bach, H. C. Barlebo, F. Ingerslev, M. Hansen, B. H. Sorensen. 2007. Leaching of estrogenic hormones from manure-treated structured soils. *Environmental Science and Technology* 41(11): 3911–3917.

Klein, K. O., J. Baron, M. J. Colli, D. P. McDonnell, and G. B. Cutler. 1994. Estrogen levels in childhood determined by an ultrasensitive recombinant cell bioassay. *Journal of Clinical Invest* 94: 2475–2480.

Kolodziej, E. P., and D. L. Sedlak. 2007. Rangeland grazing as a source of steroid hormones to surface waters. *Environmental Science and Technology* 41(10): 3514–3520.

Kolpin, D. W., E. T. Furlong, M. T. Meyer, E. M. Thurman, S. D. Zaugg, L. B. Barber, and H. T. Buxton. 2002. Pharmaceuticals, hormones, and other organic wastewater contaminants in U.S. streams, 1999–2000: A national reconnaissance. *Environmental Science and Technology* 36: 1202–1211.

Kuster, M., M. J. López de Alda, and D. Barceló. 2004. Analysis and distribution of estrogens and progestogens in sewage sludge, soils and sediments. *Trends in Analytical Chemistry* 23: 790–798.

Lange, I. G., A. Daxenberger, B. Schiffer, H. Witters, D. Ibarreta, and H. H. D. Meyer. 2002. Sex hormones originating from different livestock production systems: Fate and potential disrupting activity in the environment. *Analytica Chimica Acta* 473: 27–37.

Larsson, D. G. J., H. Hallman, and L. Forlin. 2000. More male fish embryos near a pulp mill. *Environmental Toxicology and Chemistry* 19: 2911–2917.

Larsson, D. G. J., M. Adolfsson-Erici, and P. Thomas. 2006. Characterization of putative ligands for a fish gonadal androgen receptor in a pulp mill effluent. *Environmental Toxicology and Chemistry* 25: 419–427.

Lawrence, J. D., and M. A. Ibarburur. 2008. Economic analysis of pharmaceutical technologies in modern beef production in a bioeconomy. Available at: http://www.econ.iastate.edu/faculty/lawrence. Accessed September 6, 2008.

Lee, L. S., T. J. Strock, A. K. Sarmah, and P. S. C. Rao. 2003. Sorption and dissipation of testosterone, estrogens, and their primary transformation products in soils and sediment. *Environmental Science & Technology* 37: 4098–4105.

Lobley, G. E., A. Connell, G. S. Mollison, A. Brewer, C. I. Harris, V. Buchan, and H. Galbraith. 1985. The effects of a combined implant of trenbolone acetate and oestradiol-17 beta on protein and energy metabolism in growing beef steers. *British Journal of Nutrition* 54:681–694.

Lorenzen, C. L. et al. 2000. Protein kinetics in callipyge lambs. *Journal Animal Science* 78: 78–87.

Nichols, D. J., T. C. Daniel, P. A. Moore, D. R. Edwards, and D. H. Pote. 1997. Runoff of estrogen hormone 17 beta-estradiol from poultry litter applied to pasture. *Journal of Environmental Quality* 26: 1002–1006.

O'Connell, M. B. 1995. Pharmacokinetic and pharmacologic variation between different estrogen products. *Journal of Clinical Pharmacology* 35: 18S–24S.

O'Grada, C. 1995. *The great Irish famine*. Cambridge University Press. London, England.

Organic Trade Organization. 2018. Available at: https://ota.com. Accessed 1/2/2018.

Orlando, E. F., A. S. Kolok, G. A. Binzcik, J. L. Gates, M. K. Horton, C. S. Lambright, L. E. Gray, A. M. Soto, and L. J. Guillette. 2004. Endocrine-disrupting effects of cattle feedlot effluent on an aquatic sentinel species, the fathead minnow. *Environmental Health Perspectives* 112: 353–358.

Price, E. O. 2002. *Animal domestication and behavior*. CAPI Publishing. New York.

Potau, N., L. Ibañez, M. Sentis, and A. Carrascosa. 1999. Sexual dimorphism in the maturation of the pituitary-gonadal axis, assessed by GnRH agonist challenge. *European Journal Endocrinology* 141: 27–34.

Raben, A., B. Kiens, E. A. Richter, L. B. Rasmussen, B. Svenstrup, S. Micic, and P. Bennett. 1992. Serum sex hormones and endurance performance after a lacto-ovo vegetarian and a mixed diet. *Medicine and Science in Sports and Exercise* 24: 1290–1297.

Randel, R. D., C. J. Callahan, R. E. Erb, H. A. Garverick and B. L. Brown. 1973. Effect of melengestrol acetate on rate of excretion of estrogen in urine of dairy heifers. *Journal of Animal Science* 36: 741–748.

Rollin, B. E. 1992. *Animal rights and human morality*. 2nd edition. Prometheus Books. Buffalo, NY.

Ross, R. K., A. Paganini-Hill, P. C. Wan, and M. C. Pike. 2000. Effect of hormone replacement therapy on breast cancer risk: Estrogen versus estrogen plus progestin. *Journal of the National Cancer Institute* 92: 328–332.

Rubens, R., and A. Vermeulen. 1983. Estrogen production in man. In *Anabolics in animal production* pp. 249–262. E. Meissonnier, and J. Mitchell-Vigeron Eds. Soregraph, Levalloid.

Rumsey, T. S. 1982. Effect of Synovex-S implants and kiln dust on tissue gain by feedlot beef steers. *Journal of Animal Science* 54:1030–1039.

Rumsey, T. S. 1985. Chemicals for regulating animal growth and production. In *Agricultural Chemicals of the future (BARC Symposium 8)*. Hilton J. L., Ed. 91–108. Rowman and Allanheld: Totowa, NJ.

Rumsey, T. S., and A. C. Hammond. 1990. Effect of intake level on metabolic response to estrogenic growth promoters in beef cattle. *Journal of Animal Science* 68:4310–4318.

Rumsey, T. S., A. C. Hammond, and T. H. Elsasser. 1999. Responses to an estrogenic growth promoter in beef steers fed varying nutritional regimens. *Journal of Animal Science* 77:2865–2872.

Rumsey, T. S., H. F. Tyrrell, S. A. Dinius, P. W. Moe, and H. R. Cross. 1981. Effect of diethylstilbestrol on tissue gain and carcass merit of feedlot beef steers. *Journal of Animal Science* 53:589–600.

Schairer, C., J. Lubin, R. Troisi, S. Sturgeion, L. Brinton, and R. Hoover. 2000. Menopausal estrogen and estrogen-progestin replacement therapy and breast cancer risk. *Journal of American Medical Association* 283: 485–491.

Schiffer, B., A. Daxenberger, K. Meyer, and H. H. D. Meyer. 2001. The fate of trenbolone acetate and melengestrol acetate after application as growth promoters in cattle: Environmental studies. *Environmental Health Perspectives* 109: 1145–1151.

Shappell, N. W., L. O. Billey, D. Forbes, T. A. Matheny, M. E. Poach, G. B. Reddy, P. G. Hunt. 2007. Estrogenic activity and steroid hormones in swine wastewater through a lagoon constructed-wetland system. *Environmental Science and Technology* 41: 444–450.

Shore, L. S., and M. Shemesh. 2003. Naturally produced steroid hormones and their release into the environment. *Pure and Applied Chemistry* 75: 1859–1871.

Ternes, T. A., P. Kreckel, and J. Mueller. 1999a. Behaviour and occurrence of estrogens in municipal sewage treatment plants - ii. Aerobic batch experiments with activated sludge. *Science of the Total Environment* 225: 91–99.

Ternes, T. A. et al. 1999b. Behavior and occurrence of estrogens in municipal sewage treatment plants - i. investigations in Germany, Canada and Brazil. *Science of the Total Environment* 225: 81–90.

Thorpe, K. L., T. H. Hutchinson, M. J. Hetheridge, M. Scholze, J. P. Sumpter, and C. R. Tyler. 2001. Assessing the biological potency of binary mixtures of environmental estrogens using vitellogenin induction in juvenile rainbow trout (Oncorhynchus mykiss). *Environmental Science & Technology* 35: 2476–2481.

Tyler, C. R., S. Jobling, and J. P. Sumpter. 1998. Endocrine disruption in wildlife: A critical review of the evidence. *Critical Reviews in Toxicology* 28: 319–361.

United Nations, Department of Economic and Social Affairs, Population Division. 2016. World population prospects: The 2015 revision, Volume I: Comprehensive tables (ST/ESA/SER.A/379).

USDA. 2000. Part I: Baseline reference of feedlot management practices, 1999. USDA:APHIS:VS, CEAH, National animal health monitoring system. Fort Collins, CO. #N327.0500.

USDA. 2017. National agricultural statistics service. Available at: www.nass.usda.gov/. Accessed 9/28/2017.

USDA NASS. 2006. Available at https://www.nass.usda.gov. Accessed 9/12/2006.

Whole Foods Market. 2017. Available at www.wholefoodsmarket.com. Accessed 9/28/2017.

Wilson, V. S., M. C. Cardon, J. Thornton, J. J. Korte, G. T. Ankley, J. Welch, L. E. Gray, and P. C. Hartig. 2004. Cloning and in vitro expression and characterization of the androgen receptor and isolation of estrogen receptor alpha from the fathead minnow (Pimephales promelas). *Environmental Science & Technology* 38: 6314–6321.

Woodling, J. D., E. M. Lopez, T. A. Maldonado, D. O. Norris, and A. M. Vajda. 2006. Intersex and other reproductive disruption of fish in wastewater effluent dominated Colorado streams. *Comparative Biochemistry and Physiology Part C: Toxicology & Pharmacology* 144: 10–15.

World Health Organization. 2002. Global assessment of the state-of-the-science of endocrine disruptors. Available at http:// http://www.who.int/ipcs/publications/new_issues/endocrine_disruptors/en/. Accessed 8/29/2018.

Zeuner, F. E. 1963. A history of domesticated animals. Harper and Row. New York, NY.

PART IV

Dairy Cattle

CHAPTER 22

To Meet the Ethical Imperative of *telos* in Modern Dairy Production: Societal Concern for Naturalness, Animal Welfare, and Opportunities for Resolution through Science

Beth Ventura
University of Minnesota

Candace Croney
Purdue University

CONTENTS

THE RELATIONSHIP BETWEEN DAIRY COW ETHOLOGY, *TELOS*, AND ANIMAL WELFARE

Any discussion of the extent to which an animal industry appropriately addresses the *telos* of the respective species it governs (i.e., the animal's genetically encoded "nature" which also reflects the function for which it was intended [as per Rollin, 1993]) should include at minimum a brief review of the ethology of the species in question. Today's cattle are thought to be descended from several different subspecies of aurochs domesticated around 9,000 years ago in western Asia, Africa, China, and India (Clutton-Brock, 1999). Cattle have been altered significantly from their wild progenitors through both artificial and natural selection, and the evolving demands of modern-dairy production place additional adaptation pressures upon them. However, commonalities still exist between the working dairy cow, feral cattle, and her ancestors, suggesting that despite selection for production traits, the domestic dairy cow retains the drive to engage in certain key behaviors, as reflected through her ethogram. For example, cattle will typically spend the majority of their days alternating between feeding and rumination (Hall, 2002). Grazing usually begins at sunrise, followed by rumination during the mid-morning period, after which they continue to graze sporadically through the afternoon and evening hours. In addition, like her wild counterparts, the modern-dairy cow will tend to form and strive to maintain relatively stable dominance hierarchies that are reinforced by behaviors such as lying in proximity to those of similar rank and engaging in allo-grooming. Groups of free-ranging cattle are generally observed to consist primarily of cows and their calves, although bulls join and leave the groups periodically (Hall, 2002).

The potential thwarting of behaviors inherent to the nature of the dairy cow, particularly those she remains highly motivated to perform, can be considered as a violation of her *telos*. It also appears to elicit societal concerns as indicated by studies of public perceptions of dairy cow welfare (see section "Societal Perceptions of Cow Welfare and *telos*"). Restricting such behaviors is also likely to undermine an intrinsic aspect of cattle welfare, that is, cows being allowed to live according to their *telos*, which requires "meeting of those needs and interests that matter to the animal and affect its feelings by virtue of its biological and psychological nature," (Rollin, 1993, 2007). Such restriction, particularly over any continuous length of time, may generate frustration and other negative affective states and therefore undermine psychological and emotional well-being as well. The concurrent physiological stress responses that are evoked may also ultimately undermine the cow's biological "fitness," impacting her reproductive, productive, and survival capabilities as resources are mobilized in an attempt to help her cope (Broom and Johnson, 1993; von Borell, 1995; Moberg, 2000).

Consideration of the dairy cow's *telos*—together with her evolutionary history and modern conceptions of animal welfare—is thus integral to evaluations of her capacity to cope successfully with environmental and production demands. Coincident societal pressures for the dairy industries to better accommodate animal natures into production systems and practices provide additional impetus to evaluate the role of *telos* in helping to inform best practice for modern-dairy farms, enhance public trust, and allow farmers to maintain their social license to operate (Croney and Anthony, 2011).

Conceptual Framework

Before examining public views of the more contentious animal welfare challenges faced in the dairy industry, it is useful to first establish a conceptual framework from which to understand these challenges and how they may relate to the *telos* of the dairy cow. Others (Moberg, 1985; Dawkins, 1988; Broom, 1991; Duncan, 1993; Gonyou, 1993; McGlone, 1993) have written extensively on the differing definitions of animal welfare, but for the purposes of this chapter, we will rely on the conception of animal welfare put forth by David Fraser, Daniel Weary, Ed Pajor, and Barry Milligan (1997), which is to say that animal welfare can be thought of as three distinct but interrelated concepts. First, that welfare has to do with how an animal functions *physically*, for example, in order to have good welfare a dairy cow must be free from disease and injury and consume biologically appropriate feed; second, that welfare has to do with how an animal *feels*, for example, a cow should experience positive mental states like pleasure and comfort and avoid negative states like pain, frustration, and fear; and third, that animal welfare must involve an animal being able to live *naturally*, that is, a cow should be able to express the behaviors intrinsic to her natural repertoire and live within a natural social structure and physical environment. The latter concept perhaps relates most closely to *telos*, though it should be noted that the interrelatedness of each of these three concepts dictates that a cow's ability to physically function and feel well are also critically important to fulfilling her *telos*.

SOCIETAL PERCEPTIONS OF COW WELFARE AND *TELOS*

The public has become increasingly interested in the quality of life that cows are afforded in modern production systems and may become particularly critical relative to potential mismatches between cow ethology and living conditions. Some have suggested that members of the public may be more attuned to dairy cattle welfare issues than those of other livestock species, due in part to strong associations between dairy product consumption and childhood nutrition (Widmar et al., 2017). The idea of the dairy cow as "foster mother" to the human race (Albright, 1987) likely also plays a significant role in drawing attention to cow welfare. Bucolic images of cattle grazing near public roadways, as well as commercials featuring "Happy Cows" (see inset on page 257), may also serve to keep the dairy cow more visible in the public eye than other livestock species (Glenn, 2004; Croney and Reynnells, 2008). This level of attentiveness to dairy cattle holds significance for public and consumer support of today's dairy industries. Findings from Wolf et al. (2016), who indicated that over 63% of respondents expressed concerns about dairy cattle welfare in a national study of 1201 US residents' perceptions about dairy industry practices, appear to illustrate this point. Further to this point, Widmar et al. (2017) reported that in a study of over 1200 US residents surveyed about their purchasing behaviors and perceptions of dairy production practices, 12% had altered their consumption of dairy products over the past 3 years specifically because of animal welfare concerns.

When public concerns about cow welfare are analyzed, several consistent themes emerge that reflect underlying expectations that cows should be cared for and managed in a manner consistent with their ethology, that is, in a way that affords some semblance of "natural living." For example, the quality of the animals' living environments, the amount of space provided (Ellis et al., 2009), the extent to which they are able to express natural behaviors (Te Velde et al., 2002; Vanhonacker et al., 2008), move around freely (Lassen et al., 2006; Maria, 2006; Vanhonacker et al., 2008; Ellis et al., 2009), and access the outdoors (Ellis et al., 2009) have all been identified as points of contention in various studies of public perceptions of dairy cow welfare. For example, McKendree et al. (2012) found that in a national survey of US residents, confinement of cows indoors generated the most concern of all the production practices investigated. Similarly, Schuppli et al. (2014) reported that although most North American dairy farms provide little to no access to pasture, Canadian and US participants with and without affiliations to dairy production considered cow access to more natural, pastoral living conditions to be important.

Pasture access seems to be particularly important to public stakeholders, who associate it with access to fresh air, social living, freedom of movement, better cow health, and more healthful milk products. Cardoso et al. (2016) found that public respondents in an online survey were concerned about the quality of life experienced by dairy cows, with a special focus on natural living and its perceived relationship to (improved) milk quality. Responses to questions aimed at identifying people's conceptions of ideal dairy farm characteristics included the importance of cows being able to roam freely and have access to open space, the outdoors, and a diet that included grass. A survey by Ventura et al. (2016) similarly found that public views on what constitutes a good life for cows included several elements typical of a natural living conception of animal welfare (see Fraser et al., 1997), such as pasture, space, fresh air and sunshine, social interactions with companions, and feedstuffs deemed to be natural for cows—preferences that held even after respondents toured a dairy farm in person.

These sorts of characteristics appear to significantly influence people's perceptions of dairy farming practices. However, the body of research on perceptions of dairy cow welfare indicates several areas of growing disconnect between the public's view of what constitutes a good life for a cow and standard dairy industry practices. For example, Ventura et al. (2016) observed that some participants' initially positive attitudes about the welfare of cows on dairy farms became more negative after a dairy farm visit. Moreover, those who did become more critical about dairy farming after touring a farm did so in part because they learned of practices (e.g., zero grazing, early cow–calf separation) that conflicted with their expectations that cattle should be managed in more natural conditions (Ventura et al., 2016). Te Velde et al. (2002) observed that people with no connection to farming often conceptualize animal welfare in terms of how "natural" the animal's living experience is, while those associated with animal production typically prioritize the biological functioning conception of animal welfare, focusing on factors such as animal health, reproductive success, and productivity metrics (Te Velde et al., 2002; Verbeke, 2009). It is important to stress that neither of these value-based conceptions of cow welfare are "wrong;" from a holistic conception of animal welfare, it may be most useful to seek systems that maximize welfare in *multiple* spheres. However, these differing conceptions of welfare often lead to conflict between concerned citizens and the livestock industries. For example, nonindustry groups may decry that standard practices are abusive or inhumane, while industry stakeholders maintain that those same practices conform to high standards of animal care. Ultimately, such disagreement, with each side firmly maintaining their own perspectives, is certainly frustrating for all, and profoundly unhelpful if we are to develop and implement standards that work for farmers but also improve the lives of their animals (see Ventura et al., 2015).

In short, the majority of published studies suggest that many interested members of the public may be apprehensive as to whether modern production offers a fair deal to cows or whether it represents a violation of "the ancient contract" between animals and society, a notion discussed in detail by Rollin (2008). The dissonance between public beliefs about appropriate quality of life for animals and the conception of a "good life" offered under the aegis of standard industry practice must be resolved if sustainable, socially acceptable forms of dairy production are to be achieved—and by that we mean that solutions must be achievable for the farmer while also meeting the evolving societal consensus about what it means to provide a farm animal with a good life. Given the recurrent theme from survey respondents that cows' living conditions and experiences should ideally incorporate certain basic aspects of "natural living," we review the extent to which dairy cow production and management practices currently reflect cow ethology and discuss ways in which cows might be permitted to live in a manner more consistent with their *telos*.

ANIMAL WELFARE AND *TELOS* IN THE DAIRY INDUSTRY

Contemporary dairy production presents numerous challenges to the welfare of the cow, though the vast variation among farms due to various management factors means that some farms certainly manage these challenges more successfully than others. Nonetheless, comprehensive discussions of animal welfare in the dairy industry should include, but are not limited to: diseases such as mastitis and metritis, lameness, painful procedures such as disbudding and dehorning, housing and cow comfort, early separation of calves from their dams, end-of-life and transport decisions and operating procedures (particularly in relation to downed cattle), and management of male calves (see von Keyserlingk et al., 2009 and Barkema et al., 2015 for comprehensive reviews; Ventura et al., 2015;). Given their significance, thorough discussion of all of these topics would warrant multiple chapters. This chapter will focus by necessity on those issues which are especially salient to the natural living conception of animal welfare, the *telos* of the cow and the present understanding of public/consumer concerns associated with dairy production to date.

Cow Housing

Zero grazing. The most common housing systems (see inset on page 255 for description) for lactating dairy cattle in the United States are tie stalls or stanchions (38.9%), freestalls *without* access to open or dry lots (20%), freestalls *with* access to open or dry lots (19.7%), pasture (7.5%), and open or dry lots with or without shed or barn access (7.3%; USDA, 2016a). In general, pasture access falls as herd size increases. What this means in practice is that the majority of both lactating and dry cows do not receive regular pasture access in the United States.

Dairy Cattle Housing Definitions

Dry lot	Open dirt lot with no vegetative cover used for housing cows in arid climates; common in California and American southwest.
Freestall	Housing consisting of resting cubicles or "beds" in which dairy cows are free to enter and leave at will.
Pasture	Area with vegetation suitable for grazing.
Stanchion	Housing in which a cow is restrained to a particular stall in a device with two rails that close around the cow's neck after she enters the stall. Cows are not able to enter and leave the stalls at will.
Tie stall	Housing in which a cow is restrained to a particular stall by a neck collar attached to the stall by a chain. Cows are not able to enter and leave the stalls at will.

Source: Reproduced from USDA (2016a).

Of course, from a scientific perspective, high animal welfare can be achieved in housing systems that do not provide access to the outdoors, and it is not the intent of this chapter to suggest otherwise. However, the lack of outdoor and pasture access for the majority of US dairy cattle may lead to ethical concerns over cattle being able to fulfill their *telos* among large portions of the public; it may also pose some welfare challenges if cattle are denied the opportunity to engage in motivated behaviors. As summarized by Charlton and Rutter (2017), "for cattle, pasture is a natural environment, allowing them to express normal behaviors. It can provide ample comfortable lying space, allowing cows to lie in stretched positions..." To reflect on the natural ethogram of the cow is to acknowledge that, left to her own devices, a cow would opt to spend most of her day engaging in grazing and ruminating behaviors (Hall, 2002). Cattle housed on pasture spend their time engaged in up to 40 different categories of behavior (see Kilgour, 2012 for the ethogram of behaviors exhibited by cattle on pasture). Of these ~40 behaviors, a substantial number, including grazing (the most frequently-expressed behavior, at anywhere between 8 and 12 hours of the 24-hour cycle), are much more easily expressed in an open or grassy environment than they can be in most indoor types of housing.

Providing cattle with the opportunity to access pasture also appears to provide a number of health benefits. Decades of research indicate that pasture access can positively impact cow health, for example, through reduction of lameness (Leaver, 1988; Smits et al., 1992; Gitau et al., 1996; Hernandez-Mendo et al., 2007) and mastitis (see Charlton and Rutter [2017] for a brief review). While some studies suggest that pasture housing may come with a production cost (Fontaneli et al., 2005; Hernandez-Mendo et al., 2007), others have demonstrated that it is possible to graze cattle on pasture with minimal to no production losses (e.g., in feed intake and milk yield, Chapinal et al., 2010; Motupalli et al., 2014).

Looking to some of the research techniques used in animal welfare science may help further resolve sticky questions about *telos* and dairy farming by providing more information about the cow's basic behavioral needs. For example, preference and motivation tests help scientists understand the relative importance of different resources and environments to the cow, by quite literally allowing her to vote with her feet. Cattle appear to be highly motivated (i.e., willing to work) to gain access to pasture, but both preference and motivation for pasture depend on a number of factors, including the individual, time of day, temperature, and location of feed (Legrand et al., 2009; Falk et al., 2012; Charlton et al., 2013; Charlton and Rutter, 2017). For example, when given the choice between pasture or indoors, dairy cows tend to spend their time inside the barn (where it is cooler and feed is freely available) during the daytime before moving onto pasture at night (at least during warmer months; Legrand et al., 2009). They will also work harder to access pasture during the evening hours than they do during the day (Charlton et al., 2013).

Incorporating these types of research results into the design and operations of new dairying facilities may help to resolve some of the ethical conflicts that arise around the naturalness (or lack thereof) of dairy farming, while maintaining and even improving cow productivity. From the cow's perspective, the ideal (and thus, most ethical, if we are to emphasize *telos*) solution might be to allow her to *choose* to flow between an indoor environment and an outdoor pasture. However, logistical and practical constraints may prevent individual farms from being able to provide choice-based access to pasture. These constraints include, but are certainly not limited to: farmers' access to and/or ownership of suitable land for grazing, environmental considerations pertaining to land use, weather-related conflicts, predator control, and persistent economic barriers. Future compromises between practical limitations and fulfillment of natural living for cows with relation to outdoor access may yet exist, however. For example, within the last few years researchers have begun to explore ways to provide alternative outdoor access to allow cattle a greater range of movement and socialization in ways that are feasible for farmers (e.g., small adjoining sand or deep-bedded bark mulch packs; Smid et al. 2018).

Postural and behavioral restriction. A recent review (Barkema et al., 2015) indicated that zero grazing is not the only contentious housing system used in the dairy industry. It is likely that as research continues to explore stakeholder concerns about dairying, those external to the livestock industries will increasingly voice objections to basic restrictions to the cows' opportunities for movement and social interaction (Boogaard et al., 2011; Popescu et al., 2013), rendering systems like tie stalls or stanchions, wherein cattle are tethered for long periods of time without the ability to turn around or engage in some important social behaviors, questionable from a social sustainability perspective. Furthermore, there is evidence that at least some stakeholders working within or for the dairy industry also object to such housing on the basis that it impedes a cow's nature, see, for example, the comments of an interviewed animal scientist that "the tie stall, as opposed to the free stall, is an issue...where the ability of the animal to move within that space is really hindered...I'm talking about the ability to make postural changes, to explore her environment and to choose to move away from other cattle..." (Ventura et al., 2015). Reiterating this concern, a veterinarian in that same study remarked, "It's about respect [for] the nature of the animal!" (Ventura et al., 2015).

As with many other contentious practices in dairying, the science on cattle health in tethered systems (vs. other indoor systems like the free stall) is ambiguous. For instance, in some studies, cattle in tie stalls seem to exhibit poorer health and worse reproductive outcomes than those kept in free stalls (e.g., higher rates of most diseases and lower reproductive performance on Norwegian farms [Valde et al., 1997; Simensen et al., 2010]). However, lameness incidences have been reported to be lower in tie stall systems than in free stalls (on Wisconsin and Ontario farms [Cook, 2003; Cramer et al., 2009] and again in Norway [Sogstad et al., 2005]). Others have found little to no evidence of acute or chronic stress in cattle housed in tie stalls (Veissier et al., 2008). It may well be that welfare is not necessarily poor in tie stall systems *per se*, but it is important to note that researchers who reach this conclusion also specify that in order to mitigate the potential health and welfare impacts of prolonged tethering, cattle housed in tie stalls should also be allowed exercise opportunities in paddocks or pastures (Veissier et al., 2008; Popescu et al. 2013).

Calf Housing and Management

Separation from the cow and individual housing. Few issues seem to provoke as much public ire or better illustrate the divergence in views on animal welfare between industry and lay stakeholders than that of early cow–calf separation. Here again, contradictory scientific findings complicate matters. First, it is important to note that in nature, a cow will typically separate herself from

DAIRY BOOSTERISM AND THE PASTORAL IDYLL OF DAIRY FARMING

The desire to see animals outdoors and on pasture is deeply embedded in the societal consciousness of what a "good life" for farm animals should look like in the United States (and elsewhere). The industries clearly recognize this. Marketing efforts (dairy "boosterism," as it were [see DuPuis, 2002]) within the dairy industry have historically cultivated dairying as an inherently pastoral pursuit, in so doing positioning the dairy sector as relatively immune to some social and activist concerns about industrialization, intensification, and denaturing that confront other livestock sectors (DuPuis, 2002; Molloy, 2011).

As far back as the 19th century, antebellum social reformers envisioned and sought to position milk and its production as a solution to the perceived moral depravities of industrialization and social ills befalling newly formed cities, constructing milk as a perfect food that would aid city dwellers in escaping the ills plaguing urban life. Public health officials extolled milk as "the modern elixir of life," describing it as "the most nearly perfect of human foods for it is the only single article of diet which contains practically all of the elements necessary to sustain and nourish the human system," (Crumbine and Tobey, 1930). See also the writings of one William Prout, who in 1802 described milk thusly: "Of all the evidences of design in the whole order of nature, milk affords one of the most unequivocal. No one can doubt for the moment the object for which this valuable fluid is prepared." Scholars (see DuPuis, 2002) follow the positioning of milk as wholesome and clean into 20th- and 21st-century marketing—themes that are underscored with the implication that dairy is a wholly natural product. That theme of nature boosterism has marked not only the end product of dairying, but the production process itself. Enter the dairy cow, and her position in a green, utterly (udderly?) natural landscape to produce the ultimate, perfect, *natural* food.

At first glance, depicting dairy cattle on pasture makes perfect sense from an advertising perspective, as such imagery appears to fulfill a latent urge to return to nature for many people both within and external to the dairy industry itself (see section "Societal Perceptions of Cow Welfare and *telos*"). Even dairy farmers, when surveyed, indicate that they too would love to see more cattle on pasture (Schuppli et al., 2014). This collective desire for the pastoral carries through even to the labels adorning dairy products, for example, through label design (e.g., rolling hills and bucolic landscapes) and even the brand name itself (e.g., Organic Valley, Meadowbrook). Such messages seem to subconsciously reinforce collective societal and agrarian ideals relative to cows living their lives outdoors on verdant pastures. Television commercials have done likewise: see, for example, the enormously successful "Happy Cows" campaign for California Milk in the early 2000s, in which anthropomorphized talking cows conducted intimate social lives on Californian rolling hills. That notion of "the cute cow" is, as DuPuis writes, "an American cultural phenomenon in itself...all of these cute cows represent a friendly, controllable, yet natural provision system, a sort of identity-based pastoral ideal" (DuPuis, 2002, p. 235). Agrarian imagery likewise abounds in dairy advertising, through depictions of quaint red barns situated against rolling hills, more realistic and perhaps depersonalized silhouettes of cattle embedded into the landscape alongside the farmer, and farming families portrayed as working in close partnership with their cows. These portrayals are "reminiscent of the romantic pastoral images of the mid-19th century, in which the tending milkmaid represented the care of nature [by which such] discourses emphasize agrarian values and cooperation between the farmer and the consumer, and farmer and nature" (DuPuis, 2002).

However, to depict dairying as a wholly pastoral pursuit becomes increasingly risky as members of society learn about current dairy production practices, and inevitably begin to question the integrity of dairy industry communications and marketing efforts. In California, for example, dairy farms are much more likely to house cattle in open dirt lots with not a blade of green in sight. It is of course important to clarify that this contrast does not in itself necessarily or even directly impact the welfare of dairy cattle, as high welfare (at least from the perspective of most animal welfare scientists) is absolutely achievable in a range of dairy housing systems. However, that conflict can and does negatively impact public trust and confidence when lay people must question why farmers house cattle in ways that look dramatically different from the ways in which advertising has led consumers to believe dairy cattle are housed. Indeed, California Milk's advertising has since transitioned to broader depictions of the types of dairy farms that supply their milk.

her herd to calve, hide her calf for the first few weeks of life, nurse frequently for months and wean gradually (Kilgour and Dalton, 1984; Vitale et al., 1986; Lidfors et al., 1994; Langbein and Raasch, 2000). Research indicates that modern cows housed indoors retain the motivation to separate and hide before calving (Proudfoot et al., 2014a, b). In contrast, typical practice on the vast majority of farms is to separate the calf from the cow within a few hours to a day or two after birth, after which the cow rejoins the production cycle and the calf is housed separately, fed milk artificially and weaned to solid feed some time between 6 and 8 weeks of age (von Keyserlingk and Weary, 2007).

There are a number of issues to tease out here which may provoke ethical concern on the basis of conflicts with the *telos* of the dairy calf and cow. First, the act of separation itself, which does indeed elicit strong objection among nondairy farming persons when they learn about the practice, based on concerns about the unnaturalness of the practice and subsequent negative impacts on both cow and calf (Boogaard et al., 2010; Ventura et al., 2013, 2015). Farmers separate the cow and calf for various reasons that are scientifically supported. For one, it is thought that the cow–calf bond does not develop *immediately* after birth (Hall, 2002), though the bond likely develops rapidly and strengthens over time (as reviewed in Flower and Weary, 2003). It is therefore thought that separating calves from cows immediately may prevent the greater distress that could come from separating later after the bond is allowed to develop (Weary and Chua, 2000; Flower and Weary, 2001, 2003). For some farmers, early separation is also justified to minimize disease transmission between mother and offspring (Ridge et al., 2005), reduce exposure to environmental pathogens (Windsor and Whittington, 2009), and to help facilitate individual care, including monitoring and managing colostrum intake (Vasseur et al., 2010). However, recent research suggests that there are also certain advantages conferred by allowing the calf to remain with the dam for a period of time, including increased learning and social flexibility for those calves, which may improve their success in the herd later in life (Costa et al., 2014; Gaillard et al., 2014; Meagher et al., 2014). Calves permitted contact with their dams for 2–12 weeks not only had greater social activity, they also demonstrated fewer undesirable or abnormal oral behaviors, such as cross-sucking (Flower and Weary, 2001; Fröberg and Lidfors, 2009; Wagner et al., 2015). As for health effects of separation on the calf, the science again is mixed. Calves remaining with the dam to nurse have been shown to have improved digestive health and overall lower morbidity and mortality (Selman et al., 1970; Metz and Metz, 1986; Metz, 1987; Weary and Chua, 2000; Flower and Weary, 2001; EFSA, 2006). Yet other studies have indicated that calves left with their dams may be at greater risk of failed passive transfer, diarrhea, and Johne's disease (Wesselink et al., 1999; Svensson et al., 2003; Marcé et al., 2011).

Calf housing. While researchers are beginning to re-examine whether rearing the calf with the dam may be possible on dairy farms (see Johnsen et al., 2016 for a review), it is unlikely that the industry's reliance on early cow–calf separation will cease any time in the near future. Thus, it may be beneficial to examine whether compromises between ethics, animal welfare, and practicality may exist with regard to the calf's management once separated from the cow. Many farmers report a preference for individual calf housing due to management ease and perceived health benefits for the calf, and the majority of dairy operations in the United States report housing pre-weaning heifer calves individually (37.9% in outside hutches or pens, with an additional 25.1% in indoor hutches or pens; USDA, 2016a). While in nature the neonate calf would indeed be separated from other herd mates in those first days of life, she would eventually integrate into the herd and form strong, lifelong bonds with her dam and with other cows and calves (Kilgour and Dalton, 1984; Vitale et al., 1986).

In contrast, the eventual transition from individual to group housing, occurring much later and more abruptly on most dairy farms, is neither natural nor optimal for heifer success in the herd later in life. Here again science may help bridge the gap between preferences for more natural calf housing and practical constraints for dairy farmers. Rather than housing calves individually from birth to weaning, farmers might instead consider housing calves in small groups of 6–8 or even in pairs after a few days to a week after birth. Doing so has been shown to confer abundant benefits to calves, including improved cognitive development, reduced weaning distress, and improved post-weaning growth performance (Jensen et al., 1997; Faerevik et al., 2006; De Paula Vieira et al., 2010, 2012a, b; Gaillard et al., 2014; Pempek et al., 2016).

Calffeeding protocols. We may seek further resolution of ethical concerns pertaining to permitting calves to fulfill their *telos* in the management of calf nutrition and feeding protocols. While calves would naturally choose to suckle frequently and decrease the number of bouts with age, the majority of dairy farms (88.9%; USDA, 2016a) feed calves twice per day (an additional 6.8% report thrice-per-day feeding) at approximately 10% body weight (BW) or approximately half what they would consume voluntarily (Appleby et al., 2001). Not surprisingly perhaps, calves raised on such farms often struggle with gaining weight during their first few days of life (Hammon et al., 2002) and vocalize out of hunger when deprived of adequate amounts of milk (Thomas et al., 2001). Feeding *more* milk, on the other hand, can improve BW gain and feed conversion in calves (Diaz et al., 2001; Shamay et al., 2005) and reduce hunger signals (De Paula Vieira et al., 2008) without compromising calf health (e.g., no increase in scouring, Appleby et al., 2001; De Paula Vieira et al., 2008).

Rethinking milk delivery is also in order; calves are often bucket-fed, depriving them of the opportunity to consume nutrition through suckling. Preventing calves from suckling, a behavior that they are highly motivated to perform, is thought to contribute to undesirable behaviors, such as cross-sucking or nonnutritive sucking of the bodies of other calves (Jensen, 2003; de Passillé and Rushen, 2006). Shifting to nipple/teat delivery systems, which allow calves to express their natural suckling behavior, can improve their physiology, promote relaxation (de Passillé et al., 1993; Hänninen et al., 2008) and essentially eliminate cross-sucking behavior (de Passillé and Rushen, 2006).

Routine Alterations

A number of dairy management procedures involve the removal or alteration of a body part (the horns or the tail) in an attempt to improve some aspect of animal management. Society may object to these practices on the basis that modifying an animal to fit its environment is a misguided and deeply unethical approach that violates the animal's bodily integrity and so its *telos* (Bovenkerk et al., 2002; Gavrell-Ortiz, 2004). Rather, stakeholders external to the dairy industry (and many within it) may instead wish farmers to amend the environment to fit the animal's nature. Indeed, altering the environments in which cows are housed and refining the technologies, equipment and management practices used in dairy production may be a far more palatable means of improving welfare than altering the animals themselves.

Disbudding/dehorning. Dairy farms commonly remove a calf's horns, most often through appli-cation of a hot-iron or caustic paste (to disbud) or via surgical means (dehorning), in order to prevent horn growth and avoid injury for both cattle and handlers (AVMA, 2012). Though earlier interven-tion is considered less invasive and hence less painful (AVMA, 2012), all of these procedures result in varying levels of acute and chronic pain and distress for the animals involved, particularly when performed without analgesia or anesthesia (Heinrich et al., 2010; Stafford and Mellor, 2011).

The proportion of American farms reporting use of pain control for disbudding or dehorning, however, is generally low (<18%, Hoe and Ruegg, 2006; Fulwider et al., 2008). Research in the early 2000s indicated that livestock farmers may ascribe lower concern to pain arising from short-term procedures compared to other stakeholders (Vanhonacker et al., 2008; Phillips et al., 2009; Spooner et al., 2012) perhaps partly due to the perception of the necessity of such procedures in mitigating other management problems (Kjaernes et al., 2007; Spooner et al., 2012).

However, the lack of pain control for disbudding and dehorning stands in stark opposition to values even among stakeholders within the dairy industry (Robbins et al., 2015; Ventura et al., 2015) and certainly contrasts with broader societal imperatives that farm animals not be subjected to unnecessary pain (Rutgers, 2003; Spooner et al., 2014; Robbins et al., 2015). Decades of research indicate that the pain response to these procedures is greatly attenuated through provision of comprehensive, multi-modal pain management (McMeekan et al., 1999; Heinrich et al., 2010; Stilwell et al., 2012; Huber et al., 2013; Coetzee, 2013a; Winder et al., 2017). Furthermore, provision of analgesia for calves undergoing disbudding or dehorning is beneficial for production parameters as well as welfare indicators of importance to farmers. For instance, calves disbudded without analgesia have slower growth rates than do calves receiving pain relief (Bates et al., 2016). It is esti-mated that lidocaine provision costs under $0.50 per calf (Misch et al., 2007), and comprehensive management is estimated at less than $4.00 per head, or approximately 0.004% of the total estimated cost of raising a replacement heifer (Gabler et al., 2000; see Robbins et al., 2015). Thus, increasing the application of comprehensive pain management may help resolve some social concerns about pain control while also remaining feasible for farmers.

In keeping with recent research findings, the recently revised animal care standards for the National Milk Producers Federation (under the Farmers Assuring Responsible Management, or FARM, program) encourage farmers to develop comprehensive pain management protocols together with their veterinarians (NMPF, 2016). Others have suggested more long-term solutions to replace the need for dehorning in the first place, through introduction of polled genetics into dairy herds (Long and Gregory, 1978; Hoeschele, 1990). Genetic engineering provides a viable means by which to accomplish horn removal by introducing an allele that knocks out the gene that produces horns (Carlson et al., 2016) without causing direct negative effects (such as pain). Thus, it warrants greater consideration for implementation into dairy production and management decision-making. Unlike other methods of altering cows and calves, genetic alteration does not compromise *telos* given that there are naturally occurring alleles for polled cattle. In line with Sandøe et al.'s (2014) premise, genetically altered calves would have a different *telos* than horned animals, and consequently, dif-ferent interests (see Rollin's [1995] suggestion) that would require attention.

Robbins et al. (2015) note that that use of polled genetics has long been suggested (Long and Gregory, 1978; Hoeschele, 1990) and has already been adopted by the US beef industry, which reported that in 2007, over 85% of beef calves were born without horns—an increase of 17% from 1992 (USDA, 2008). As Robbins et al. (2015) conclude, "Given the obvious benefit of this approach for both dairy producers (i.e., reduced labor and improved public image) and dairy cattle (i.e., reduced pain), greater investment of this option seems prudent." The NMPF animal care standards now suggest the potential of polled dairy genetics to supplant dehorning (NMPF, 2016).

Tail docking. It is thought that farmers in New Zealand began docking cows' tails in an effort to control transmission of leptospirosis, which was believed to be connected with milkers coming into contact with the pathogen, which can be shed in urine found on cows' tails (Tucker et al, 2001).

As the practice became more widespread, farmers began to cite numerous other reasons for docking, namely improved milk hygiene, udder cleanliness and health, and enhanced milker comfort, health, and hygiene (Tucker et al., 2001; Croney and Anthony, 2011). Consequently, by 2007, docking was reportedly performed by 48% of US dairy farms (USDA, 2007; Barkema et al., 2015).

Despite the claimed benefits, tail docking raises significant ethical and scientific issues, especially related to infringements on the *telos* of the cow. Tails by design provide functional benefits to cows. They facilitate social communication and fly deterrence, both of which are impaired when they are docked (Croney and Anthony, 2011). Not surprisingly, the obvious alteration of cows is viewed quite negatively in public perception studies (Widmar et al., 2017). Weary et al. (2011) noted that survey respondents objected to tail docking in part because they viewed the practice as "unnatural" or because it interfered with the "natural" behavior of the cow (and her ability to disperse flies). Here, the question of violating the integrity of the cow appears to be a central, underlying concern.

Tail docking is also unjustified based on the prevailing scientific evidence. The nature and extent of pain caused by docking raises immediate concern, although most studies to date suggest that cows may experience only minor levels of acute pain upon docking (Tom et al, 2002). There is still debate about whether docking causes chronic pain (Eicher et al., 2006; von Keyserlingk et al., 2009). Additionally, the stated rationales for docking do not withstand scientific scrutiny. For example, no differences are observed in milk production or milk hygiene in docked cows; docked cows have, however, been observed to have higher fly loads likely due to the loss of a tail switch that assists with ridding themselves of flies (Ladewig and Matthews, 1992; Mathews et al., 1995; Eicher et al., 2001; Tucker et al., 2001; Eicher and Dailey, 2002; Schreiner and Ruegg, 2002; von Keyserlingk et al., 2009). The primary benefit appears to be worker comfort, which can be accommodated by management of the cow's environment as well as by trimming the hairy switch of the tail, further weakening the argument for the alteration. Moreover, unlike some of the other welfare issues previously discussed, tail docking can be easily and economically abandoned, as no infrastructural or financial investment is needed—farmers can simply elect to simply stop the practice immediately.

In light of the current state of scientific evidence, the National Milk Producers Federation set January 1, 2017 as the date by which farmers participating in the industry's FARM Animal Care Program to phase out routine tail docking. Additionally, the AVMA opposes routine tail docking of cattle, noting that "current scientific literature indicates that routine tail docking provides no benefit to the animal, and that tail docking can lead to distress during fly seasons," (AVMA, 2017). Thus, the issue of tail docking may come to represent a critical success story in terms of the dairy industry integrating both scientific and ethical consensus into standard management procedures.

In summary, the dairy industry has an opportunity to address a major source of ethical concern by taking steps to avoid procedures that cause cattle pain by exploring viable, cost-effective alternatives or through making management decisions that better incorporate pain control. Given that a major challenge relative to achieving the latter goal is the lack of approved analgesics for cattle in the United States (Coetzee, 2013b), it is critical for the industries to invest further in research that may lead to development of approved analgesics and best practice protocols for pain management in dairy cows, while revisiting the necessity and implications of painful alterations of animals.

Longevity

As animal agriculture strives to meet global food demands, the impetus for increased production has correspondingly grown. Predictably, the production demands placed on livestock animals have subsequently increased. Nowhere is this paradigm better exemplified than in contemporary dairy production. Over the past 40 years, milk yield per cow has more than doubled (Oltenacu and Broom, 2010). However, the dramatic increase in productivity has potentially come at a cost to the cow, with the reported incidence of production-related diseases dramatically increasing over the past few decades. Despite evidence that onset of reproduction at earlier ages is correlated with reduced longevity, the dairy industry has pushed to lower the age at first calving for economic reasons (Knaus, 2009). The tremendous increase in cow production along with early first calving age now manifests in metabolic disorders, foot and leg problems, and fertility issues, all of which contribute to reduced longevity (Oltenacu and Broom, 2010).

Lameness and mastitis have become particularly troublesome issues, routinely ranking at the top of the most serious and costly welfare problems in dairy production (Ventura et al., 2015). In a review of the welfare impacts of lameness in dairy cows, Whay and Shearer (2017) noted that a study conducted in Minnesota of 50 cow herds revealed that 15% of the cows were clinically lame (lameness score ≥3) and 2.5% were severely lame. However, detection of lameness, particularly early in its development, remains a challenge for herd managers who underreport it compared to trained evaluators (Espejo et al., 2006; von Keyserlingk et al., 2009). Lameness reduces cow comfort; in more advanced cases, the pain and suffering associated with failure to detect and treat the condition in cows presents even greater challenges to cow welfare, while also increasing costs for farmers who are subsequently forced to cull animals whose productivity is compromised.

Here, considerations of permitting cows to live according to their *telos* or natures become relevant, as existing science suggests that cow management and housing impacts lameness (Espejo and Endres, 2007) and in turn, longevity (Whay et al., 2003). Furthermore, lame cows given short-term access to *well-maintained* pasture showed greater improvements in gait than those continuously maintained in free stalls (Hernandez-Mendo et al., 2007). Thus, while pasture access should not be considered a panacea for welfare issues in dairy cows, accommodation of natural living via even temporary pasture may prove to be meaningful and beneficial for both cows and farmers.

Mastitis likewise poses a serious challenge for dairy production. As is the case for lameness, mastitis is not only costly, but significantly impactful on cow welfare as it is associated with pain, suffering and ultimately, reduced cow longevity. A survey of US dairy producers indicated that over 26% of dairy cows culled were lost because of udder health or mastitis problems, and mastitis was also a major cause of cow deaths (USDA, 2002); in 2007, 23% of cows were culled for udder health issues (USDA, 2008). In 2002, clinical mastitis was found to be the most prevalent of all diseases impacting dairy cows; the percentage of clinical mastitis cases remained unchanged in 2013 (USDA, 2016b), with almost all milk producers (99.7%) reporting at least one mastitis case.

In addition to raising questions pertaining to the "naturalness" of early and repeated calving, the roles that environmental management, hygiene, and housing play in the prevalence of mastitis cases must be considered. The complex interplay of contributing factors complicates resolution of this welfare problem. However, pasture access again may offer some benefits relative to yielding smaller numbers of cows with mastitis. For example, Washburn et al. (2002) found that cows maintained in confinement had almost twice the clinical mastitis numbers and eight times the culling rate compared to those kept on pasture. It has been suggested that these differences may be due to pastured cows potentially being exposed to fewer environmental pathogens than those housed indoors (Smith and Hogan, 1994). It may also help that cows on pasture spend more time standing, and less time lying down (Hernandez-Mendo et al., 2007), especially in areas that may have higher pathogen loads. Coincidentally, the changes in lying versus standing time on pasture may also facilitate recovery from lameness (Hernandez-Mendo et al., 2007). Thus, the case for incorporating some aspects of natural living, such as pasture access, becomes stronger, as doing so may not only better align with the *telos* of the cow, but may also improve her health and longevity.

Exclusively focusing on single-trait selection (such as exceptionally high milk production) has been shown to impair cow health, reproductive performance, increase veterinary costs, and decrease longevity (Phuong et al, 2016). These outcomes should not be surprising as it is well documented that animals selected for very high production efficiency are at greater risk of developing behavioral, physiological, and immunological problems (Rauw et al., 1998). Course-correction will likely require a multifaceted approach. To that end, management approaches must be devised to reduce metabolic stress on the cow (Oltenacu and Broom, 2010). Selection strategies are needed for multiple traits that favor overall cow welfare while maintaining acceptable levels of productivity. On this front, the dairy industries are already making significant progress. For example, Zoetis has introduced a commercially available product for genetic evaluation of wellness traits that allow farmers to assess risk factors for economically relevant diseases and conditions such as mastitis, metritis, ketosis and lameness that specifically designed for wellness traits in US cattle (www.zoetisus.com/animal-genetics/dairy/clarifide/clarifide-plus.aspx). Likewise, the Council on Dairy Cattle Breeding has recently completed efforts to develop genetic and genomic evaluations for health conditions of dairy cattle (including common problems such as displaced abomasum, mastitis, metritis, ketosis) that can potentially inform decision-making.

Finally, continued deliberation is needed about reasonable, economically feasible ways in which to better address the nature of the cow in contemporary production so as to identify win–win solutions for the dairy industries.

CONCLUSIONS

Societal pressures require the US dairy industries to better and more comprehensively address animal welfare in keeping with the expectations of all stakeholders. As consumers and members of the public increasingly attend to natural living as an indicator of animal quality of life, and by extension, the quality of products derived from the animals (Harper and Makatouni, 2002; Cardoso et al., 2016), the notion of *telos* and its application to dairy cow management and production becomes increasingly relevant. The idea that cows have natures of their own and interests that flow from those (Rollin, 1993, 2016) suggests that their preferences should be paramount in the systems in which we choose to raise them. The extent to which cows are afforded their *telos* by way of concessions toward natural living appears to have tangible impacts on various aspects of their welfare, impacts that also matter to farmers and others within the dairy industries. Bruijnis et al. (2013) call for an integrated perspective wherein "more than only the functioning and feelings of the animals in the present is of importance… and where flourishing is worthwhile striving for in itself." Broader consideration of the cow's welfare in the context of flourishing and living in accordance with her *telos*—rather than simply surviving, reproducing and producing—is therefore needed, and may be critical in ensuring the sustainability and social acceptability of dairy production.

REFERENCES

Albright, J. 1987. Dairy animal welfare: Current and needed research. *J. Dairy Sci.* 70:2711–2731.

American Veterinary Medical Association (AVMA). 2012. AVMA Policy: Castration and Dehorning of Cattle. Accessed June 18, 2018. https://www.avma.org/KB/Policies/Pages/Castration-and-Dehorning-of-Cattle.aspx

American Veterinary Medical Association (AVMA) 2017. AVMA Policy: Tail Docking of Cattle. Accessed June 18, 2018. https://www.avma.org/KB/Policies/Pages/Tail-Docking-of-Cattle.aspx

Appleby, M. C., D. M. Weary, and B. Chua. 2001. Performance and feeding behaviour of calves on ad-libitum milk from artificial teats. *Appl. Anim. Behav. Sci.* 74:91–201.

Barkema, H. W., M. A. G. von Keyserlingk, J. P. Kastelic, T. J. Lam, C. Luby, J.-P. Roy, S. J. LeBlanc, G. P. Keefe, and D. F. Kelton. 2015. *Invited review*: Changes in the dairy industry affecting dairy cattle health and welfare. *J. Dairy Sci.* 98:7426–7445.

Bates, A. J., R. A. Laven, F. Chapple, and D. S. Weeks. 2016. The effect of different combinations of local anaesthesia, sedative and non-steroidal anti-inflammatory drugs on daily growth rates of dairy calves after disbudding. *NZ Vet. J.* 64(5): 282–287.

Boogaard, B. K., B. B. Bock, S. J. Oosting, and E. Krogh. 2010. Visiting a farm: An exploratory study of the social construction of animal farming in Norway and the Netherlands based on sensory perception. *Int. J. Soc. Agric. Food* 17:24–50.

Boogaard, B. K., B. B. Bock, S. J. Oosting, J. S. C. Wiskerke, and A. J. van der Zijpp. 2011. Social acceptance of dairy farming: The ambivalence between the two faces of modernity. *J. Agric. Environ. Ethics* 24:259–282.

Bovenkerk, B., F. W. A. Brom, and B. J. Van Den Bergh. 2002. Brave new birds: The use of "animal integrity" in animal ethics. *Hastings Cent. Rep.* doi: 10.2307/3528292.

Broom, D. M. 1991. Animal welfare: Concepts and measurement. *J. Anim. Sci.* 69:4167–4175.

Broom, D. M., and K. G. Johnson. 1993. *Stress and animal welfare*. London: Chapman & Hall.

Bruijnis, M. R. N., F. L. B. Meijboom, and E. N. J. Stassen. 2013. Longevity as an animal welfare issues applied to the case of foot disorders in dairy cattle. *Agric. Environ. Ethics* 26:191–205.

Cardoso C. S., M. J. Hötzel, D. M. Weary, J. Robbins, and M. G. von Keyserlingk. 2016. Imagining the ideal dairy farm. *J. Dairy Sci.* 99(2):1–9.

Carlson, D. F., C. A. Lancto, B. Zang, E. S., Kim, M. Walton, et al. 2016. Production of hornless dairy cattle from genome-edited cell lines. *Nat. Biotechnol.* 34:479–481.

Chapinal, N., C. Goldhawk, A. M. de Passillé, M. A. G von Keyserlingk, D. M. Weary, and J. Rushen, 2010. Overnight access to pasture does not reduce milk production or feed intake in dairy cattle. *Livest. Sci.* 129:104–110.

Charlton, G. L., S. M. Rutter, M. East, and L. A. Sinclair. 2013. The motivation of dairy cows for access to pasture. *J. Dairy Sci.* 96:4387–4396.

Charlton, G. L., and S. M. Rutter. 2017. The behaviour of housed dairy cattle with and without pasture access: A review. *J. Appl. Anim. Behav. Sci.* 192:2–9.

Clutton-Brock, J. 1999. *A natural history of domesticated mammals*, 2nd edn. Cambridge: Cambridge University Press.

Coetzee, J. F. 2013a. Assessment and management of pain associated with castration in cattle. *Vet. Clin. North Am. Food Anim. Pract.* 29(1):75–101.

Coetzee, J. F. 2013b. A review of analgesic compounds used in food animals in the United States. *Vet. Clin. North Am. Food Anim. Pract.* 29(1):11–28.

Cook, N. B. 2003. Prevalence of lameness among dairy cattle in Wisconsin as a function of housing type and stall surface. *J. Am. Vet. Med. Assoc.* 223:1324–1328.

Costa, J. H. C., R. R. Daros, M. A. G. von Keyserlingk, and D. M. Weary. 2014. Complex social housing reduces food neophobia in dairy calves. *J. Dairy Sci.* 97(12):7804–7810.

Cramer, G., K. D. Lissemore, C. L. Guard, K. E. Leslie, and D. F. Kelton. 2009. Herd-level risk factors for seven different foot lesions in Ontario Holstein cattle housed in tie stalls or free stalls. *J. Dairy Sci.* 92:1404–1411.

Croney, C. C., and R. D. Reynnells. 2008. The ethics of semantics: Do we clarify or obfuscate reality to influence perceptions of farm animal production? *Poult. Sci.* 87(1):387–391.

Croney, C., and R. Anthony. 2011. Ruminating conscientiously: Scientific and socio-ethical challenges for US dairy production. *J. Dairy Sci.* 94(2):539–546.

Crumbine, S. J., and J. A. Tobey. 1930. *The most nearly perfect food: The story of milk*. Baltimore: Williams & Wilkin, p. 17.

Dawkins, M. S. 1988. Behavioural deprivation: A central problem in animal welfare. *Appl. Anim. Behav. Sci.* 20:209–225.

de Passillé, A. M., R. J. Christopherson, and J. Rushen. 1993. Nonnutritive sucking and the postprandial secretion of insulin, CCK and gastrin in the calf. *Physiol. Behav.* 54:1069–1073.

de Passillé, A. M. B., and J. Rushen. 2006. Calves' behaviour during nursing is affected by feeding motivation and milk availability. *Appl. Anim. Behav. Sci.* 101:264–275.

De Paula Vieira, A., V. Guesdon, A. M. B. de Passillé, M. A. G. von Keyserlingk, and D. M. Weary. 2008. Behavioural indicators of hunger in dairy calves. *Appl. Anim. Behav. Sci.* 109:180–189.

De Paula Vieira, A., M. A. G. von Keyserlingk, and D. M. Weary. 2010. Effects of pair versus single housing on performance and behavior of dairy calves before and after weaning from milk. *J. Dairy Sci.* 93:3079–3085.

De Paula Vieira A., A. M. de Passillé, and D. M. Weary. 2012a. Effects of the early social environment on behavioral responses of dairy calves to novel events. *J. Dairy Sci.* 95:5149–5155.

De Paula Vieira, A., M. A. G. von Keyserlingk, and D. M. Weary. 2012b. Presence of an older weaned companion influences feeding behavior and improves performance of dairy calves before and after weaning from milk. *J. Dairy Sci.* 95:3218–3224.

Diaz, M. C., M. E. Van Amburgh, J. M. Smith, J. M. Kelsey, and E. L. Hutten. 2001. Composition of growth of Holstein calves fed milk replacer from birth to 105-kilogram body weight. *J. Dairy Sci.* 84:830–842.

Duncan, I. J. H. 1993 Welfare is to do with what animals feel. *J. Agric. Environ. Ethics* 6(2):8–14.

DuPuis, E. M. 2002. *Nature's perfect food: How milk became America's drink.* New York and London: New York University Press.

Eicher, S. D., J. L. Morrow-Tesch, J. L. Albright, and R. E. Williams. 2001. Tail-docking alters fly numbers, fly-avoidance behaviours, and cleanliness, but not physiological measures. *J. Dairy. Sci.* 84:1822–1828.

Eicher, S. D., and J. W. Dailey. 2002. Indicators of acute pain and fly avoidance behaviors in Holstein calves following tail-docking. *J. Dairy Sci.* 85(11)2850–2858.

Eicher, S. D., H. W. Cheng, A. D. Sorrells, and M. M. Schutz. 2006. Behavioral and physiological indicators of sensitivity or chronic pain following tail docking. *J. Dairy Sci.* 89(8):3047–3051.

Ellis, K. A., K. Billington, B. McNeil, and D. E. F. McKeegan. 2009. Public opinion on UK milk marketing and dairy cow welfare. *Anim. Welf.* 18:267–282.

Espejo, L. A., M. I. Endres, and J. A. Salfer. 2006. Prevalence of lameness in high-producing Holstein cows housed in freestall barns in Minnesota. *J. Dairy Sci.* 89:3052–3058.

Espejo, L. A., and M. I. Endres. 2007. Herd-level risk factors for lameness in high-producing Holstein cows housed in freestall barns. *J. Dairy Sci.* 90:306–314.

European Food Safety Authority (EFSA). 2006. *The risks of poor welfare in intensive calf farming systems. An updated of the scientific veterinary committee report on the welfare of calves.* EFSA-Q-2005-014. European Food Safety Authority, Parma, Italy.

Faerevik, G., M. B. Jensen, and K. E. Boe. 2006. Dairy calves social preferences and the significance of a companion animal during separation from the group. *Appl. Anim. Behav. Sci.* 99:205–221.

Falk, A. C., D. M. Weary, C. Winkler, and M. A. G. von Keyserlingk. 2012. Preference for pasture versus freestall housing by dairy cattle when stall availability indoors is reduced. *J. Dairy Sci.* 95:6409–6415.

Flower, F. C., and D. M. Weary. 2001. Effects of early separation on the dairy cow and calf. II: Separation at 1 day and 2 weeks after birth. *Appl. Anim. Behav. Sci.* 70:275–284.

Flower, F. C., and D. M. Weary. 2003. The effects of early separation on the dairy cow and calf. *Anim. Welf.* 12:339–348.

Fontaneli, R. S., L. E. Sollenberger, R. C, Littell, and C. R. Staples. 2005. Performance of lactating dairy cows managed on pasture-based or in free stall barn-feeding systems. *J. Dairy Sci.* 88:1264–1276.

Fraser, D., D. M. Weary, E. A. Pajor, and B. N. Milligan. 1997. A scientific conception of animal welfare that reflects ethical concerns. *Anim. Welf.* 6:187–205.

Fröberg, S., and L. Lidfors. 2009. Behaviour of dairy calves suckling the dam in a barn with automatic milking or being fed milk substitute from an automatic feeder in a group pen. *Appl. Anim. Behav. Sci.* 117:150–158.

Fulwider, W. K., T. Grandin, B. E. Rollin, T. E. Engle, N. L. Dalsted, and W. D. Lamm. 2008. Survey of dairy management practices on one hundred thirteen North Central and Northeastern United States dairies. *J. Dairy Sci.* 91:1686–1692.

Gabler M, P. R. Tozer, and A. J. Heinrichs. 2000. Development of a cost analysis spreadsheet for calculating the costs to raise a replacement dairy heifer. *J. Dairy Sci.* 83:1104–1109.

Gaillard, C., R. K. Meagher, M. A. G. von Keyserlingk, and D. M. Weary. 2014. Social housing improves dairy calves' performance in two cognitive tests. *PLoS One.* doi:10.1371/journal.pone.0090205.

Gavrell-Ortiz, S. E. 2004. Beyond welfare: Animal integrity, animal dignity, and genetic engineering. *Ethics Environ.* 9(1):94–120.

Gitau, T., J. J. McDermott, and S. M. Mbiuki. 1996. Prevalence, incidence and risk factors for lameness in dairy cattle in small scale farms in Kikuyu Division, Kenya. *Prev. Vet. Med.* 28:101–115.

Glenn, C. B. 2004. Constructing consumables and consent: A critical analysis of factory farm industry discourse. *J. Commun. Inq.* 28:63–81.

Gonyou, H. W. 1993 Animal welfare: Definitions and assessment. *J. Agric. Ethics* 6(2):37–43.

Hall, S. J. G. 2002. Behaviour of cattle. In P. Jensen (Ed.). *The ethology of domestic animals: An introductory text*. Wallingford, Oxon: CABI Publishing, pp. 131–142.

Hammon, H. M., G. Schiessler, A. Nussbaum, and J. W. Blum. 2002. Feed intake patterns, growth performance, and metabolic and endocrine traits in calves fed unlimited amounts of colostrum and milk by automate, starting in the neonatal period. *J. Dairy Sci.* 85:3352–3362.

Hänninen, L., H. Hepola, S. Raussi, and H. Saloniemi. 2008. Effect of colostrum feeding method and presence of dam on the sleep, rest and sucking behaviour of newborn calves. *Appl. Anim. Behav. Sci.* 112:213–222.

Harper, G. C., and A. Makatouni. 2002. Consumer perception of organic food production and farm animal welfare. *Br. Food J.* 104 (3/4/5):287–299.

Heinrich, A., T. F Duffield, K. D. Lissemore, and S. T Millman. 2010. The effect of meloxicam on behavior and pain sensitivity of dairy calves following cautery dehorning with a local anesthetic. *J. Dairy Sci.* 93(6):2450–2457.

Hernandez-Mendo, O., M. A. G von Keyserlingk, D. M., Veira, and D. M Weary. 2007. Effects of pasture on lameness of dairy cows. *J. Dairy Sci.* 90:1209–1214.

Hoe, F. G. H., and P. L. Ruegg. 2006. Opinions and practices of Wisconsin dairy producers about biosecurity and animal well-being. *J. Dairy Sci.* 89:2297–2308.

Hoeschele, I. 1990. Potential gain from insertion of major genes into dairy cattle. *J. Dairy Sci.* 73:2601.

Huber, J., T. Arnholdt, E. Möstl, C. C. Gelfert, and M. Drillich. 2013. Pain management with flunixin meglumine at dehorning of calves. *J. Dairy Sci.* 96:132–140.

Jensen, M. B., K. S. Vestergaard, C. C. Krohn, and L. Munksgaard. 1997. Effect of single versus group housing and space allowance on responses of calves during open-field tests. *Appl. Anim. Behav. Sci.* 54:109–121.

M. B. Jensen. 2003. The effects of feeding method, milk allowance and social factors on milk feeding behaviour and cross-sucking in group housed dairy calves. *Appl. Anim. Behav. Sci.* 80(3):191–206.

Johnsen, J. F., K.A. Zipp, T. Kälber, A.M. de Passillé, U. Knierim, K. Barth, and C.M. Mejdell. 2016. Is rearing calves with the dam a feasible option for dairy farms?—Current and future research. *Appl. Anim. Behav. Sci.* 181:1–11.

Kilgour, R., and C. Dalton. 1984. *Livestock Behaviour: A Practical Guide*. London: Granada Publishing Ltd.

Kilgour, R. J. 2012. In pursuit of "normal": A review of the behaviour of cattle at pasture. *J. Appl. Anim. Behav. Sci.* 138:1–11.

Kjaernes, U., M. Miele, and J. Roex. 2007. Attitudes of consumers, retailers and producers to farm animal welfare. Welfare Quality Reports No. 2. Cardiff: Cardiff University.

Knaus, W. 2009. Dairy cows trapped between performance demands and adaptability. *J. Sci. Food Agric.* 89:1107–1114.

Ladewig, J., and L. R. Matthews. 1992. The importance of physiological measurements in farm animal stress research. *Proc. NZ Soc. Anim. Prod.* 52:77–79.

Langbein, J., and M. L. Raasch. 2000. Investigations on the hiding behaviour of calves at pasture. *Arch. Tierzucht.* 43:203–210.

Lassen, J., P. Sandøe, and B. Forkman. 2006. Happy pigs are dirty!—Conflicting perspectives on animal welfare. *Livest. Sci.* 103:221–230.

Leaver, J. D. 1988. *Management and welfare of animals*, 3rd edn. London: Bailliere Tindall.

Legrand, A. L., M. A. G. von Keyserlingk, and D. M. Weary. 2009. Preference and usage of pasture versus free-stall housing by lactating dairy cattle. *J. Dairy Sci.* 92:3651–3658.

Lidfors, L. M., D. Moran, J. Jung, P. Jensen, and H. Castren. 1994. Behaviour at calving and choice of calving place in cattle kept in different environments. *Appl. Anim. Behav. Sci.* 42:11–28.

Long, C. R., and K. E. Gregory. 1978. Inheritance of the horned, scurred, and polled condition in cattle. *J. Hered.* 69(6):395–400.

Marcé, C., P. Ezanno, H. Seegers, D. U. Pfeiffer, and C. Fourichon. 2011. Within-herd contact structure and transmission of *Mycobacterium avium* subspecies *paratuberculosis* in a persistently infected dairy cattle herd. *Prev. Vet. Med.* 100:116–125.

Maria, G. A. 2006. Public perception of farm animal welfare in Spain. *Livest. Sci.* 103:250–256.

Mathews, L. R., R. A. Phipps, and D. Verkerk. 1995. The effects of taildocking and trimming on milker comfort and dairy cattle health, welfare and production. *Anim. Behav. Welf. Res. Cent.* 1–25.

McGlone, J. J. 1993. What is animal welfare? *J. Agric. Environ. Ethics* 6(2):26–36.

McKendree, M., N. J. Olynk, and D. L. Ortega. 2012. Consumer preferences and perceptions of food safety, production practices and food product labeling: A spotlight on dairy product purchasing behavior in 2011. Purdue University, CAB RP 12.1 http://agribusiness.purdue.edu/files/resources/r-1-2012-mckendree-olynk-ortega.pdf. Accessed October 12, 2017.

McMeekan, C. M., K. J., Stafford, D. J. Mellor, R. A. Bruce, R. N. Ward, and N. G. Gregory. 1999. Effects of a local anaesthetic and a nonsteroidal anti-inflammatory analgesic on the behavioural responses of calves to dehorning. *NZ Vet. J.* 47:92–96.

Meagher, R. K., R. D. Rolnei, J. H. C. Costa, M. A. G. Von Keyserlingk, and D. M. Weary. 2014. Social housing reduces fear of novelty and improves reversal learning performance in dairy calves. In I. Estevez, X. Manteca, R. H. Marin, X. Averos (Eds.). *48th Congress of the International society for applied ethology.* Vitoria-Gasteiz: Wageningen Academic Publishers, p. 96.

Metz, J., and J. Metz. 1986. Maternal influence on defecation and urination in the newborn calf. *Appl. Anim. Behav. Sci.* 16:325–333.

Metz, J. 1987. Productivity aspects of keeping dairy cow and calf together in the post-partum period. *Livest. Prod. Sci.* 16:385–394.

Misch L. J., T. F. Duffield, S. T. Millman, and K. D. Lissemore. 2007. An investigation into the practices of dairy producers and veterinarians in dehorning dairy calves in Ontario. *Can. Vet. J.* 48:1249–1254.

Moberg G. P. 1985. Biological response to stress: key to assessment of animal well-being? In Moberg G. P. (Ed.). *Animal stress.* Bethesda: American Physiological Society, pp 27–49.

Moberg G. P. 2000. Biological response to stress: Implications for animal welfare. In Moberg G. P., Mench J. A. (Ed.). *The biology of animal stress.* Wallingford, Oxon: CABI Publishing, pp. 123–146.

Molloy, C. 2011. *Popular media and animals.* Basingstoke: Palgrave Macmillan.

Motupalli, P. R., L. A. Sinclair, G. L. Charlton, E. C., Bleach, and S. M. Rutter. 2014. Pasture access increases dairy cow milk yield but preference for pasture is not affected by herbage allowance. *J. Anim. Sci.* 92:5175–5184.

National Milk Producers Federation (NMPF). 2016. Farmer's assuring responsible management (FARM) animal care reference manual. Available online at www.nationaldairyfarm.com/sites/default/files/Version-3-Manual.pdf.

Oltenacu, P. A., and D. M. Broom. 2010. The impact of genetic selection for increased milk yield on the welfare of dairy cows. *Anim. Welf.* 19:39–49.

Pempek, J. A., J. A. Eastridge, M. L. Swartzwelder, S. S. Daniels, and K. M. Yohe. 2016. Housing system may affect behavior and growth performance of Jersey heifer calves. *J. Dairy Sci.* 99:569–578.

Phillips, C. J., J. Wojciechowska, J. Meng, and N. Cross. 2009. Perceptions of the importance of different welfare issues in livestock production. *Anim.: Int. J. Anim. Biosci.* 3(8):1152–1166.

Phuong, H., P. Blavy, O. Martin., P. Schmidely, and N. Friggens. 2016. Modelling impacts of performance on the probability of reproducing, and thereby on productive lifespan, allow prediction of lifetime efficiency in dairy cows. *Animal* 10(1), 106–116.

Popescu, S., C. Borda, E. A. Diugan, M. Spinu, I. S. Groza, and C. D. Sandru. 2013. Dairy cows welfare quality in tie-stall housing system with or without access to exercise. *Acta Vet. Scand.* 55:43.

Proudfoot, K. L., D. M. Weary, and M. A. G. von Keyserlingk. 2014a. Maternal isolation behavior of Holstein dairy cows kept indoors. *J. Anim. Sci.* 92:277–281.

Proudfoot, K. L., M. B. Jensen, D. M. Weary, and M. A. G. von Keyserlingk. 2014b. Dairy cows seek isolation at calving and when ill. *J. Dairy Sci.* 97:2731–2739.

Rauw W. M., E. Kanis, E. N Noordhuizen-Stassen, and F. J Grommers. 1998. Undesirable side effects of selection for high production efficiency in farm animals: A review. *Livest. Prod. Sci.* 56:15–33.

Ridge, S. E, I. M. Baker, and M. Hannah. 2005. Effect of compliance with recommended calf-rearing practices on control of bovine Johne's disease. *Aus. Vet. J.* 83(1–2):85–90.

Robbins, J. A., D. M. Weary, C. A. Schuppli, and M. A. G. von Keyserlingk. 2015. Stakeholder views on treating pain due to dehorning dairy calves. *Anim. Welf.* 24:399–406.

Rollin, B. E. 1993 Animal welfare, science, and values. *J. Agric. Environ. Ethics* 6(2):44–50.

Rollin, B. E. 1995. The Frankenstein Syndrome. Ethical and Social Issues in the Genetic Engineering of Animals. Cambridge: Cambridge University Press.

Rollin, B. E. 2007. Cultural variation, animal welfare and telos. *Anim. Welf.* 16(S):129–133.

Rollin, B. E. 2008. The ethics of agriculture: the end of true husbandry. In: Dawkins, M. S., Bonney, R. (Eds.), The Future of Animal Farming: Renewing the Ancient Contract. Blackwell Publishing, Oxford, UK, pp. 7–19.

Rollin, B. E. 2016. A New Basis for Animal Ethics: Telos and Common Sense. Columbia, MO. University of Missouri Press.

Rutgers, 2003. New Jerseyans' opinions on humane standards for treatment of livestock. Conducted for Farm Sanctuary. Eagleton Institute of Politics Center for Public Interest Polling, Rutgers, NJ.

Sandøe, P., P. M. Hocking, B. Förkman, K. Haldane, H. H. Kristensen, and C. Palmer. 2014. The blind hens' challenge: Does it undermine the view that only welfare matters in our dealings with animals? *Environ. Values* 23(6):727–742.

Schreiner, D. A., and P. L. Ruegg. 2002. Effects of tail docking on milk quality and cow cleanliness. *J. Dairy Sci.* 85:2503–2511.

Schuppli, C. A., M. A. G. von Keyserlingk, and D. M. Weary. 2014. Access to pasture for dairy cows: Responses from an online engagement. *J. Anim. Sci.* 92(11):5185–5192.

Selman, M., A. McEwan, and E. Fisher. 1970. Studies on natural suckling in cattle during the first eight hours post partum. I. Behavioural studies (dams). *Anim. Behav.* 18:276–289.

Shamay, A., D. Werner, U. Moallem, H. Barash, and I. Bruckental. 2005. Effect of nursing management and skeletal size at weaning on puberty, skeletal growth rate and milk production during first lactation of dairy heifers. *J. Dairy Sci.* 88:1460–1469.

Simensen, E., O. Østerås, K. E. Bøe, C. Kielland, L. E. Ruud, and G. Naess. 2010. Housing system and herd size interactions in Norwegian dairy herds; associations with performance and disease incidence. *Acta Vet. Scand.* 52:14.

Smid, A. M. C., D. M. Weary, J. Costa, and M. A. G. von Keyserlingk. 2018. Dairy cow preference for different types of outdoor access. *J. Dairy Sci.* 101(2):1448–1455.

Smith, K. L., and J. S. Hogan. 1994. Mastitis control. Page 11 in Intensive Grazing Seasonal Dairying: The Mahoning County Dairy Program. Ohio Agricultural Research and Development Center Research Bulletin 1190, The Ohio State University, D. L. Zartman, Editor.

Smits, M. C. J., K. Frankena, J. H. M. Metz, and J. P. T. M. Noordhuizen. 1992. Prevalence of digital disorders in zero-grazing dairy farms. *Livest. Prod. Sci.* 32:231–244.

Sogstad, Å. M., T. Fjeldaas, O. Østerås, and K. Plym Forshell. 2005. Prevalence of claw lesions in Norwegian dairy cattle housed in tie stalls and free stalls. *Prev. Vet. Med.* 70:191–209.

Spooner, J. M., C. A. Schuppli, and D. Fraser. 2012. Attitudes of Canadian beef producers toward animal welfare. *Anim. Welf.* 21(2):273–283.

Spooner, J. M., C. A. Schuppli, and D. Fraser. 2014. Attitudes of Canadian citizens toward farm animal welfare: A qualitative study. *Livest. Sci.* 163:150–158.

Stafford, K. J., and D. J. Mellor. 2011. Addressing the pain associated with disbudding and dehorning in cattle. *Appl. Anim. Behav. Sci.* 135(3):226–231.

Stilwell G., M. S. Lima, R. C. Carvalho, and D. M. Broom. 2012. Effects of hot-iron disbudding, using regional anaesthesia with and without carprofen, on cortisol and behaviour of calves. *Res. Vet. Sci.* 92:338–341.

Svensson, C., K. Lundborg, U. Emanuelson, and S. O. Olsson. 2003. Morbidity in Swedish dairy calves from birth to 90 days of age and individual calf-level risk factors for infectious diseases. *Prev. Vet. Med.* 58:179–197.

Te Velde, H., N. Aarts, and C. Van Woerkum. 2002. Dealing with ambivalence: Farmers' and consumers' perceptions of animal welfare in livestock breeding. *J. Agric. Environ. Ethics* 15:203–219.

Thomas, T. J., D. M. Weary, and M. C. Appleby. 2001. Newborn and 5-week-old calves vocalize in response to milk deprivation. *Appl. Anim. Behav. Sci.* 74:165–173.

Tom, E. M., I. J. H. Duncan, T. M. Widowski, K. G. Bateman, and K. E. Leslie. 2002. Effects of tail docking using a rubber ring with or without anesthetic on behavior and production of lactating cows. *J. Dairy Sci.* 85:2257–2265.

Tucker, C. B., D. Fraser, and D. M. Weary. 2001. Tail docking dairy cattle: Effects on cow cleanliness and udder health. *Dairy Sci.* 84:84–87.

USDA. 2002. Part I: Reference of Dairy Health and Management in the United States, 2002. USDA:APHIS:VS:CEAH. #N377.1202. National Animal Health Monitoring System, Fort Collins, CO.

USDA. 2007. Dairy 2007, Part IV Reference of dairy cattle health and management practices in the United States, 2007, USDA-APHIS-VS, CEAH. Fort Collins, CO, USA (2007) #N480.1007. Accessed August12, 2017. www.aphis.usda.gov/animal_health/nahms/dairy/downloads/dairy07/Dairy07_dr_PartIV.pdf.

USDA. 2008. Dairy 2007, Part II: Changes in the U.S. Dairy Cattle Industry, 1991–2007 USDA-APHIS-VS, CEAH. Fort Collins, CO #N481.0308. www.aphis.usda.gov/animal_health/nahms/dairy/downloads/dairy07/Dairy07_dr_PartII_rev.pdf. Accessed October 1, 2017.

USDA. 2016a. Dairy 2014, "Dairy Cattle Management Practices in the United States, 2014" USDA–APHIS–VS–CEAH–NAHMS. Fort Collins, CO #692.0216. www.aphis.usda.gov/animal_health/nahms/dairy/downloads/dairy14/Dairy14_dr_PartI.pdf.

USDA. 2016b. "Dairy 2014, Milk Quality, Milking Procedures, and Mastitis in the United States, 2014" USDA–APHIS–VS–CEAH–NAHMS. Fort Collins, CO #704.0916. www.aphis.usda.gov/animal_health/nahms/dairy/downloads/dairy14/Dairy14_dr_Mastitis.pdf. Accessed October 1, 2017.

Valde, J. P., D. W. Hird, M. C. Thurmond, and O. Østerås. 1997. Comparison of ketosis, clinical mastitis, somatic cell count, and reproductive performance between free stall and tie stall barns in Norwegian dairy herds with automatic feeding. *Acta Vet. Scand.* 38:181–192.

Vanhonacker, F., W. Verbeke, E. Van Poucke, and F. A. Tuyttens. 2008. Do citizens and farmers interpret the concept of farm animal welfare differently? *Livest. Sci.* 116(1–3):126–136.

Vasseur, E., F. Borderas, R.I. Cue, D. Lefebvre, D. Pellerin, J. Rushen, K.M. Wade, and A.M. de Passillé. 2010. A survey of dairy calf management practices in Canada that affect animal welfare. *J. Dairy. Sci.* 93:1307–1315.

Veissier, I., S. Andanson, H. Dubroeucq, and D. Pomiès. 2008. The motivation of cows to walk as thwarted by tethering. *J. Anim. Sci.* 86:2723–2729.

Ventura, B. A, M. A. G. von Keyserlingk, C. A. Schuppl, and D. M. Weary. 2013. Views on contentious practices in dairy farming: The case of early cow-calf separation. *J. Dairy Sci.* 96:6105–6116.

Ventura, B. A, M. A. G. von Keyserlingk, and D. M. Weary. 2015. Animal welfare concerns and values of stakeholders within the dairy industry. *J. Agric. Environ. Ethics* 28:109–126.

Ventura, B. A., M. A. G. von Keyserlingk, H. Wittman, and D. M. Weary. 2016. What difference does a visit make? Changes in animal welfare perceptions after interested citizens tour a dairy farm. *PLoS One* 11:e0154733.

Verbeke, W. 2009. Stakeholder, citizen and consumer interests in farm animal welfare. *Anim. Welf.* 18:325–333.

Vitale, A. F., M. Tenucci, M. Papino, and S. Lovari. 1986. Social behavior of the calves of semi-wild Maremma cattle, Bos primigenious taurus. *Appl. Anim. Behav. Sci.* 16:217–231.

Von Borell, E. 1995. Neurocrine integration of stress and significance of stress for the performances of farm animals. *Appl. Anim. Behav. Sci.* 44: 219–227.

Von Keyserlingk, M. A. G., and D. M. Weary. 2007. Maternal behaviour in cattle. *Horm. Behav.* 52:106–113.

Von Keyserlingk, M. A. G., D. M. Weary, J. Rushen, and A. M. de Passillé. 2009. Invited review: the welfare of dairy cattle—Key concepts and the role of science. *J. Dairy Sci.* 92:4101–4111.

Wagner, K., D. Seitner, K. Barth, R. Palme, A. Futschik, and S. Waiblinger, 2015. Effects of mother versus artificial rearing during the first 12 weeks of life on challenge responses of dairy cows. *Appl. Anim. Behav. Sci.* 164, 1–11.

Washburn,S.P. S. L. White, J. T. Green Jr., and G. A. Benson. 2002. Reproduction, mastitis, and body condition of sseasonally calved Holstein and Jersey cows in confinement or pasture systems. *J. Dairy Sci.* 85(2002):105–111.

Weary, D. M. and B. Chua. 2000. Effects of early separation on the dairy cow and calf. 1: Separation at 6h, 1 day and 4 days after birth. *Appl. Anim. Behav. Sci.* 69:177–188.

Weary, D. M., C. A. Schuppli, and M. A. G. von Keyserlingk. 2011. Tail docking dairy cattle: Responses from an online engagement. *J. Anim. Sci.* 89(11):3831–3837.

Wesselink, R., K. J. Stafford, D. J. Mellor, S. Todd, and N. G. Gregory. 1999. Colostrum intake by dairy calves. *NZ Vet. J.* 47:31–34.

Whay, H. R., D. C. J. Main, L. E. Green, and A. J. F. Webster. 2003. Assessment of the welfare of dairy cattle using animal-based measurements: Direct observations and investigation of farm records. *Vet. Rec.* 153:197–202.

Whay, H. R. and J. K. Shearer. 2017. The impact of lameness on welfare of the dairy cow. *Vet. Clin. North Am. Food Anim. Pract.* 33(2):153–164.

Widmar, N. O., C. J. Morgan, C. A. Wolf, E. A. Yeager, S. R. Dominick and C. C. Croney. 2017. US resident perceptions of dairy cattle management practices. *Agric. Sci.* 8:645–656.

Winder, C. B., S. J. LeBlanc, D. B. Haley, K. D. Lissemore, M. A. Godkin, and T. F. Duffield. 2017. Clinical trial of local anesthetic protocols for acute pain associated with caustic paste disbudding in dairy calves. *J. Dairy Sci.* 100:6429–6441.

Windsor, P. A., and R. J. Whittington. 2009. Evidence for age susceptibility of cattle to Johne's disease. *Vet. J.* doi:10.1016/j.tvjl.2009.01.007.

Wolf, C. A., G. T. Tonsor, and N. J. Olynk. 2011. Understanding US consumer demand for milk production attributes. *J. Agric. Res. Econ.* 36:326–342.

Wolf, C.A., G.T. Tonsor, M.G.S. McKendree, D.U. Thomson, and J.C. Swanson. 2016. Public and farmer perceptions of dairy cattle welfare in the United States. *J. Dairy Sci.* 99(7):5892–5903.

CHAPTER 23

Herdsmanship and Human Interaction with Dairy Cattle

Kurt D. Vogel
University of Wisconsin

CONTENTS

INTRODUCTION

Importance of Stockmanship

A recent survey of dairy industry stakeholders, predominantly from North America with minimal representation from European countries, identified concerns about poor handling and stockmanship as one of the pressing animal welfare concerns for the dairy industry (Ventura et al., 2015). The authors of the study reported that four out of five stakeholder focus groups within the study shared this sentiment while the dairy producer focus group did not comment on stockmanship. Animal handling was identified as a welfare issue for dairy cattle in the contexts of inducing stress and other trickle-down effects on animals (Ventura et al., 2015). The duration of stockmanship's impact on animal welfare was specifically identified by one focus group member as a key factor in identifying stockmanship as a welfare concern. The focus group member stated, "When it comes to animal handling and stockmanship, you know that's a whole herd, so 100% of a herd is affected by poor stockmanship…and that exists for the lifetime of the cow" (Ventura et al., 2015).

The quality of stockmanship matters to the animals that receive it. In addition, stockmanship appears to have a relationship with handler safety. Sorge et al. (2014) reported that 73.3% of injuries on Minnesota dairy farms were a result of interactions with cattle. Stockmanship training is becoming increasingly common for dairy producers. In the previously mentioned study, nearly 30% of dairy farmers reported that they had participated in previous stockmanship training. The development of stockmanship skills often starts with initiation by family members as 42.6% of Minnesota dairy farmers reported that they learned cattle handling techniques from family members (Sorge et al., 2014). In the same study, 62.3% of dairy farms reported that they provide training regarding handling techniques for dairy cattle.

Defining Stockmanship

A wide variety of definitions have been established in the scientific literature to describe the term "stockmanship." At its core, stockmanship appears to involve the direct handling and management of animals by their caretakers and the knowledge of the caretakers regarding the animals they care for (Fukasawa et al., 2017; Ventura et al., 2015). One focus group member described stockmanship as, "…how we handle cows, how we work around them, how we treat them… on an individual basis and work with them as a group" (Ventura et al., 2015). Sorge et al. (2014) defined stockmanship simply as, "proper cattle handling techniques." In general, it seems that the term "stockmanship" applies to the general care and handling of a specific type of animal. Since stockmanship refers to the interaction with animals in general, it is possible that stockmanship could be excellent, poor, or somewhere within the continuum that separates those two poles.

DAIRY CATTLE HANDLING

The handling of dairy cattle is best facilitated by patience and calm, confident interaction. In general, dairy cows are highly routinized animals. The willingness of dairy cows to accept routine is valuable because positive human interaction over time will reduce the effect of fear on handling. The ability of a handler to apply flight zone pressure is heavily reliant on the ability of the animal to experience fear. As an animal loses the ability to fear handlers to some extent, they become more challenging to handle with conventional methods that rely on flight zone pressure. As a result, it is most advisable to train cows to perform specific activities in a predictable pattern throughout the day. When cattle become accustomed to moving to a specific location at the same time each day, they tend to be less fearful and safer to handle during the time they are moving (Lindahl et al., 2016). For less common handling events, such as movement to and through hoof trimming equipment and initial introduction of heifers to a milking parlor, dairy cattle tend to experience greater physiological stress and behavioral unpredictability (Lindahl et al., 2016). When dairy cows were moved to hoof trimming, handler safety became compromised as the handler began to forcefully apply a tactile object to the animals. Application of a tactile object in a forceful manner resulted in increased occurrence of incidents in which the handler was kicked by a cow. It is important to take the time to train animals to willingly move through the unfamiliar or rarely used handling facility on their own terms before they are forced into the facility.

It appears that allowing animals that are unfamiliar with new facilities and stockpeople to explore a new facility without time pressure may be most appropriate and possibly just as effective as offering a feed reward. Pajor et al. (2000) reported that heifers that had no prior experience with a single file race were presented with either no interaction, gentle petting, or food at the end of the race. The treatment at the end of the race had no effect on the time spent in the race or the length of time it took for the heifers to enter the race. However, the heifers entered the race more quickly, spent less time in the race, and required less force to move through the race over nine successive trips through the race. The results of this study suggested that exposure to a facility without activation of fear may be more valuable than the provision of a reward at the end of the facility. The authors postulated that human presence could have been considered aversive by the heifers if they did not have substantial prior exposure, or negative exposure in general, to stockpeople.

Dairy cattle typically require less physical restraint than beef cattle. In most dairy operations, cows are restrained for the majority of veterinary procedures in a locking stanchion in the location where the herd is fed. This is commonly referred to as a head lock. Most head locks are self-locking after the operator sets the system to allow the stanchions to lock after they are closed during feeding.

IMPACT OF HUMAN INTERACTION ON ANIMAL WELFARE

In the early 2010s, I received a phone call one day from a herdsman from a large dairy that was recently the focal point of an undercover exposé that resulted in the release of a video of his facility by an activist group. He was looking for guidance on how he and his staff should proceed following the video release and associated increase in scrutiny from his customers and the public in general. At one point in the conversation, I asked him to watch the video with me so we could discuss specific events that occurred. He played the video on his computer and I played it on mine as we talked on the phone. During this exercise, we agreed that all of the major issues that we observed in the video involved human interaction with animals. In all cases, lack of appropriate training and supervision were identified as the primary causative factors for the events that we observed.

There were two key concepts that we identified as we reviewed the handling of non-ambulatory cows. First, we observed one scene where an employee unleashed a profanity-laden tirade on a cow that refused to stand. The use of such language toward another sentient being has the potential to escalate to physical abuse. The use of yelling and demeaning language toward animals appeared to be culturally driven within that facility. Second, it appeared that the handlers of non-ambulatory cows lacked empathy for the animals they were trying to move.

To remediate the previously described issues, a cultural shift was necessary. The employees that were responsible for caring for the cattle were formally trained in animal handling. Specifically, they were no longer allowed to use derogatory language toward the cows they were handling. This change was simple to implement, but vigilance in supervision was needed to prevent relapse to old behaviors.

The herdsman had a plan to address the apparent lack of empathy for the non-ambulatory cows his staff were tasked with handling. His simple—but powerful—suggestion was that all handlers involved in moving a non-ambulatory cow remove a glove and touch or pet the cow. He shared that his herd veterinarian laughed at the suggestion, but he proceeded to implement it. His rationale was that his employees felt considerable time pressure to move non-ambulatory cows so they could get back to their normal job duties. By prompting the workers to take time to pet the cow before working on her, he thought the action would give the workers an opportunity to redirect their focus to the cow instead of all of the other tasks they had to do. In essence, he was giving his workers an opportunity to develop empathy.

Empathy can be defined as, "…the capacity to vicariously experience the emotions of another" (Coleman et al., 1998). Furthermore, it has been argued that a specific bond between the individual experiencing an emotion and an empathetic observer is not required. What is required is for an observer to experience at least some indication of the emotion displayed by the individual they are observing.

Physiological Responses

When animals experience fear-based interactions with people, their bodies respond with activation of biological pathways that are intended to help the individual survive during periods of stress. At their core, these responses are centered on increasing energy availability throughout the body and getting that energy to the tissues that will need it to overcome a stressor.

In cattle, the impact of positive human interaction can be profound. One example can be found in meat science research. The dark cutting condition—defined by dark lean muscle color and elevated postmortem pH—is an area of focus for many meat scientists. The condition does not occur regularly throughout the year, but seems to occur in "rashes" during the times of year when major shifts in environmental conditions occur. Since beef from dark cutting carcasses may be difficult for researchers to procure on a consistent basis due to the relative rarity of its occurrence, some researchers have worked to establish models to create the dark cutting condition in cattle so they would have a means of producing dark cutting beef when it was needed for study.

In one study of methods to induce dark cutting beef, researchers used castrated male Holstein calves in an attempt to induce dark cutting beef through social isolation stress (Apple et al., 2005). The investigators exposed the calves to either 0, 2, 4, or 6 hours of restraint and isolation stress prior to slaughter. The restraint and isolation stress involved taping all four legs of each animal together and laying them alone on their side on a padded surface. The rationale behind this approach of inducing stress was centered on the idea that cattle were gregarious by nature and would experience considerable distress if they were not allowed to stay with a group of conspecifics during a stressful event (being tied up and denied the ability to move). The experiment was replicated four times over a 4 week period with two calves per social isolation and restraint stress treatment per week. The 6-hour social isolation and restraint stress treatment consistently produced dark cutting meat in the carcasses of the calves with the exception of 1 week (two calves).

The only explanation that the authors of the study could produce to explain the difference during 1 week provided support for the concept that positive human interaction can have a substantial calming effect on cattle. During that week of the study, the ambient temperature declined suddenly and drastically. This sudden drop in temperature is typically associated with an increase in the occurrence of dark cutting. This was puzzling at first for the authors. As they reviewed their methods, they found a difference in how the calves were handled during the cold week compared to other weeks—the students that were working on the project felt badly for the calves since they had to be tied up and left alone in the cold, so they sat with the calves intermittently and petted them. The authors concluded that human presence and gentle contact helped the calves to cope with the psychological distress associated with the treatment and the environmental stress of the sudden temperature change. This study suggested that human presence and gentle contact has the potential to improve the ability of Holstein calves to cope with stressors.

Rushen et al. (1999) reported that the presence of either an aversive or gentle handler resulted in an increase in heart rate in dairy cows during milking. Control cows that were milked without either the presence of a handler that had used gentle methods to move them or a handler that used aversive methods to handle them displayed no change in heart rate from the pre-milking baseline. However, the presence of the aversive handler caused a greater increase (5.94 beats/minute) in heart rate than the gentle handler (3.42 beats/minute).

Behavioral Responses

Multiple studies have been conducted to improve understanding of the behavioral implications of human interaction and indicators of both positive and negative stockmanship styles (Ellingsen et al., 2014). In one study, researchers worked to determine if gentle interactions—being stroked gently for 3 minutes per day for the first 14 days of life—with stock people would impact the flight distance of dairy heifers after a year had passed (Lürzel et al., 2016). Ultimately, the heifers that were stroked by stock people as calves did not display a difference in avoidance distance compared to heifers that were not stroked as calves. However, all yearling heifers in the study displayed a reduction in avoidance distance from handlers for a minimum of 5 weeks following being stroked and petted 14 times for 3 minutes each time. The reduced avoidance distance was also displayed toward a handler that was blind to the treatment and not directly involved in the study. This study suggested that positive interaction throughout the life of the animal is important to maintain reduced avoidance distance toward handlers. In essence, positive interaction during calfhood is not enough positive interaction to influence dairy cattle behavior for their entire lives.

In cows, Rushen et al. (1999) reported fewer kicks per minute from cows during milking preparation when a handler that had used aversive methods to move the cow was present. The authors attributed the difference in kicking to fear of the handler. After the milking unit was attached, the same cows displayed more movements per minute while an aversive handler was present but fewer movements per minute than when no handler was present.

IMPACT OF HUMAN INTERACTION ON PRODUCTIVITY

The impact of negative experiences with people on dairy cow productivity has been described by multiple researchers, with increasing interest over the past 20 years. The relationship between human interaction and animal productivity operates on the core tenet that negative interactions between humans and animals are detrimental to animal productivity. In dairy cattle, the source of reduced milk production due to negative experiences with their handlers can be traced to the physiological mechanisms that are activated when they are afraid.

As much as 19% of variation in milk yield between dairy farms has been attributed to the level of fear experienced by the cows (Breuer et al., 2000). It has been suggested that a sequential relationship exists between stockperson attitudes and behaviors and dairy cattle behavior and productivity (Breuer et al., 2000). In essence, a link between stockperson attitude and milk yield has been isolated but stockperson behavior toward the animals is the conduit through which the effects of attitudes toward animals flow. Research focused on the behavior of pig caretakers identified stockperson attitude as the most reliable predictor of behavioral approach to handling animals, but other job-related variables, such as satisfaction and interest, were indirectly involved in the development of stockperson behavior by impacting stockperson attitude (Coleman et al., 1998).

The behavior of dairy cows has been identified as an indicator of the quality of stockmanship afforded to them (Breuer et al., 2000). Cows that were willing to spend a greater amount of time within three meters distance of a human handler achieved greater milk yield, milk protein concentration, and milk fat concentration. Highly negative interaction between handlers and cows, including such behaviors as "forceful hits, slaps, pushes, and tail-twists" had a negative correlation with milk yield, protein concentration, and fat concentration. Loud or harsh vocalization from handlers during interaction with dairy cows also contributed to the same effects on milk yield and composition (Breuer et al., 2000). Handlers that consistently moved cows at greater speeds over the final 50 meters between pasture and milking facilities were identified as a contributing factor in reduced milk yield (Breuer et al., 2000). Cows that were milked in proximity to a handler who used aversive techniques to move them experienced 70% greater residual milk than when a gentle handler was present (Rushen et al., 1999). Interestingly, the difference in residual milk when a gentle versus aversive handler was present during milking only occurred for cows that were considered to be good at discriminating between handlers based on appearance. Cows that were not able to discriminate experienced the same quantity of residual milk regardless of the type of handler that was present.

HERDSMANSHIP ASSURANCE

There are multiple aspects of animal care that can be assessed through audit data collection. However, human interaction and animal fear have proven to be difficult to meaningfully quantify. A substantial component in this challenge lies in the telos of humans in general. As a species we tend to display behavior that matches expectations when we are aware that we are being monitored and evaluated.

In general, the direct interaction between handlers and cows is challenging to assess in quantifiable terms through in-person assessments. The potential exists for the implementation of remote video auditing systems in larger-scale dairies where animal handling and stockmanship activities occur for several hours each day. Monitoring becomes more difficult at smaller farms where the contact time between handlers and animals may be much more limited to very concentrated events at specific times of the day. Since the amount of time available to observe animal handling is so small in these cases, repeated observations are needed to gain an accurate portrayal of normal handling practices for a single farm.

Challenges exist regarding the identification and selection of means of quantifying the quality of stockmanship at the farm level through means that are reliable and efficient to perform (de Passillé and Rushen, 2005). As a result, animal welfare assessment programs will generally focus less or not at all on the interactions between people and animals and more on applied indicators of an animal's relationship with the environment, such as body condition or lameness score.

One approach that has been used in multiple studies to quantify the response of dairy cows to previous handling experiences is the distance the animals maintain from people. This type of test often occurs in the form of an avoidance test, which measures the distance animals will maintain as a handler approaches. An additional type of test in this category is the approach distance test. The approach distance test is conducted by measuring the duration of time that animals spend within predetermined threshold distances of people or the distance that an animal will willingly approach a handler. One challenge with the "distance measures" described above is the difficulty of comparing results across different studies due to the lack of uniformity in testing at this time (de Passillé and Rushen, 2005). An additional issue that arises with distance tests—either active approach by a handler or allowing the animal to approach the handler—is a limited consistency of repeatability between handlers (Windschnurer et al., 2008).

An additional category of tests that have been used to quantify animal responses to human presence and handling are the handling tests. Handling tests typically focus on the length of time required to complete a handling task or measures that indicate fear or restlessness of the animal (Breuer et al., 2000; de Passillé and Rushen, 2005; Lanier et al., 2001). A wide array of handling tests have been used in published studies of animal response to human interaction with limited standardization of methods between studies. It is likely that standardized approaches to handling tests will emerge as common methods are shared between studies.

The third category of tests of responses to human presence and handling are the rating scales. These tests typically include the assignment of specific definitions to a numerical scale that allows an observer to describe animal behavior through the assignment of numerical scores (Lanier et al., 2001; Voisinet et al., 1997). The concept behind these scoring systems is similar to the scoring systems used for body condition and lameness assessment in multiple species in the approach to making a subjective visual assessment more objective through the use of clearly defined numerical categories.

Other research regarding stockmanship has applied a set of descriptive terms that observers of human and animal behavior assign to both the stockperson and the animals they handle after observing the handling interaction for a specific period of time (Ellingsen et al., 2014). In one particular study, observers were asked to assign descriptive terms to stock people and the animals they handled immediately after observing their interaction during a standard husbandry procedure (weight estimation by heart girth measurement). Seventeen descriptors were available for observers to select to describe stockperson behavior. They were as follows: quick, dominating, aggressive, fearful, patient, careful, calm, determined, focused, insecure, careless, talks to the animals, cuddles the animals, inventive, nervous, boisterous, and including. The observers completed a visual assessment scale for each of the 17 variables to create numerical scores based on the location of the mark they drew on the scale. This is one method of converting subjective descriptions to more objective terms. In the same study, the researchers also had observers assess the behavior of the calves during stockperson interaction. Thirty-one descriptors were included in the list to describe the behavior of the calves during interaction with the stockperson. They were as follows: nervous, frustrated, fearful, enjoying, distressed, uncomfortable, friendly, content, sociable, uneasy, calm, confident, agitated, unwell, happy, scared, positively occupied, relaxed, boisterous, inquisitive, playful, tense, aggressive, bored, depressed, active, lively, irritable, vigilant, apathetic, and indifferent. After observing the interaction of a stockperson and calf, the observer employed the same assessment method described above to create numerical values describing each of the 31 characteristics. This approach is inherently subjective as each individual observer will vary to some extent regarding specific level of any single descriptor they observe. However, the introduction of a means of converting a discrete outcome—was the descriptor observed versus not?—to a continuous outcome—what level of each descriptor was observed?—facilitates the calculation of statistics that can better describe the differences between observers.

An additional challenge that requires attention as methods of testing human-animal interaction are developed and implemented is controlling for the ability of animals to recognize handlers and respond to handlers in different ways (de Passillé and Rushen, 2005). It is likely that common methods of quantifying animal responses to human interaction will emerge as the field of animal welfare science continues to mature. As these common methods arise, the following factors must be considered: (1) How well are the assessment methods applied in field versus laboratory settings?, (2) Do the assessment methods accurately capture and quantify animal behavior when applied over time?, (3) What are acceptable levels of repeatability and reliability for tests of human-animal interaction?, and (4) What is an appropriate threshold between acceptable and unacceptable test results?.

It is important to note that behavioral assessments may not be the only set of measures that are necessary to fully understand the quality of stockmanship (de Passillé and Rushen, 2005). The early debate over the relative importance behavioral versus physiological assessments of animal welfare states was discussed in detail by Duncan (1997). He described two primary schools of thought regarding animal welfare assessment, namely, Biological Function and Feelings. He described the Biological Function school of thought as an approach that was focused on assessing the operative and physiological states of the body as the primary means of understanding welfare states. The Feelings school of thought was focused on understanding the affective state of the animal. This approach did not consider the physiological state of the animal, but relied heavily on interpreting behavioral displays instead. A reasonable argument can be made that a well-informed assessment of animal welfare should include a blend of both schools of thought. Such an approach aligns with a point made by de Passillé and Rushen (2005); that a full understanding of the implications of good or poor stockmanship should weigh the behaviors an animal displays, but also consider other outcomes such as physiological responses, animal productivity, and final product quality attributes if possible.

IMPORTANCE OF MANAGEMENT

Prevention of Abusive Behavior

Abusive behavior is difficult to detect in animal welfare assessments for multiple reasons. First, people tend to display their best behavior when the feel that they are being monitored. In addition, a formal animal welfare assessment captures a limited view of the state of animal welfare at a farm.

This is often described as a "snapshot" because it allows the viewer of the assessment results to see the state of animal welfare within a limited window of time. Since abusive behavior is often manifested in the form of short duration and variable frequency events, the limited amount of time that observation occurs during an audit is not conducive to catching abusive behavior.

It is very difficult to capture an accurate representation of the existence or frequency of abusive behaviors toward animals during an animal welfare audit. The inability of an in-person animal welfare audit is not the result of asking the wrong questions or looking in the wrong places. It is the fact that in-person audits of animal handling and welfare occur for very limited periods of time (Vogel, 2015). In fact, a single, half-day animal welfare audit represents less than 0.05% of the available animal handling time in a year (Vogel, 2015).

Since it is extremely difficult to police the handling of animals through the use of audits, an alternative approach must be taken to ensure that dairy cattle are treated humanely and not abused. It appears that one of the most important factors in the prevention of animal abuse is the culture within each animal facility (Vogel, 2015). If the standard practices on the farm clearly reflect a high regard for animal care and a complete lack of tolerance for abuse, it is likely that new employees that join the workforce at a specific farm will assimilate with their new cohorts and avoid abusive behavior. This approach is not a complete guarantee that all animal handlers would avoid being abusive on a farm that has a culture of care and concern for animal welfare. However, it reduces the likelihood of abuse becoming a systemic problem.

Hemsworth (2003, 2007) discussed the value of developing cognitive behavioral training to assist in retraining stock people to ensure that animals are handled and cared for appropriately. Cognitive behavioral training involves addressing the core beliefs that a person holds that underlie their behavior. If a meaningful shift in underlying beliefs regarding animals or the tasks associated with caring for the animals can be achieved, the behaviors that stock people display toward animals improves as well. However, one must acknowledge that a shift in the quality of human-animal interaction requires considerable effort at the onset because of pre-existing behaviors and attitudes that take time to adjust. In addition, it is important to focus not just on the mechanics of improving animal handling when an improvement in stockmanship is desired. Instead the long-term focus should be placed on developing a "cowshed culture" in which positive behaviors are developed by stockpeople and positive behaviors are developed by animals in turn (Burton et al., 2012). Such a culture is self-sustaining as long as it is maintained.

Impact of Personality Types

There appear to be some core character attributes that make individuals more likely to be successful animal handlers. Patience and empathy have been specifically described as important traits of good stock people in general (Fukasawa et al., 2017; Ward and Melfi, 2015).

It may be possible to classify the general approach that stock people take to handling their animals based on the grouping of behavioral descriptors mentioned earlier in this chapter (Ellingsen et al., 2014). Four primary handling styles were isolated among the 110 Norwegian dairy farmers in the study. They were as follows: (1) calm/patient, (2) dominating/aggressive, (3) positive interactions, and (4) insecure/nervous. Of the four styles, the calm/patient and positive interactions groups were considered to be positive handling styles while dominating/aggressive and insecure/nervous were negative. The primary difference in the positive handling styles was the level of interaction with the calves. The positive interactions style was more interactive and passionate toward the calves. Ultimately, the researchers reported that stock- people that engaged in positive interaction, which included gently petting and talking calmly to their calves, as well calm and patient handlers, had calves that were more friendly, content, and sociable. Calves were classified as being more nervous, frustrated, and fearful when their stock people displayed handling styles that were more consistent with the dominating/aggressive and insecure/nervous classifications.

Impact of Attitudes

Multiple studies have investigated the relationships between the attitude of stockpersons toward animals, animal care activities, and the ultimate interaction and care afforded to animals (Fukasawa et al., 2017; Waiblinger et al., 2002; Ward and Melfi, 2015). Attitudes of animal handlers toward the animals they work with as well as extensive experience and knowledge of the species they work with are conducive to good animal handling and welfare (Ward and Melfi, 2015).

Dairy stockpersons with greater composite attitude scores, defined through a series of survey questionnaire items regarding petting and talking to cows, the ability of cows to recognize people, and the ease of moving cows, had a positive impact on milk yield and protein concentration (Breuer et al., 2000). In the same study, reduction in composite attitude score—indicative of increasingly negative attitude toward dairy cows—had a negative correlation with the occurrence of flinch, step, and kick behaviors by cows during milking. This means that as stockperson attitude toward the cows they care for becomes increasingly negative, the cows display a greater incidence of flinches, steps, and kicks during milking. Such behaviors have been identified as indicators of a cow's restlessness (Breuer et al., 2000). Although the authors in the previously mentioned study attributed the restlessness to human interaction, they acknowledged that dairy cattle restlessness is also impacted by other factors such as stray voltage within their environment, nutrient deficiency, social pressures from other cows, and lameness.

CONCLUSIONS

Herdsmanship and Human Interaction Are Very Important

The intimacy and duration of dairy cattle handling is unique to animal agriculture in the sense that other species are not handled on a daily basis to carry out standard routines. As any animal becomes acclimated to routine handling, it is likely that fear will subside and a reliance on routinized behavior will be necessary to achieve desired handling outcomes. It is important for dairy stock people to understand and embrace this substantial difference between dairy cattle and other farm animal species. Researchers that have worked to better understand the relationship between stockmanship and dairy cattle welfare and productivity have indicated that improvement of the attitude and behavior of stockpeople has the potential to improve the welfare and productivity of dairy cows. There is still much to learn regarding the development, remediation, and maintenance of good stockmanship. In the meantime, it is imperative that stockpeople are consistently reminded that it is not just acceptable to care and express empathy for dairy cattle—it is the very core of being a good stockperson.

REFERENCES

Apple, J. K., E. B. Kegley, D. L. Galloway, T. J. Wistuba, and L. K. Rakes. 2005. Duration of restraint and isolation stress as a model to study the dark-cutting condition in cattle. *J. Anim. Sci.* 83:1202–1214.

Breuer, K., P. H. Hemsworth, J. L. Barnett, L. R. Matthews, and G. J. Coleman. 2000. Behavioural response to humans and the productivity of commercial dairy cows. *Appl. Anim. Behav. Sci.* 66:273–288.

Burton, R. J. F., S. Peoples, and M. H. Cooper. 2012. Building "cowshed cultures": A cultural perspective on the promotion of stockmanship and animal welfare on dairy farms. *J. Rural Stud.* 28:174–187.

Coleman, G. J., P. H. Hemsworth, and M. Hay. 1998. Predicting stockperson behavior towards pigs from attitudinal and job-related variables and empathy. *Appl. Anim. Behav. Sci.* 58:63–75.

de Passillé, A. M. and J. Rushen. 2005. Can we measure human-animal interactions in on-farm animal welfare assessment? Some unresolved issues. *Appl. Anim. Behav. Sci.* 92:193–209.

Duncan, I. J. H. 1997. A concept of welfare based on feelings. In *Farm animal behavior and welfare* pp. 85–101. A. F. Fraser and D. M. Broom, eds. CABI Publishing, Wallingford, Oxfordshire.

Ellingsen, K., G. J. Coleman, V. Lund, and C. M. Mejdell. 2014. Using qualitative behaviour assessment to explore the link between stockperson behaviour and dairy calf behaviour. *Appl. Anim Behav. Sci.* 153:10–17.

Fukasawa, M., M. Kawahata, Y. Higashiyama, and T. Komatsu. 2017. Relationship between the stockperson's attitudes and dairy productivity in Japan. *Anim. Sci. J.* 88:394–400.

Hemsworth, P. H. 2003. Human-animal interactions in livestock production. *Appl. Anim. Behav. Sci.* 81:185–198.

Hemsworth, P. H. 2007. Ethical Stockmanship. *Aust. Vet. J.* 85:194–200.

Lanier, J. L., T. Grandin, R. Green, D. Avery, and K. McGee. 2001. A note on hair whorl position and cattle temperament in the auction ring. *Appl. Anim. Behav. Sci.* 73:93–101.

Lindahl, C., S. Pinzke, A. Herlin, and L. J. Keeling. 2016. Human-animal interactions and safety during dairy cattle handling—comparing moving cows to milking and hoof trimming. *J. Dairy. Sci.* 99:2131–2141.

Lürzel, S., I Windschnurer, A. Futschik, and S. Waiblinger. 2016. Gentle interactions decrease the fear of humans in dairy heifers independently of early experience of stroking. *Appl. Anim. Behav. Sci.* 178:16–22.

Pajor, E. A., J. Rushen, and A. M. B de Passillé. 2000. Aversion learning techniques to evaluate dairy cattle handling practices. *Appl. Anim. Behav. Sci.* 69:89–102.

Rushen, J., A. M. B de Passillé, and L. Munksgaard. 1999. Fear of people by cows and effects on milk yield, behavior, and heart rate at milking. *J. Dairy Sci.* 82:720–727.

Sorge, U. S., C. Cherry, and J. B. Bender. 2014. Perception of the importance of human-animal interactions on cattle flow and worker safety on Minnesota dairy farms. *J. Dairy Sci.* 97:4632–4638.

Ventura, B. A., M. A. G. von Keyserlingk, and D. M. Weary. 2015. Animal welfare concerns and values of stakeholders within the dairy industry. *J. Agric. Environ. Ethics.* 28:109–126.

Vogel, K. D. October 2015. Farm to plate: Culture sets the tone: The best tool we have to prevent abuse is culture. *National Provisioner.* 229:30–32.

Voisinet, B. D., T. Grandin, J. D. Tatum, S. F. O'Connor, and J. J. Struthers. 1997. Feedlot cattle with calm temperaments have higher average daily gains than cattle with excitable temperaments. *J. Anim. Sci.* 75:892–896.

Waiblinger, S., C. Menke, and G. Coleman. 2002. The relationship between attitudes, personal characteristics and behaviour of stockpeople and subsequent behaviour and production of dairy cows. *Appl. Anim. Behav. Sci.* 79:195–219.

Ward S., and V. Melfi. 2015. Keeper-animal interactions: Differences between the behaviour of zoo animals affect stockmanship. *PLoS One.* 10:e0140237.

Windschnurer, I., C. Schmied, X. Bovin, and S. Waiblinger. 2008. Reliability and inter-test relationship of tests for on-farm assessment of dairy cows' relationship to humans. *Appl. Anim. Behav. Sci.* 114:37–53.

CHAPTER 24

Health, Disease, and Animal Welfare Perspectives for Dairy Cattle

Jerry D. Olson
Colorado State University

CONTENTS

ANIMAL HUSBANDRY/STOCKMANSHIP

"Stockmanship has been defined as the knowledgeable and skillful handling of livestock in a safe, efficient, effective, and low-stress manner and denotes a low-stress, integrated, comprehensive, holistic approach to livestock handling (*Stockmanship Journal*)." However, stockmanship is more than just handling. It is concerned with the whole life of the animal in our care. In the past, we have called Stockmanship animal *husbandry* or *stewardship*.

There are three essential elements of good stockmanship: an environment that provides protection and comfort appropriate for the species; adequate, well-designed facilities that enables low-stress handling; and a comprehensive, herd health management program. We as veterinarians are often exposed to some very insightful and effective observations on animal handling by some producers. A friend relayed how an animal trainer had trained beef bulls to hop on trailers out in the pasture. If the bulls moved away from the trailer, the bulls were chased and harassed by a rider on horse-back. If the bulls moved toward the trailer, they were left alone. They quickly learned the trailer was a place of solace from being chased and harassed and would hop in to get away from horse and rider.

THREE PERSPECTIVES ON HEALTH AND WELFARE OF DAIRY CATTLE

Societal Expectations

The future sustainability of agriculture depends on establishing trust between the producer and the consumer. Today, <1% of the population in the US is involved in production agriculture. Consumers have very little contact with the people that care for the animals and produce the food they ultimately consume. Furthermore, they have little understanding about the life of the animal, or the care of these animals. Whatever understanding and trust that may have been established with the consumer can be easily and quickly destroyed by one video that goes viral of one incident of an animal being mishandled or abused. Arnot (2008) states that agriculture needs to address changing perceptions about the environment that affects the animals' lives on the farm and how to incorporate the best practices that are ethically based, scientifically sound and economically viable for the welfare of the animal.

PRODUCER'S CONTRACT AND OBLIGATIONS TO SOCIETY AND ANIMALS

The consumer has an expectation that the dairy products and meat that they consume are safe and wholesome and that the animals that produce their food should be cared for in a humane manner. FDA and USDA regulations provide guidance for testing of milk for drug residue to assure the safety and wholesomeness of the food. Every tanker of milk is tested for beta-lactam antibiotics upon arrival at the processor. In addition, processors may randomly test for other drug residues. As an industry, just testing for beta-lactam antibiotics will not be adequate in the future with increased consumer concerns about drug residue, antibiotic resistance and safety of milk. It is likely that additional drugs will be added to the routine list of mandatory tests. In the case of meat safety, the carcass of any cull dairy cow that has any indication of injection site lesions is pulled off the line for further testing for drug residues. The owner of animal has a legal obligation to send residue –free animals to slaughter. The veterinarian has an obligation to the owner/producer that treatment protocols are compliant with the label for the drug including the dose, route of administration, duration of treatment, milk withhold following cessation of treatment and stated slaughter withhold and that only healthy animals are sent to slaughter. If there is any deviation from the label guidance, the veterinarian of record can alter the protocol under the guidance of AMDUCA (Animal Medical Drug Use Clarification Act) but assumes responsibility and liability for any residue along with the producer.

The three factors that are necessary for building and sustaining trust in this social contract between the producer and the consumer are that the treatment of the animals is ethically grounded, scientifically verified and is economically viable. Arnot (2008), states that to be ethically grounded, there should be shared values including compassion, responsibility, respect, fairness and truth. Scientific verification uses specific, measurable, and repeatable observations to provide data for objective decisions. Scientific verification can provide insight in how food systems should be managed but needs to maintain balance with ethical consideration, not only justification for a scientific process for the care and treatment of the animal. The process of producing food must be economically viable for without a profit margin for producer, the producer cannot sustain the business. The variations of input prices of commodities and meat and milk in the market place can and at times reduce the price of product below the cost of production.

Webster (2001) has described the relationship between the animal care giver and the animal with respect to animal welfare in terms of five freedoms.

1. Freedom from hunger and thirst.
2. Freedom from discomfort.
3. Freedom from pain, injury, and discomfort.
4. Freedom of expression of normal behavior by providing sufficient space, proper facilities with animals of its own kind.
5. Freedom from fear and distress by providing appropriate treatment and conditions to avoid anxiety.

The first freedom is freedom from hunger and thirst. The animal should have ready access to fresh water and a diet to maintain full health and vigor. This a fairly straight forward obligation of good husbandry/stockman ship by the owner/care giver of the dairy animals. Anything less is animal abuse. For the dairy owner, it is in their best interest to meet these needs as failure to do so will compromise productivity and violate consumer trust.

The second freedom is freedom from discomfort. By providing an appropriate environment including shelter and a comfortable resting area. The second freedom is closely associated with the third freedom—freedom from pain, injury or disease—and the fourth freedom—expression of normal behavior by providing sufficient space, proper facilities and company of the animal's own kind. Poorly designed and managed facilities can be responsible for discomfort in a number of ways. In overcrowed facilities, submissive cows may not have access to bunk space and to free stalls. Cows that spend more of their time standing without access to free stalls results in an increased incidence of lameness. Overcrowding of facilities leads to stress and disease. Nordlund (2009) has shown that transition cows that don't have adequate bunk space have a higher incidence of disease associated with transition from dry cows to lactation including displaced abomasum, ketosis, and metritis.

The third freedom is the freedom from pain, injury, or disease. By prevention of pain during routine surgical procedures such as dehorning and minimizing discomfort by rapid diagnosis and treatment of disease.

Dehorning and tail-docking are two procedures that should be seriously evaluated in relation to the social contract. Most veterinarians, consumers, and producers feel that tail-docking should be discontinued. Consumers feel that tail-docking is mutilation of the animal and prevents the animal from displaying their natural means of fly control. The primary reasons given for tail-docking are improved cow cleanliness and better udder health. There seems to be little scientific justification for the tail-docking for either of these reasons. According to an observation from an NHMS study, cows had better hygiene scores in herds that did not dock tails (Lombard, 2010). If cleanliness or worker safety is an issue, the switch can be clipped at times when the animal is routinely handled such as post-freshening. In addition, the practice of tail-docking dairy cattle is banned or discouraged in most industrialized countries except the US (Sutherland and Tucker, 2011). Dehorning is a common practice to prevent the potential of injury to herd mates and humans. There are several means of dehorning animals. The most important considerations are that dehorning should be done early in life, it should include some means of pain mitigation, and that person doing the dehorning be trained to do the procedure safely and effectively.

Although the polled trait in Holstein cattle is transmitted as autosomal dominant trait, researchers have demonstrated that selection for the trait would negatively affect selection for Net Merit dollars (Spurlock et al., 2014).

Many of the injuries to adult cattle are preventable. It is a good practice to develop a system of categorizing injuries "splits," the cow traffic areas should evaluated for the slipperiness of the surface. In addition, the cow handling should be evaluated to determine how the cows are moved; are the cows being moved in a quiet manner without yelling and hitting cows and allowed to move at a peaceful pace.

DISEASES OF DAIRY CATTLE

The primary disease conditions of dairy cows are mastitis, lameness, retained fetal membranes, metritis, ketosis, milk fever, and dystocia. From an animal welfare perspective, two things are important. First, early identification of the disease condition is important to minimize the suffering of the animal and the production loss and improve chance of recovery. Second, it is also important to know the incidence or prevalence of a disease. If the disease exceeds a threshold incidence or prevalence, one should determine the reasons and address causes of excess. The dairy needs to maintain a record system that records the disease condition, the treatment protocol for the individual animal, and means of assessing the treatment outcome. It should be apparent that without a method of assessing treatment outcomes, we cannot know what proportion of animals return to the herd as productive animals, what proportion have relapses, and what proportion either die or are culled from the herd. Assessing outcomes is not only an animal welfare issue but affects the rate of culling and profitability of the dairy. This also suggests that the veterinarian can have critical role as the herd epidemiologist in developing the ongoing means of record evaluation. Prevention of disease is certainly the desired method approach to minimize animal welfare and improve profitability.

Lameness is a common disease condition in dairy cows and can be *cat*egorized into either those of infectious origin or metabolic origin. The two primary causes of infectious foot lameness are foot rot and digital dermatitis. The primary consequences of metabolic disorders causing metabolic acidosis are laminitis resulting in sole ulcers, white-line disease, and abscesses. In a large Canadian study including 28,607 cows in in 156 herds (Solano, 2016), the prevalence of digital dermatitis was in 15% of the cows and 94% of the herds. Sole ulcers and white-line disease are a consequence of laminitis. The conditions were detected in 6% and 4% of the cows and 92% and 93% of herds, respectively. In a Wisconsin survey of lameness in 66 high- performing herds, the prevalence of clinical lameness was 13% and severe lameness was 2.5% (Cook, 2016). Rations that are low in functional fiber and/or high in readily degradable carbohydrates can contribute to ruminal acidosis and laminitis. Heat stress is also associated with laminitis. The consequences of lameness besides being painful to the cow are reduced milk production, poorer reproductive performance, and increased culling. Free stall design and bedding materials have a significant interaction with the cow and risk of lameness. For cows housed in confinement barns with free stalls, the odds of sole ulcers and white-line disease were twice that of those housed on deep-bedded packs (Solano et al., 2016). The prevalence of lameness can be monitored by locomotion scoring of subsets of cows by pens and/or parity.

Mastitis is the most common infectious disease of dairy cows in most herds. The prevalence of subclinical mastitis as measured by bulk tank somatic cell count (BTSCC) in a herd affects the quality of milk, the risk of individual cows developing clinical mastitis and subsequent use of antibiotics to treat clinical cases, and the risk of cows being culled or dying, and poorer reproductive performance of infected cows. The herd veterinarian should be a key participant with management in implementing an udder health program. Items to consider in prevention of udder infections include dry cow therapy, teat sealants, pre- and post-milking teat disinfectants, routine culture of fresh cows for *Staphyloccus aureaus* and mycoplasma spp., culture of clinical cases, and recommendations on culling chronic cases. In addition, other items that might be included in an udder health program are treatment protocols for clinical mastitis, evaluation of milking procedure, monitoring udder health through somatic cell counts, equipment evaluation, and environmental assessment including bedding, cow hygiene scoring, and recommendations for cow vaccination. Some authors have suggested that nonsteroidal anti-inflammatory drugs should be included in the treatment regime for severe clinical mastitis for the amelioration of pain (Suojala et al., 2013).

DISEASES AFFECTED BY CONFINEMENT
COMPARED TO ACCESS TO PASTURE

The following meta-analysis by Arnott, G., et al. (2017). addresses two of the freedoms that relate to animal welfare, freedom from disease, and appropriate facilities, i.e., pasture vs. confinement, and shows that there is reduction in diseases and mortality when dairy cattle have access to pasture compared to animals in confinement facilities.

Arnott et al. conducted a review of the literature comparing pasture-based dairy cows to continuously housed cows. The review showed significant benefits of pasture-based cattle over continuously housed systems with respect to respect to health, behavior, and physiology. The health benefits included less lameness, fewer hock lesions, less mastitis and metritis, and lower mortality. Cattle which had access to pasture showed an overall preference for pasture when given a choice between pasture and indoor housing. The 2010 National Animal Health Monitoring System (USDA, NAHMS) study reported that 63.9% of North American dairies had some type of housing system which accounted 82.2% of US dairy cows. However one third of US dairies use some combination of grazing. But in 2007, 75% of dairy cows in the US were housed in free stall or dry lots.

Lameness

The causes of lameness are multifactorial and the access to pasture is factor that can affect the incidence of lameness. Two controlled studies have shown a significant improvement in locomotion scores and reduction in lameness when cows had access to pasture (Hernandez-Mendo et al., 2007; Olmos et al., 2009) and two controlled studies showed difference (Baird et al., 2009; Chapinal et al., 2010). For cows housed in confinement barns with free stalls, the odds of sole ulcers and white-line disease were twice that of those housed on deep-bedded packs (Solano et al., 2016). Other studies have shown that hard surfaces in confinement building contribute to reduce cow comfort and increased incidence of lameness (Endres, 2017). Covering concrete slatted floors with slatted rubber floors resulted in improved hoof health and animal hygiene (Ahrens, 2011).

Mastitis

In this multiyear meta-analysis of seasonally calving cows, Holstein cows in confinement had an increased prevalence of mastitis, a greater number (clinical cases of mastitis per cow) (1.1 vs. 0.6), and greater risk of being culled due to mastitis (9.7% vs. 1.6%) compared to cows on pasture (Washburn, 2002). Many confinement dairies today have attained excellent udder health as measured by bulk tank SCCs and a low incidence of clinical mastitis through implementation of a good mastitis control program, including a clean, dry environment, and good milking procedures.

Reproductive Disease

In a comparison of cows in confinement to cows in pasture-based system, cows in confinement tended to have an increased incidence of dystocia, metritis, and endometritis on a small sample size of cows and disease (Olmos, 2009). Cows in confinement tended to fared better nutritionally in early lactation because dairies have greater control over the nutrient density of the ration than for animals on pasture.

Mortality

Several studies have shown that the mortality risk was reduced when cows had access to pasture. Thomsen et al. (2006) found the mortality risk reduced in Danish herds for cows during the first 100 days of lactation that had access to pasture during the summer compared to continuous confinement (OR = 0.78). Burow et al. (2011) compared 131 Danish herds using summer grazing to 260 herd identified as zero-grazing herds. The risk of cows dying was 4.8% vs. 6.0% ($p = 0.11$) in grazing herds compared to zero-grazing herds, a 46% reduction in grazing herd compared to zero-grazing herds.

Summary of the Meta-analysis Studies on Comparison of Pasture to Confinement

The meta-analysis showed an improvement in the cow health in a reduction of lameness in cows that access to pasture compared to continuous confinement. Other aspects of cow health were not as apparent. *Whenever possible, it appears desirable to give cows access to pasture even when part of the life-cycle includes confinement.*

The fourth of the five freedoms is the expression of normal behavior by providing sufficient space, proper facilities and company of the animal's own kind. At face value this seems like a fairly straight forward means of comparing the cow's preference for one environment over another as a viable means of evaluating preference in expression of normal behavior. There are numerous factors that confound what seems to a fairly straight forward comparison. Few studies have directly compared dairy cow behavior of cows on pasture-based systems to cows in a continuously housed in a confinement environment. Most of the studies that have that have made comparisons were done a number of years ago. The comparisons do not split out dairies where the cows in confinement have access to an outside dry lot. The understanding of the effect of stall design and bedding materials and stall surface relative cow comfort have improved over time. In addition to cow comfort, environmental and management factors affect the prevalence of lameness. Access to pasture and depth of bedding explained 50% of the variation in clinical lameness in NE and California dairy herds (Chapinal, 2013). When Canadian researchers offered cows a choice between pasture and free stalls with Total Mixed Ration (TMR) indoors, cows spent 54% of their time on pasture, mostly during the night time (Legrand, 2009) preferring the confinement during the day. In addition to stall design, dairies in areas of the country with climatic heat stress have implemented various means of cow cooling and improved ventilation to mitigate heat stress and improve cow comfort. Because of the potential extreme stress in the desert southwest, dairies have had develop sophisticated methods of cooling cows to attain high levels of milk production. Some dairies developed systems with bedded packs to provide good cow comfort. The surface on which cows walk affects cow comfort. Some dairies have implemented rubber mats in high cow traffic areas and along feed alleys to improve cow comfort.

However, there are a large range of factors affecting cow comfort in confinement facilities. These include heat abatement and ventilation, surface on which the cows walk, free stall design, bedding materials, and space allocation. Climate has a profound effect on animal preferences as result of extremes in temperature and moisture. Depending on latitude in the US, northern states have winter conditions which are not suitable for being outside in the winter month. In southern latitudes, summer may provide enough heat stress that cattle will seek means of shelter from the sun and heat abatement when available. I have seen cattle run from the milking parlor to cooling ponds in the summer on Texas dairies to mitigate heat stress. Rainy weather may result in muddy condition in the traffic areas between milk parlor and pasture. Nutrition will be different between pasture and confinement. Cows on pasture are primarily limited to the nutrient content of the grasses. In confinement facilities, TMRs are the most common means of feeding dairy cattle which will tend to have higher nutrient density. As a result, cows on pasture will usually have lower body condition scores compared cows in confinement at comparable stages of lactation. Cows fed TMRs because of the greater nutrient density will usually have high milk production for the lactation. Cows on pasture usually have seasonal calving patterns while cows in confinement will usually be calving year-around with a skewing of patterns of calving due to seasonal heat stress affecting fertility during the hotter months. As generalizations, pasture-based herds will be smaller compared to confinement herds.

FREEDOM FROM FEAR AND DISTRESS: BY ENSURING CONDITIONS AND TREATMENT THAT AVOID MENTAL SUFFERING

One of the best approaches to providing freedom from fear and distress is through good stockmanship—the best handling and movement procedures of animals. This means that everyone on the dairy should know what good stockmanship is and every new employee should be trained in good stockmanship and there should be periodic retraining of employees. *There is no reason for using cattle prods on a dairy.* With appropriate training of the employees, there should never be a video of employees abusing animals and there should never a video of animal abuse that goes viral. There should never be a need for the industry to attempt to get legislation against secret video tapping of abuse in animal handling on a dairy. Support of this kind of legislation by the industry erodes public trust and confidence in the dairy industry. One condition that all dairies should have is a protocol to deal with is "downer cows." All dairies will eventually have downer cows and some forethought needs to be given to the handling of these animals so that an onlooker doesn't perceive this as a case of animal abuse. Dairies should also have written protocols to deal with animals that need to be euthanized. These should define the criteria for euthanasia, how it will be done and who will do it.

THE ROLE OF THE VETERINARIAN

There are several key areas that the veterinarian could be involved with on the dairy operation. The herd veterinarian could be involved with management to develop, monitor, and add or change items within herd record system that affect animal welfare. The veterinarian could work with the management of the dairy to develop protocols in the following areas: vaccination, treatment, and udder health. Veterinarians are frequently involved in worker training in the areas of milking procedure, disease diagnosis and treatment, calving management, and reproductive management including semen handling and artificial insemination.

Genomic testing is not only able to identify animals with superior production traits but better fertility, longer herd life, and greater health characteristics. The goal is to have highly productive animals that are healthy (Zwald, 2004). This benefits the producer in having more profitable animals and improves the consumer perception of a better, healthier animal that needs fewer treatments.

Guatteo et al. (2012) have suggested that there are three things that can be done on the dairy to mitigate pain and improve animal welfare on the dairy: (1) suppress the source of the pain, (2) substitute with less painful alternatives, and (3) soothe the pain with an appropriate analgesic drug when possible. Examples of means of suppressing pain include genetic selection for the polled trait and selection for health traits through genomic testing and the need for fewer treatments. Means of substitution with less painful alternatives include timing of procedures such as dehorning as early as possible in life of the calf. Mitigation of pain includes appropriate use of local anesthetics for surgical procedures and use of effective nonsteroidal anti-inflammatory drugs.

REFERENCES

Ahrens, F., et al. (2011). Changes in hoof health and animal hygiene in a dairy herd after covering concrete slatted floor with slatted rubber mats: A case study. *J Dairy Sci* **94**(5): 2341–2350.

Arnot, C. (2008). Sustainability takes balance. *Feedstuffs* 80(17).

Arnott, G., et al. (2017). Review: Welfare of dairy cows in continuously housed and pasture-based production systems. *Animal* **11**(2): 261–273.

Baird, L. G., et al. (2009). Effects of breed and production system on lameness parameters in dairy cattle. *J Dairy Sci* **92**: 2174–2182.

Burow, E., et al. (2011). The effect of grazing on cow mortality in Danish dairy herds. *Prev Vet Med* **100**: 237–241.

Chapinal, N., et al. (2010). Overnight access to pasture does not reduce milk production or feed intake in dairy cattle. *Livest Sci* **129**: 104–110.

Chapinal, N., et al. (2013). Herd-level risk factors for lameness in freestall farms in the northeastern United States and California. *J Dairy Sci* **96**(1): 318–328.

Cook, N. B., et al. (2016). Management characteristics, lameness, and body injuries of dairy cattle housed in high-performance dairy herds in Wisconsin. *J Dairy Sci* **99**(7): 5879–5891.

Endres, M. I. (2017). The relationship of cow comfort and flooring to lameness disorders in dairy cattle. *Vet Clin: Food Anim Pract* **33**(2): 227–233.

Esslemont, R. J. and Kossaibati, M. A. (1996). Incidence of production diseases and other health problems in a group of dairy herds in England. *Vet Rec* **139**: 486–490.

Guatteo, R., et al. (2012). Minimising pain in farm animals: The 3S approach – 'Suppress, substitute, soothe'. *Animal* **6**(8): 1261–1274.

Hernandez-Mendo, O., et al. (2007). Effects of pasture on lameness in dairy cows. *J Dairy Sci* **90**: 1209–1214.

Legrand, A. L., et al. (2009). Preference and usage of pasture versus free-stall housing by lactating dairy cattle. *J Dairy Sci* **92**(8): 3651–3658.

Lombard, J. E., et al. (2010). Associations between cow hygiene, hock injuries, and free stall usage on US dairy farms. *J Dairy Sci* **93**(10): 4668–4676.

Nordlund, K. (2009). The five key factors in transition cow management of freestall dairy herds. *46th Florida Dairy Production Conference*: 27–32.

Olmos, G., et al. (2009). Hoof disorders, locomotion ability and lying times of cubicle-housed compared to pasture-based dairy cows. *Livest Sci* **125**: 199–207.

Solano, L., et al. (2016). Prevalence and distribution of foot lesions in dairy cattle in Alberta, Canada. *J Dairy Sci* **99**(8): 6828–6841.

Spurlock, D. M., Stock, M. L. and Coetzee, J. F. (2014). The impact of 3 strategies for incorporating polled genetics into a dairy cattle breeding program on the overall herd genetic merit. *J Dairy Sci* **97**(8): 5265–5274.

Suojala, L., et al. (2013). Treatment for bovine Escherichia coli mastitis – An evidence-based approach. *J Vet Pharm and Therap* **35**: 521–531.

Sutherland and Tucker (2011). The long and the short of it: A review of tail docking in farm animals. *Appl Anim Behav Sci* **135**(3): 179–191.

Thomsen, P. T., et al. (2006). Herd-level risk factors for the mortality of cows in Danish dairy herds. *Vet Rec* **158**: 622–625.

USDA, APHIS. (2007). Dairy 2007 Part II: Changes in the U.S. dairy cattle industry, 1991–2007. Report, Veterinary Services, Centers for Epidemiology and Animal Health: 1–100.

USDA, APHIS. (2010). Facility characteristics and cow comfort on U.S. dairy operations, 2007. Report, Veterinary Services, Centers for Epidemiology and Animal Health: 1–184.

Washburn, S. P., et al. (2002). Reproduction, mastitis, and body condition of seasonally calved Holstein and Jersey cows in confinement or pasture systems. *J Dairy Sci* **85**: 105–111.

Webster, A. J. F. (2001). Farm animal welfare: The five freedoms and the free market. *Vet J* **161**(3): 229–237.

Zwald, N. R., et al. (2004). Genetic selection for health traits using producer-recorded data. I. incidence rates, heritability estimates, and sire breeding values. *J Dairy Sci* **87**(12): 4287–4294.

CHAPTER 25

Dairy Cow Welfare and Herd Turnover Rates

Donald J. Klingborg
University of California

CONTENTS

INTRODUCTION

Society has a major role to play in how the welfare of animals is valued as a direct result of their market decisions about which product to purchase, and through their collective effort at influencing social, legislative, and moral standards. Healthy markets and society rely on consumers making their choices based on factual information rather than being influenced by fads, marketing, and special interest promotions.

Cattle have been an important part of the human experience for over 6,000 years based on attested evidence of their domestication. Across those years thousands of cattle generations have been selectively bred to create our dairy and beef breeds by influencing their size, growth rate, reproduction, disposition, diet, tendency to panic, and social structure.[1]

Today's cattle are very different than their wild ancestors, and they are still sentient beings worthy of our respect and thoughtful consideration. They should be provided access to water, nutrition able to support their metabolic needs, and an environment appropriate to their care, use, and welfare. They should be afforded consideration for their safety, health and species-specific biological needs and most of their behavioral natures. They should also be cared for in ways that minimize fear, distress, pain, and suffering.[2]

Most of the information presented in this chapter intentionally draws from larger multiyear studies from neutral and authoritative sources and published in peer-reviewed journals or from USDA published data.

COW WELFARE AND TODAY'S DAIRY PRODUCTION SYSTEM

Across time cattle have represented, and represent today, different values to individuals based on time, place and culture including: status, wealth, food, currency, power, fuel, food, fertilizer, heat, fat, and textiles.[1] These value differences likely have an influence on the principles that are the foundation of how cattle are considered and treated. What isn't different is the fact that healthy cows produce more with fewer expenses than unhealthy cows, so every producer has a vested interest in caring for their cattle.

Dairy farms today are the product of economic, market, environmental, and legislative forces that have shaped it with rapid change starting in about 1950 and continuing today. Half of US dairy herds disappeared between 1950 and 1975, and another 18% were lost between 1975 and 2000. From 1950 to 1975 milk production per cow almost doubled, and it grew another 76% between 1975 and 2000.[3] Since then production per cow has continued to climb and milk quality has improved. Technology, specialization, advances in nutrition, health, managing cows and a trained labor force, and other improvements, all contributed to these changes.

Why are herds getting bigger? The price the dairy producer receives for their milk is largely regulated by others and the economic inputs they can control include the costs of facilities, cattle, equipment, debt, labor, and feed. Figure 25.1 shows the relationship between herd size and income over operational costs. While profit margins remain low there is a clear advantage in today's economic market for herds producing more milk per cow and with more cows in the herd.

As required for economic survival, the dominant family dairy farm today has changed into a much more specialized system of production, with a greater focus on milk production as the primary cash crop of the operation, a permanent workforce with specialized training in their tasks (and greater accountability for their knowledge and performance), densely populated housing with ventilation fans, protection from the elements and, in hot climates, cow cooling systems. Employees today enjoy a more defined work day, and benefits including health care, paid vacations, housing, training and days off. These changes are a result of "right sizing" the herd with current costs of production, including land, facilities, cattle, equipment, debt, labor and feed. They also involve more technology and facilities to allow dairy cattle to use more of their nutritional support to produce milk (Table 25.1).

Figure 25.2 shows the relationship between herd size, costs of production, and the percentage of herds with income from milk sold exceeding the costs of production. Milk sales represent 90% of the typical dairy's income. In this study, herds had to grow beyond 500 cows to have more than

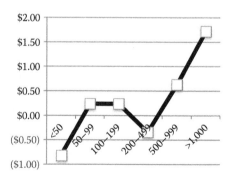

Figure 25.1   Milk income minus operating costs per hundredweight (CWT) of milk sold by herd size.

(*Source*: **USDA National Agriculture Statistics Service: Milk Production, Disposition & Income 2016 Summary (April, 2017).**)

Table 25.1 Changing % of National Dairy Herd in Small vs. Large Dairies

Year	% Share of Cows in Herds < 100 Cows	% Share of Cows in Herds > 999 Cows
1992	49	10
1997	39	18
2002	29	29
2007	21	40
2012	11	49

Source: USDA Census of Agriculture.

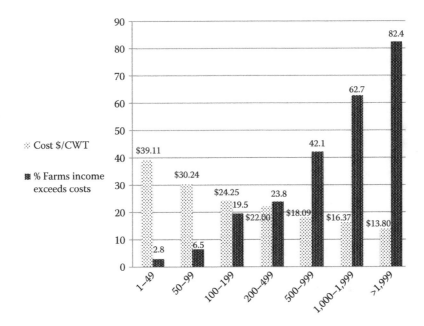

Figure 25.2 Costs of milk production and proportion of herds where sales exceeded costs of production (2012).

(*Source:* **USDA ERS using USDA, National Agricultural Statistics Service, Agricultural Resource Management Survey (ARMS) 2007.**)

a quarter of farms achieving income exceeding their costs of production. A change in feed prices, replacement heifer costs, and/or milk price will impact this graph significantly; however, the advantage in lowering costs with "right sizing" provides a huge benefit to larger herds.

As a personal observation, "right sizing" in 1970 in California, meant a minimum of about 250 cows, and that number seems to have doubled about every 10 years requiring closer to a minimum of 2,000 cows today.[4] In 2012, the midpoint herd size in the US, where half the herds had more cows and half had fewer, was 900 cows. The average herd had 144 cows. In 1982 there were 135,000 dairies with fewer than 100 cows, and in 2012 there were 50,000 dairies of that size. The number of herds with more than 1,000 cows tripled to 1,807 over the same period.[5] Smaller herds may remain in business, but a careful look at their books will show they are most likely surviving by having no pay or low pay family labor, no or low debt, access to low-cost feed (their farming is supplementing the dairy effort), low facility costs (often older and without modern technology), are subsidizing the dairy with non-dairy income, are slowly depleting assets, or some combination of all of these.

COW LONGEVITY

Key components of a dairy farm's economic health depend on more sales per cow and more income over operating costs. Operating costs include the investment in raising a calf from birth to first lactation at which time income from milk sales begins. That cost of caring for them to first lactation, and the cost associated with the dry period of 40–60 days between lactations when cattle are no longer being milked as they rest and recuperate before calving and starting their next lactation, must be paid for by milk sales. As those costs are repaid the operational cost for that animal decreases and she becomes more profitable. In general, profitability peaks at third lactation when the animal reaches their mature weight, achieves her maximum feed intake capacity, and has developed her maximum milk secreting tissues.

Cow longevity may be important to survivability of a dairy enterprise; however, economically higher production levels make average herd life beyond 48 months less important when compared to herds with lower production. A calculated optimum economic annualized removal rate of 25% results in an average herd life of about 48 months (exiting in the 3rd lactation), and the study identified milk yield, milk price, feed costs, and replacement heifer prices as the most impactful variables on removal rates.[6] Lower removal rates of first and second lactation animals were seen as important economic goals because these animals are still paying for the debt incurred while they were raised to the beginning of their first lactation.[7]

Producers usually sit down and evaluate their monthly production test results, first to identify those cows that are not producing at herd average thereby making them candidates for removal, and then to compare those animals to others based on stage of lactation, lactation number, reproductive status (non-pregnant cows are at higher risk of removal than pregnant cows), and somatic cell counts (SCCs) in milk as a proxy for mastitis. Finally, they'll likely look at other potential problems such as lameness, conformation of feet, legs and udder, family/genetic potential, what fluctuations in herd numbers are anticipated in the next month or so and the value if sold for beef rather than retained for milk production. The decision to remove a cow from the herd comes down to whether another animal replacing the candidate for exiting the herd will deliver more value to the herd both in the short and long term.

Actuarial tables, sometimes called survival tables, are used in determining what we have to pay for our life insurance by calculating how long we will live on average for our population. Social security payments are also determined from actuarial tables. Both include data on race, occupation, geographical region, and other variables as they impact survival rates. They are also useful in understanding turnover in dairy herds and these may include breed, production, lactation number, and other variables that impact longevity. Figures 25.3, 25.4, and 25.5 show data taken from a study of 13.8 million US dairy cattle by scientists at USDA.[8] These survival tables relate to how long cattle stay in a herd and the term "survival" reflects exits from a herd for all reasons (including but not limited to on-farm death). In Figure 25.3, the authors show survival rates (how long they remained in the herd) for the two most popular breeds of dairy cattle in the US across five consecutive years after their first calving. The slope of the line indicates a pretty constant rate of exits up to 4 years after first calving and more rapid removals thereafter. There are some impacts of breed on longevity.

In Figure 25.4, the same authors present the mean months an individual cow stayed in the herd, reported by the year of first calving, excluding any cows at or beyond their 9th lactation. Breed differences are evident. Average longevity declines from about 1981 until it starts increasing for unknown reasons in 1990. Holsteins "lost" ~4 months of longevity in this 15 year period, and Jersey's and Guernsey's each lost ~2 months.

Figure 25.5 shows the percentage of three breeds of cows remaining in the herd by parity. Taking Holsteins as an example, about 25% of those in the herd at first calving are gone before the end of their second lactation, and another 25% exit by the end of their third lactation. Exit rates then

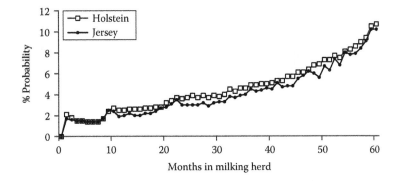

Figure 25.3   Survival analysis of Jersey & Holstein cattle first calving in 1998–1999 in US dairy herds: Probability of removal from herd by months after first calving.

(*Source*: **USDA ARS, Norman, Hare and Wright 2006.**[8])

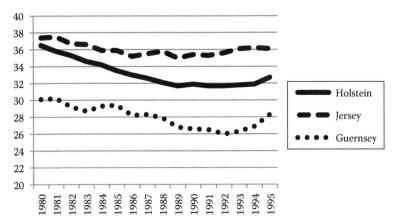

Source: Survival Rates and Productive Herd Life of Dairy Cattle in the United States
Journal of Dairy Science 89:3713–3720

Figure 25.4   Mean productive herd life in months through parity 8 by breed and year of first calving from 13.8   million US dairy cows from 1980 to 2005.

(*Source*: **Adapted from Hare, Norman & Wright, 2006[9].**)

drop to ~18%, 13%, 9%, 5%, and 3% in each subsequent lactation. The patterns are similar for all breeds with Jersey's enjoying the longest longevity and Guernsey's the shortest.

TURNOVER RATES, ANIMAL HEALTH, AND WELFARE

There are a number of forces that impact turnover rates other than the health of the animals. Turnover rates are lower when: herds are expanding so they do not remove as many animals from the herd; replacement costs are high—the replacement market in North America includes animal movement moves between Canada, the US, and Mexico and in times of high demand and high prices the producer may choose to sell replacement heifers rather than keep them in the herd, and therefore heifers are not available to replace exiting milking cows so they stay longer and, herd

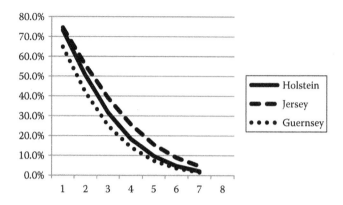

Figure 25.5   Survival rate by parity and breed from 1980 to 1995.

(Source: Adapted from Hare, Norman & Wright.[9])

removal rates will decline; feed costs are low, or feed supply is high on the dairy prompting a delay in removal; and in herds that use rBST that increases persistency of production and allows animals to have positive income over expense ratio longer into their lactation. Herds with purebred (registered) cattle often have lower removal rates because these cattle have more economic value from their genetic base without additional operational costs.

Removal rates tend to be higher when the price of slaughter beef is high, feed costs are high, the milk price is low or there is a surplus of replacement heifers in the herd or replacement heifer market price is low.

Calculating a consistent and accurate turnover rate that can be used within and between herds is a challenge. There are a number of formulas used today to calculate culling rates. Often these calculations use different information especially in the denominator and are not directly comparable.

The use of the terminology "Turnover Rates" is recommended as a more accurate description of what's been measured than the historical "culling rates" terminology. The formula used should be standardized across all data collection services and systems with the numerator equaling the number of cattle under review (herd, defined lactation, breed, etc.) leaving the herd across a specified time (month, year, etc.) and a denominator based on how long each animal under review was in the herd across the same time period, qualifying them as "at risk" of being removed. The time frame may be days, months or years with a cow resident in the herd all year counting as one cow-year, while one entering or leaving the herd with only 6 months of total residency counting as one-half of a cow-year. Using cow-days would provide the most accurate denominator but requires more effort to determine. Doing a monthly calculation is easier and can be compiled into an annual turnover rate that is very similar to one calculated using cow-days.[10]

Changes are urgently needed in standardizing and improving the turnover rates calculation making it more useful, including: using four mutually exclusive destinations for an exiting cow to include: (1) sale—animals going to other farms to produce milk or for reproductive reasons such as embryo transfer, etc. (2) Slaughter—noncompetitive animals unable to remain in the herd and are sold for immediate or eventual (after a period of fattening) slaughter for meat and valuable by-products via sales. This would include those animals that are slaughtered on-farm slaughter for employee or owner consumption. (3) Salvage—animals that are not sufficiently competitive to remain in the herd and their meat is not usable for human consumption so they go to rendering. (4) On-farm death.[10]

Following the destination there should be opportunities to provide at a minimum primary and secondary reason for the removal, and adding a tertiary or even more reasons will improve the usability of the information. Reasons need to be specific with no or minimal overlap that may result in errors of assignment and make interpretation more difficult. The goal is to compare a dairy to itself across time, and to other dairies at a given point in time, to identify where to focus management attention for improvement.

The majority of animals removed from dairies are for economic reasons—where the decision is made to replace that animal with another that has more economic promise or requires fewer farm resources while in the herd. The replacements, animals that enter the herd to assume the space of those departing, are usually younger animals who have greater potential due to the accumulated genetic improvement between the sire's genetics used at the time of conception of the replacement vs. the genetics of the sire used to conceive the animal "to be removed". Animals removed with a milking herd life of 48 months means the replacement will be 2+ generations "better" than the exiting animal.

The following five tables and figures summarize the live-removal and on-farm death results from one or both of the two large published studies (Pinedo, DeVries & Webb[11] who looked at records from 3.6 million Holstein lactation records from 2,054 herds from 2001 to 2006, and Hadley, Wolf & Harsh[12] who looked at 1.5 million lactation records from an average of 17,979 herds per year between 1993 and 1999).

They found as annualized removal rates increase, on-farm death rates decline, and as annualized death rates increase, on-farm removal rates decline. While the data captured in DHIA records fail to provide a clear understanding as to why these animals died, and provides only general information about why they were removed from the herd, there does appear to be a relationship between herds that remove more live animals having lower on-farm death rates, and herds with higher on-farm death rates having lower live-removal rates (Table 25.2).

Figure 25.6 demonstrates the rate of live removals and on-farm deaths are associated with more lactations. The risk of on-farm death appears to increase at almost a constant rate across lactations, while the risk of live removal accelerates after the third lactation.

Figure 25.7 shows a remarkable association with the risk of removal and failing to be pregnant. There is also a smaller but still significant risk of on-farm death for cows that are not pregnant. A part of the on-farm death increase in non-pregnant animals will be related to the period of highest risk of on-farm death, from calving to about 40 days post calving, when cows are not being inseminated and not pregnant. Other data show delayed conception, conception failure, and conception–abortion as extremely high risks for live removal and a lack of longevity.

Figure 25.8 shows in Pinedo's study removals by production levels per cow are highly associated with low milk production. Both Pinedo's and Hadley's studies showed medium-producing herds had lower live-removal rates, and high-producing herds had the lowest removal rates. On-farm deaths were similar in low- and medium-producing cows and very slightly higher in high-producing cows. Looking at production on a herd basis, there were no significant differences in live removals or on-farm deaths across low and medium herds and a very slight increase in high-producing herds.

Table 25.2 Low-, Medium-, and High-Producing Herds and Percent Annualized Live Removal and On-Farm Deaths

| | % Annualized Herd Removal Rates | | % Annualized Herd Death Rates | |
	Removed	Died	Died	Removed
Low	13.6	7.2	2	27.3
Medium	25.3	6.6	6.1	25.3
High	36.2	5.6	5.6	22.7

Source: Pinedo et al.[11] and Hadley et al.[12]

In relation to herd size, Pinedo's study shows a slight increase in removals associated with herd size, with the lowest death rates in small herds and highest in the largest herds. Hadley's study shows an increase in increase in live removals and death rates as herd sizes grew from low to medium and from medium to high (Figure 25.9).

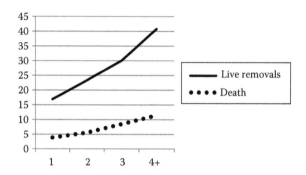

Figure 25.6 Percent live removal and % death by parity.

(*Source:* **Pinedo et al.[11]**)

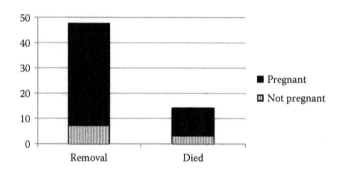

Figure 25.7 Percentage removals vs. on-farm death by pregnancy status.

(*Source:* **Pinedo et al.[11]**)

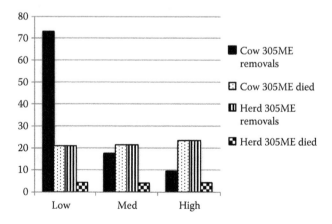

Figure 25.8 Percentage removals and died by 305ME production.

(*Source:* **After Pinedo et al.[11] and Hadley et al.[12]**)

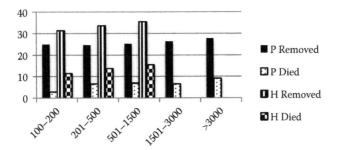

Figure 25.9 Percentage removal and died in Pinedo et al. (P) and Hadley et al. (H) by size of herd. (*Source:* **Pinedo et al.[11] and Hadley et al.[12]**)

REASONS FOR REMOVALS

Pinedo's study reported an annualized live-removal rate 25.1% and an annualized on-farm death rate of 6.6%. Hadley's study reported an annualized live-removal rate of 28.5% and an annualized on-farm death rate of 4.4%. The table below reports the reasons for removal using frequency rates. Frequency rates are useful in identifying the relative risk of removal by reason but do not include the number of animals affected. As an example, a herd of 1,000 cows with one cow each exiting the herd in the time period under consideration and because of: low production; delayed reproduction; injury; mastitis; and death would report a frequency distribution of 20% for each of those five reasons to sum to 100%. It would not tell you the total number of animals removed from the herd was very small (total removal rate is 5/1000=0.5% in this example). Pinedo's cattle were at a 20.6% risk of on-farm death in comparison to the other risks reported for those cattle, but the actual percent of those dying of all cows counted was 6.6%. Similarly, Hadley's cattle were at a 13.9% risk of on-farm death in comparison to the other risks of those cattle, but had a 4.4% on-farm death rate for all cows counted (Table 25.3).

The Pinedo's study used records from DHIA services that provide producers the option to select from up to eight reasons for removals. Surprisingly 43% of the herds only selected one or two reasons. Those herds reported much higher death rates than herds using three or more reasons demonstrating how a recording system may misrepresent the reporting of live removal or on-farm death rates.

Table 25.3 Frequency Distribution of Cows Removed from Herd According to Dairy Herd Improvement (DHI) Codes

Reason	Pinedo et al. %	Hadley et al. %
Low production	12.1	12.8
Died	20.6	13.9
Reproduction	17.7	20.5
Injury/other	14.3	29.1
Mastitis	12.1	13.1
Feet and legs	8.1	4.8
Disease	6.9	3.2
Udder	3.2	3.9
Reason not reported	5.0	–
Sum	100	100

Source: Pinedo, DeVries and Webb[11] and Hadley, Wolf and Harsh.[12]

The reasons reported for live removals and on-farm deaths represent "buckets" each holding many possible problems rather than distinctly different reasons. If reasons were more accurately and consistently reported within a herd and between herds it would be possible to better focus treatment and, more importantly, prevention programs. Persistent Bovine Virus Diarrhea, Bovine Leukosis, Johne's disease and neospora-associated abortions remain challenges in the dairy industry and are likely significant "silent" contributors to live removals and on-farm deaths.

It's been recommended that DHI record centers adopt USDA's Animal Improvement Program Laboratory (AIPL) listing of health trait terms to improve the usefulness of the reasons captured. The terms include: Cystic Ovary, Diarrhea/Scours, Digestive Problem/Off Feed, Displaced Abomasum, Downer Cow, Dystocia, Johne's Disease (Clinical), Ketosis/Acetonemia, Lameness, Mastitis (Clinical), Metritis, Milk Fever/Hypocalcemia, Nervous System Problem, Other Reproductive Problem, Respiratory Problem, Retained Placenta, Stillbirth, Teat Injury and Udder Edema. Other non-disease traits recommended for adoption include: Body Condition Score, Milking Speed and Temperament.[13] Collection of this kind of detail is a necessary step toward fostering better management and prevention health programs.

Many published studies report annualized death rates of around 6% of the herd. Risk factors reported vary widely and include larger herd size, higher milk production per cow, and problems associated with calving including retained placenta, dystocia, and metritis. The 2013 USDA NAHMS[14] report notes similar death rates across small, medium, and large dairies, averaging 4.8%. This study included data from 17 states, 76% of the nation's dairies, and 80% of the nation's cows. Calving's reported as stillbirths had the highest rates in the smallest herds and the lowest rates in the largest herds. This likely is associated with dedicated employees trained and responsible for managing deliveries and more frequent maternity pen checks in larger herds with day and night shifts caring for the cattle.

Pinedo et al., Hadley et al., and USDA[15] report higher on-farm death risks for cows in their later lactations. Annualized live removal and on-farm death rates were lower in higher producing cows in the USDA study.

Cow deaths in a 2007 USDA Dairy Survey, as reported by producers, listed from highest to lowest frequency, were: lameness or injury (20.0%); mastitis (16.5%); calving problems (15.2%); unknown (15.0%); respiratory problems (11.3%); diarrhea or other digestive problems (10.4%); other known reasons (10.2%); nervous condition or lack of coordination (1.0%); and poison (0.4%). These results represented data from 13.6% herds with <100 cows, 32.6% herds with 100–499 cows, and 53.8% from herds with 500 or more cows. Only about 4% of deaths were necropsied to determine a more definitive cause of death.[22] All deaths are animal welfare concerns, but combining all these conditions into one figure reported as "on-farm death" makes interpretation very difficult.

Table 25.4 shows death by herd size for four different years selected from 1996 to 2014 USDA data with the average percent of 4.08% in herds less than 100 cows, 4.85% in herds from 100 to 499 cows, and 4.95% in herds with 500 and more cows. The 2014 data suggest no significant differences in on-farm deaths based on herd size.

Table 25.4   Number of Dairy Cows that Died as a Percentage of U.S. Dairy Cow Inventories by Herd Size Categories

Herd Size (Number Dairy Cows)	Dairy 1996	Std. Error	Dairy 2002	Std. Error	Dairy 2007	Std. Error	Dairy 2014	Std. Error
Less than 100	3.6	(0.1)	4.4	(0.1)	4.8	(0.1)	4.9	(0.2)
100–499	3.9	(0.1)	5.0	(0.1)	5.8	(0.2)	4.7	(0.3)
500 or more	4.0	(0.2)	4.9	(0.1)	6.1	(0.2)	4.8	(0.2)
All operations	3.8	(0.1)	4.8	(0.1)	5.7	(0.1)	4.8	(0.1)

Source: USDA 2008[16] and USDA 2016.[17]

Reproductive failure is a huge risk factor for exiting the herd. To maintain production levels across a cow's lifetime requires her to calve at a regular interval. Lower producing cows need to calve more often than higher producing cows to remain in the herd because they reach lower levels of production later in lactation and don't maintain income over operational costs. Reproduction as a reason for removal include: cattle that got pregnant too long after the previous calving to sustain the required production until 60 days before the next calving; cattle that got pregnant within the required time period but aborted the calf; and cattle that didn't get pregnant. Pinedo shows first and second lactation animals had higher frequency distributions for removals associated with reproduction, indicating an area in need of management's attention. It's interesting that for pregnant cows in this study reproduction was the second highest frequency reported for exiting the herd. This only makes senses if these cows aborted and were open but the abortion was not reported, or they got pregnant very late in their lactation. Hadley found cows that were pregnant had one half to one fourth the risk of exiting the herd compared to open cows.

Pinedo et al. reported reproduction as the most often used disposal code noted for intermediate- and high-producing cows. Hadley's study conflicts with Pinedo's showing cows enjoying lower risk of removal if they delivered higher than average milk production, higher than average protein content, and higher than average milk persistency. High-producing cows do need extra time to get into positive energy balance following calving and that can impact reproductive performance. They also have a longer time from calving to when their production reaches the level where expenses exceed costs of production. This means they can calve less frequently and still be of economic value in the herd. More information is needed to understand the reasons for the conflicting reports.

Mastitis is commonly discussed welfare issue brought to the forefront from the use of rBST to promote milk yields. Mastitis was reported by 99.7% of producers surveyed in 2013 by USDA NAHMS survey, and about 25% of cows were reported to have a case of clinical mastitis with no difference in the percentage of cases reported in small vs. large herds.[14] Mastitis is a general term that reflects environmental causes associated sanitation, especially in bedding areas and prior to milking, and contagious causes associated with an infected cow transmitting the bacterial agent to a noninfected cow usually during the milking process. Mastitis can be painful and is a welfare concern. An SCC is a laboratory test that can provide some insight into the level of inflammation in the udder and serves as a proxy for subclinical mastitis. Many studies show animals with high SCC's are at higher risk of removal, produce less milk, and have a greater incidence of clinical mastitis. Bedding with non-organic materials that don't support microbial growth and excellent milking hygiene are important strategies for preventing mastitis. Dry cow treatment and bedding along with investments in milking equipment offer some additional preventative benefits. While only a small percentage of mastitis pathogens cause severe and even life-threatening illness, because the frequency of mastitis occurrence is high this disease accounts a high percentage of on-farm deaths. A study of SCC's in 3,000 dairies in Wisconsin showed similar mastitis levels across all herd sizes.[18] Pinedo et al. and Hadley et al. reported similar risks of live removal for low production and mastitis in their studies.

Foot disorders are among the easiest identified welfare concern in today's dairy enterprise. Unfortunately, the terms "lameness" and "feet and legs" lack specificity, combining many potential problems all with different causes and preventions. The prevalence of lameness has been associated with lactation number, hoof-trimming frequency, walking surfaces, and stall design/comfort, and has been growing across time. In 2002, the reported prevalence of lameness in heifers was 36.5%, and in 2007 it was 58.7%.[19] Flooring is an important factor and efforts to find kinder and safer footing is ongoing. The average dairy facility in the country was constructed in 1976 so we have older technology still dominating our physical plants.[20] A 2010 study showed no association between feet and leg problems and high production.[21]

Pinedo et al. reported a risk of removal for feet and legs reasons about twice that of Hadley, while removals for udder reasons were similar. Pinedo identified a relationship between removal for feet and legs and removal for udder problems and speculates these live removals may be for poor conformation rather than observed health problems. The absolute risk for removal for feet and leg reasons was twice as high for open vs. pregnant cows indicating the reasons declared are only part of the story.

Table 25.5 shows a frequency distribution of reasons for removal vs. herd size. Removals for production were similarly distributed across all herd sizes. Increasing herd size resulted in a significantly lower frequency of reproduction removals. Injuries decreased in frequency as herd size increased, mastitis risk was about equal across all herd sizes, and live removals due to feet and leg reasons increased in larger herds. The risk of removal due to disease was highest in the middle-sized herds, and smaller herds were at a greater risk of removal for udder reasons. Health removals as a percentage of all removals were lowest in the smallest and largest herds.

Figure 25.10 shows time period of risk for live removal and on-farm death for each reason by days in milk (DIM). The graph also uses frequency rates so it tells us when the risk of removal changed across a lactation. The percentages in this figure reflect the causes as a proportion of the daily loss of cattle, with each day adding to 100%. On the first day (DIM 1) 58% of those animals leaving the herd were from on-farm deaths, ~15% were from injury, and the remaining ~27% were spread across all other reasons, adding to 100%. Each day does the same, totaling 100%.

This graph gives us an idea about when the risks occur across the lactation. There is a very high risk period for on-farm deaths from birthing to about 60 DIM, during the time the cow is adjusting to a higher nutrient-dense ration, her metabolism is increasing, she's losing weight as her rumen capacity grows, and recovering from her pregnancy and delivery. Another high-risk time is late in lactation, from about 300 DIM and beyond, for live removals due to reproductive failure. These animals have had their milk production decline with time to a point where their milk isn't paying the costs of keeping the cow in the herd, and her next calving is too far away to keep her so she is transformed from a milk cow to a beef cow and sent to market.

Hadley et al. also noted that removal rates were impacted by the availability of replacement heifers in the herd. If more replacements are available then removal rates climb, and if fewer are available removal rates decline. There is an association between higher removal rates and: higher SCCs, higher services per conception, breed, higher lactation number, and season of calving. Having more

Table 25.5 Frequency Distribution of Cows that Died or were Removed According to Reason across Herd Size for Dairy Herd Improvement (DHI) Herds in 10 States from 1993 to 1999[a]

Reason	Herd Size (Cows)				
	1–150	151–300	301–450	451–600	601+
Died	9.7	13.1	14.4	14.5	16.4
Sold for dairy	8.3	5.0	4.3	3.4	6.5
Low production	12.8	12.7	12.9	13.6	13.4
Reproduction	19.5	18.1	16.9	15.7	13.4
Injury/other	27.4	25.8	24.4	22.9	23.3
Mastitis	11.7	13.3	14.4	14.7	13.4
Feet and legs	3.6	7.0	7.2	8.7	8.2
Disease	2.8	3.4	4.2	5.0	3.5
Udder	4.1	1.6	1.4	1.6	1.9
Health removals as a percentage of all removals[b]	78.9	82.3	82.8	83.1	80.0

Source: Adapted from Hadley et al.[12]

[a] The 10 states with DHI records analyzed included five Upper Midwest states (Illinois, Indiana, Iowa, Michigan, and Wisconsin) and five Northeast states (Maine, New Hampshire, New York, Pennsylvania, and Vermont).

[b] Health removals included those due to udder and mastitis problems, lameness and injury, disease, and reproduction problems but not death.

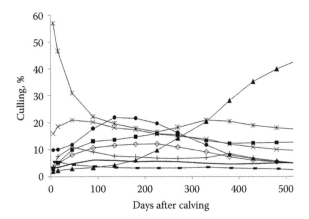

Figure 25.10 Distribution of removed cows (%) by disposal codes at different stages of lactation. Distribution of culled cows (%) by disposal codes at different stages of lactation (each day is 100%). Disposal codes: feet and legs (◊), low production (■), reproduction (▲), injury/other (×), died (*), mastitis (•), disease (+), udder problems (■), and reason not reported (clean line). Disposal codes: feet and legs (◊), low production (■), reproduction (▲), injury/other (×), died (*), mastitis (•), disease (+), udder problems (■), and reason not reported (clean line).

(*Source*: Adapted from Pinedo, P., Daniels, A., Shumaker, J., & DeVries, A., 2010.[23] Dynamics of culling risk with disposal codes reported by Dairy Herd Improvement dairy herds. *Journal of Dairy Science* 93, 2250–2261.)

of breed-registered cows in the herd or current or recent herd expansion lowered live-removal rates as did having a low proportion of replacement heifers to milk cows.

In Figure 25.11, the authors report on Holstein cows that first calved in 1988–1989 and the percentage that exited the herd by their DIM reported by their lactation number. The graph is divided into two figures to simplify the information presented. The removal rates and timing are similar in the upper graph with the expected fewer removals for first lactation animals. Removal rates by DIM are almost identical for lactations 2–4. The bottom graph shows the impact on removals lactations 4, 5, and >5, with lactation 4 included to provide a visual comparison to the upper graph because lactations 2–4 were almost identical. Lactation 5 has earlier removals compared to lactations 1–4, and Lactation >5 show removals at a much higher rate and much earlier in the lactation.[8] An average cow in her 5th lactation will be between 7 and 8 years old.

Breed differences are shown in Figure 25.12. Jersey cows remained in the herd longer and had a consistent rate of decline from 1980 to 1996. During that time period Holstein cows remained in the herd at a rate that also declined consistently but more rapidly than the Jersey's rate. Guernsey's were very

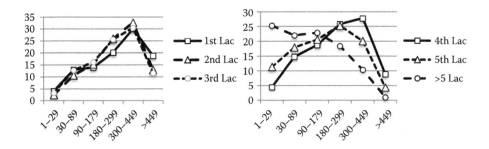

Figure 25.11 DIM at removal for Holsteins first calving in 1998–1999 and for their subsequent lactations.

(*Source*: USDA ARS, Norman, Hare & Wright 1994 ADSA/ASAS/CSAS meeting presentation.)

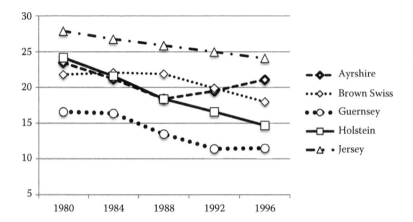

Figure 25.12   Percent survival rates to parity 5 by breed and year.

(*Source*: **USDA ARS, Norman, Hare & Wright 1994 ADSA/ASAS/CSAS meeting presentation.[8])**

consistent with no change in decline from 1980 to 1984, then declined sharply from 1984 to 1992 where is leveled off again until 1996. In 1988 Brown Swiss started declining more rapidly while Ayrshire's declined from 1980 to 1988 and then reversed that decline through 1996. Holsteins comprise the majority of the nation's dairy herd. The authors reported no association of herd size on survival rates by parity.

CONCLUSIONS

Today's larger dairy production systems have evolved in response to economic, social, and environmental forces that have shaped it, especially over the last 50 years. Modern dairy farms require a huge investment in land, facilities, cattle, equipment, debt, labor, and feed. Dairy producers have little influence over the price they receive for their milk and milk sales comprise about 90% of their income.

Those producers who have survived in business have responded to the economic realities they've faced by "right sizing" facilities, labor, and equipment to increase efficiently and lower the costs of operation, including: regularly sorting animals into groups based on the animals nutritional needs (feeding the highest producers the most nutrient-dense and expensive rations and lower producing animals less expensive rations); keeping those animals that produce or are likely to produce income over expenses while removing those that do not; and with other management changes. As a consequence, herds have grown in size, production per cow has increased, more cows find themselves housed in higher density and on harder surfaces, and longevity has decreased.

While dairy cattle represent many values to their owner, including pride, responsibility, and status, economically dairy cattle provide three potential values to the producer: the value of the milk she produces that is sold (including milk components such as protein, fat, etc.), the value of her genetics as a registered breed or potential embryo supplier for implantation into surrogates, and value she returns as a beef animal and through her by-products when slaughtered. The highest value beef cattle go to market before 2 years of age with a lower value market for older animals that are removed for reproduction failure, lameness, mastitis, and other similar reasons as dairy cattle. Dairy cattle are valued for milk and when production is no longer sufficient to cover operational costs they go to market as beef and for their by-products (leather, fat, etc.).

A dairy animal going to beef is a normal and important part of the economic equation of how a dairy survives. There is a tendency for some to regard slaughter as failure in dairy management, but it is a fundamental component of a successful dairy enterprise and may represent a significant portion of a dairy's cash flow. In the US about 20% of the beef consumed by humans comes from dairy breeds.

Producers recognize that premature removal of a cow from the herd is an economic liability and it has an emotional impact on most owners and employees. Retaining an animal in the milking herd when she is no longer delivering income over expenses also impacts owners and employees. Turnover is a normal part of dairying, and while very high turnover rates for health reasons may represent a management problem, so, too, do very low turnover.

SUMMARY

The studies presented in this chapter show dairy cattle welfare: In relation to herd size

- Had no association between herd size and production per cow
- Larger herds had better reproductive performance than smaller herds
- Mastitis prevalence was similar across all herd sizes
- Removals for injuries were higher in small herds and lower in large herds
- Live removals for feet and leg reasons were higher in larger herds
- There was no association found between herd size and survival by parity
- On-farm death rates increases slightly as herd size increases

With respect to production per cow

- Feet and leg problems were not associated with higher production
- The lowest risk of removal was associated with cows producing more milk

Other reported associations

- Feet and leg problems are a serious welfare issue especially for animals on concrete. Footbaths and regular trimming help but more work needs to be done on flooring to address this issue
- Removals for feet and leg problems were twice as high in open vs. pregnant animals suggesting a more complex issue exists than just lameness
- The lowest risks of removal were for pregnant cows
- Cows in and beyond lactation 5 are at a higher risk of on-farm death

FINAL THOUGHTS

As the concern about the welfare of dairy cattle grows, becomes more emotional, and more value laden, we all need to wrestle with how we foster production systems that meet the cow's needs while also meeting the economic realities of the marketplace. To lower a cow's risk of removal in today's dairy she must outperform the average cow in the herd throughout her life. Does maintaining the cow's health, safety and nutritional support allowing her to "win" that competition translate to good welfare? If not, what does?

Longevity is a central theme expressed by those concerned with dairy cow welfare. Cows would remain in the dairy herd longer if milk prices were higher in relation to operational costs, and if the slaughter market remained higher for older animals. Without those two changes economic survival of the dairy enterprise necessitates higher turnover rates and shorter longevity.

For a long time the emphasis on sire selection was skewed toward milk yield and component traits. Thankfully, sire selection now includes more robust and balanced criteria taking into account conformation and other longevity traits as important goals of a desirable breeding program. With changes in record keeping it's possible we'll understand more about why cattle leave the herd early in their lives and be able to create preventive programs to change the current reality. Just as the remarkable increases in production are related to improvement in both genetics and management, it's expected that genetics and management will contribute to lower live removals and on-farm deaths in the future.

Dairy producers and those interested in animal welfare share a common interest in promoting production systems that improve a cow's well-being. It is currently not economically possible for dairy producers to survive the changes needed to extend longevity without receiving higher prices for their milk.

ACKNOWLEDGMENTS

The author acknowledges and appreciates the contributions made to this chapter by Dr. Terry Lehenbauer. He did the original literature search and then circumstances prevented him from continuing his involvement.

REFERENCES

1. Diamond, J., 1999. *Zeberas, Unhappy Marriages, and the Anna Karenina Principle. Guns, Germs and Steel*, New York: W.W. Norton & Company, pp 157–175.
2. After California Veterinary Medical Association Eight Principles of Animal Care and Use. www.cvma.net.
3. USDA Census for Agriculture and Economic Reporting Services.
4. Klingborg, D.J. Personal Observation as a Dairy Practitioner from 1972 to 1993 and a Clinical Professor of Dairy Preventive Medicine from 1993 to 2012.
5. USDA ERS from USDA National Agriculture Statistics Service and Census of Agriculture data, 2013.
6. Rogers, G.W., Van Arendonk, J.A.M., McDaniel, B.T., 1988. *Journal of Dairy Science* 71(12):3453–3462.
7. Rogers, G.W., Van Arendonk, J.A.M., McDaniel, B.T., 1988. *Journal of Dairy Science* 71(12):3463–3469.
8. Norman, H.D., Hare, E., Wright, J.R. 1994. ADSA/ASAS/CSAS Meeting Presentation 2006.
9. Hare, E., Norman, H.D., Wright, J.R., 2006. Survival Rates and Productive Herd Life of Dairy Cattle in the United States. *Journal of Dairy Science* 89:3713–3720.
10. Fetrow, J., Nordland, K.V, Norman, H.D., 2006. *Journal of Dairy Science* 89(6):1896–2006.
11. Pinedo, P., DeVries, A., Webb, D., 2014. *Journal of Dairy Science* 97:2886–2895.
12. Hadley, G.L. Wolf, C.A., Harsh, S.B., 2006. *Journal of Dairy Science* 89:2286–2296.
13. USDA NAHMS Report, 2013.
14. Blayney, D.P., 2002. The Changing Landscape of US Milk Production, USDA Statistical Bulletin 978, June, 2002.
15. USDA, 2008. Dairy 2007, Part II: Changes in the US Dairy Cattle Industry 1991–2007.
16. USDA, 2016. Dairy Cattle Management Practices in the US, 2014.
17. Ruegg, P.I., 2015. Dairy Cow Welfare & Udder Health, www.milkquality.wisc.edu/animal-welfare/.
18. USDA Dairy, 2007. Part V: Changes in Dairy Cattle Health and Management Practices in the United States 1996–2007.
19. USDA, 2007. Dairy 2007.
20. Eicher, S., 2010. Dairy Cow Welfare Fact Sheet USDA ARS Livestock Behavior Research Unit, Fall.
21. USDA Dairy, 2007. Part 1: Reference of Dairy Cattle Health and Management Practices in the United States.
22. Pinedo, P., De Vries, A., Webb, D., 2014. Dynamics of Culling for Jersey, Holstein and Jersey X Holstein Crossbred Cows in Large Multibreed Dairy Herds. *Journal of Dairy Science* 97:2886–2895.

CHAPTER 26

Calf Welfare

Munashe Chigerwe
University of California

CONTENTS

SUMMARY

Several factors affect the well-being of dairy and beef calves. These factors include: housing, the environment, nutrition, handling, transportation, health management such as vaccinations and treatment of sick calves, management procedures such as tail docking, dehorning, and removal of supernumerary teats, weaning, morbidity and mortality, and euthanasia. There are several peer-reviewed literature and reviews addressing the above-mentioned factors. This chapter focuses on the following specific factors affecting calf welfare:

1. Maternity sanitation
2. Colostrum management
3. Separation of calf immediately after birth
4. Single-calf housing vs. group housing
5. Weaning

Maternity Pen Sanitation

The calf contacts the environment for the first time during delivery. In dairy operations, designated group maternity pens are usually available for close-up cows. Within the maternity pen, individual cow calving pens might be available. Thus, cows showing signs of impending labor are moved from group maternity pens to individual maternity pens. The individual maternity pen allows close supervision of parturition and the new born calf. In cow-calf operations, producers may designate a calving pasture for close monitoring of cows during the calving season. The management of the calving pen influences the level of exposure to infectious pathogens.[1] Pathogens that can be transmitted to the new born calf in the maternity pens include *Mycobacterium avium* subsp. *paratuberculosis* (MAP), *Escherichia coli*, and *Salmonella* sp. The association between calving site and incidence of calf morbidity has been reported in several studies. A study conducted in New York dairy herds reported that calves born in herds using calving pens were less likely to experience diarrhea compared to those born in herds using stanchions or loose housing areas.[2] In a study that enrolled 122 Swedish dairies, calves born in herds using single-cow calving pens were at a lower risk for respiratory disease compared to those born in herds using cubicles or group calving pens.[3] In Minnesota, herds with 50–90 cows, use of single-cow pens and removal of bedding from pens between calvings were associated with lower incidence of diarrhea.[4] Furthermore, studies by Waltner-Toews[5] and others reported a lower overall mortality in calves born in single-cow pens compared to those born in unexpected or unusual environmental sites on the farm. In contrast, a controlled field trial reported no differences in risk for calf-hood diseases (diarrhea and respiratory diseases) during the first 90 days of life between calves born in single-cow calving pens compared to calves born in multiple-cow calving pens.[6] However, a longitudinal study based on the same field trial by Pithua and others[6] later reported that that using individual calf pen for calving delayed exposure to MAP in calves and provided an effective strategy for reducing peripartum MAP transmission risks in herds focusing on reducing the impact of paratuberculosis.[7] Consequently, single-cow calving pens, which are cleaned after each, are recommended to dairy producers to reduce exposure of calves to pathogens.

Colostrum Management

Negligible transfer of antibodies occurs across the ruminant cotyledonary placenta. As a result, calves rely on the ingestion of colostrum and absorption of maternal antibodies across the intestines via pinocytosis during the first 24–48 hours after birth.[8] Calves that fail to ingest and absorb sufficient colostrum experience increased risk of morbidity and mortality due to diarrhea, enteritis, septicemia, arthritis, omphalitis, and pneumonia.[9,10] Despite vast information on the importance of feeding colostrum, prevalence of failure of passive immunity (FPI) in calves remains high, ranging at 19.2% in the United States.[11] Calves fail to ingest and absorb sufficient colostral components because of several reasons, including shortage of colostrum, poor-quality colostrum, delay in collecting or feeding colostrum, storage conditions, lactation number, and dystocia. In other instances, calves are not fed sufficient colostrum due to their low perceived economic value, for instance, dairy bull calves. While beef calves are left to nurse from the cow on their own, intervention through feeding colostrum by a nipple bottle or oroesophageal tube is necessary for dairy calves. Studies reported that prevalence of FPI of colostral immunoglobulins when dairy calves were allowed to nurse from the dam was 61.4% compared to 19.3% and 10% when calves were fed colostrum by a nipple bottle and oroesophageal tubing, respectively.[12] In addition to conferring passive immunity, colostrum is a highly digestible source of carbohydrates, protein, and fat for the calf. Quality of colostrum prior to feeding to calves can be improved by assessing the immunoglobulin concentrations using several instruments such as refractometers and hydrometers. In instances where transmission of infectious agents such as MAP or *Salmonella* through colostrum is a concern, pasteurization of colostrum should be considered. Furthermore, standard operating procedures on dairy farms should include monitoring of colostral feeding practices in calves by using farm adapted tests such as serum total protein determination by refractometry. From a calf welfare standpoint, colostrum is necessary for all new born calves for long-term health and production.

Separation of Calf after Birth

While beef calves remain with the cow until 6–7 months of age, dairy calves can be separated at 1–24 hours after parturition. Dairy calves are fed colostrum within the first 24 hours followed by milk or milk replacer. Reasons for separating calves from cows immediately include prevention of transmission of infectious pathogens from the cow to the calf, and efficient monitoring of colostrum, milk, and solid intake by the calf. Additionally, early separation of the calf from the cow is anticipated to reduce distress for both the cow and calf. In contrast, other studies reported advantages for delayed separation of the cow from the calf. Calves separated from cows at 14 days of age gained weight at three times the rate compared to those separated on the day of birth, and maintained the weight even after separation.[13] However, cows and calves separated later responded more strongly to the separation compared to separation at birth.[13] Additionally, cows kept with calves had lower milk yields, although the total lactation milk yields recovered to expected levels after separation.[13] While there are advantages associated with average daily gain by allowing calves to nurse from the cow for longer periods after parturition, such approach may have practical limitations in large dairies, and may hinder efficient individual monitoring of intake by calves. Perhaps recommendations to allow dairy calves to nurse from the cow for longer periods of time may depend on the size of the herd, space, and goals of the producer.

Single Housing vs. Group Housing

In North America, housing dairy calves in single-calf hutches has been recommended.[14] The advantages of raising calves in individual calf hutches include reducing nose–nose contact, thereby reducing transmission of diseases, efficient monitoring of feed intake and medical treatments, and reduction of behavioral vices such as excessive cross-sucking. Cattle are considered a social species and interact through visual and/or audio mechanisms. Group housing of calves allows normal social behavior during growth, provides access to space,[14] and reduces labor of cleaning pens and feeding.[15] Another approach to enhance normal social behavior has been pairing of calves. Previous studies reported no difference in weight gain, except during the week of weaning between calves raised individually and pair-housed calves.[14] During the week of weaning, individually raised calves lost weight whereas pair-housed calves maintained daily weight gain at pre-weaning levels.[14] Incidence of diarrhea was not different between individually housed calves and pair-housed calves.[14] It has been postulated that the advantages of individual housing in reducing disease transmission might be overestimated as contact of calves may occur[14] probably through fomites, calf handling, feeding utensils, and aerosols. It has been suggested that the hygiene of the calf housing and ventilation might be more important in preventing transmission of diseases rather than pairing or non-pairing of calves in hutches.[14] New technologies are now available that can assess individual milk, and solid feed intake in group housed calves. Thus, given the enhancement of development of normal social behavior and performance among calves, group housing of calves prior to weaning should be considered.

Weaning

In beef cattle, calves continue to nurse from the cow until 6–7 months of age at which they are physically separated from the cow. Assuming that the calf's rumen development is advanced by 6–7 weeks of age, beef calves will be consuming significant quantities of solid feed (forages and concentrates if provided) by the time they are weaned. Additionally, by the time the beef calf is weaned, the cow is producing smaller volumes of milk suggesting that milk will constitute a smaller proportion of the calf's diet. In contrast, dairy calves are offered solid feed as early as the first week of life and are commonly weaned from milk or milk replacer at approximately 8–12 weeks age. At the time of weaning, dairy calves are still consuming significant volumes of milk or milk replacer. Thus, the abrupt change in diet for both beef and dairy calves can cause distress. To reduce distress associated with weaning in dairy calves, several methods have been employed by producers. One method is to gradually reduce the frequency of milk or milk replacer, for instance, from three, to two, and then one feeding per day. Another method is to maintain the same frequency of feeding but dilute the milk with water. In one study, gradually weaned dairy calves (by diluting milk with water) consumed more solid feed than abruptly weaned calves during weeks 5–8.[14] However, there was no difference in weight gain between calves gradually weaned and abruptly weaned calves.[15] Surprisingly, calves weaned gradually by diluting milk with water continued to drink milk even at 100% dilution with no behavioral signs such as increased vocalization associated with abrupt weaning.[14] These studies suggest gradual weaning using various methods might be effective in reducing distress associated with weaning in dairy calves.

Weaning in beef calves is commonly abrupt. Some strategies have been recommended to reduce the stress associated with weaning in beef calves. One method involves separating the cow–calf pairs through a fence for several days before final separation to allow partial contact while preventing nursing by the calf.[16] Another method is the use of nose flaps which prevent the calf from nursing while the calf remains with the cow. In one study, calves with nose flaps exhibited less vocalization, spent more time eating and laying down than calves that were separated abruptly.[17] However, calves with nose flaps had lower weight gain and potentially had undesirable welfare concerns due to frustration as they unsuccessfully made attempts to nurse with the nose flaps.[18] In summary, the weaning methods in beef calves might depend on the goals of the producers, space, and labor availability.

REFERENCES

1. Smith BP, Oliver DG, Singh P, et al. 1989. Detection of Salmonella Dublin in mammary gland infection in carrier cows using an enzyme linked immunosorbent assay for antibody in milk or serum. *Am J Vet Res* 50: 1352–1360.
2. Curtis CR, Scarlett JM, Erb HN, et al. 1988. Path model for individual-calf risk factors for calfhood morbidity and mortality in New York Holstein herds. *Prev Vet Med* 6: 43–62.
3. Svensson C, Lundborg K, Emanuelson U, et al. 2003. Morbidity in Swedish dairy calves from birth to 90 days of age and individual calf-level risk factors for infectious diseases. *Prev Vet Med* 58: 179–197.
4. Frank NA, Kaneene JB. 1993. Management risk factors associated with calf diarrhea in Michigan dairy herds. *J Dairy Sci* 76: 1313–1323.
5. Waltner-Toews D, Martin SW, Meek AH. 1986. Dairy calf management, morbidity and mortality in Ontario Holstein herds IV. Association of management with mortality. *Prev Vet Med* 4: 159–171.
6. Pithua P, Wells SJ, Godden SM, et al. Clinical trial on type of calving pen and the risk of disease in Holstein calves during the first 90 d of life. *Prev Vet Med* 89: 8–15.
7. Pithua P, Espejo LA, Godden SM, et al. 2013. Is an individual calving pen better than a group calving pen for preventing transmission of Mycobacterium avium subsp paratuberculosis in calves? Results from a field trial. *Res Vet Sci* 95: 398–404.
8. Weaver DM, Tyler JW, VanMetre DC. 2000. Passive transfer of colostral immunoglobulins in calves. *J Vet Intern Med* 14:569–577.
9. Besser TE, Gay CC. 1994. The importance of colostrum to the health of the neonatal calf. *Vet Clin North Am Food Anim Pract* 10:107–117.
10. Tyler JW, Hancock DD, Wiksie SE, et al. 1998. Use of protein concentration to predict mortality in mixed-source dairy replacement heifers. *J Vet Intern Med* 12:79–83.
11. Beam AL, Lombard JE, Kopral CA, et al. 2009. Prevalence of failure of passive transfer of immunity in newborn heifer calves and associated management practices on US dairy operations. *J. Dairy Sci* 92:3973–3980.
12. Besser TE, Gay CC, Pritchett L. 1991. Comparison of three methods of feeding colostrum to dairy calves. *J Am Vet Med Assoc* 198:419–422.
13. Flower F, Weary DM. 2001. Effects of early separation on the dairy cow and calf: 2. Separation at 1 day and 2 weeks after birth. *Appl Anim Behav Sci* 70:275–284.

14. Weary DM. 2001. Calf management: Improving calf welfare and production. *Adv Dairy Technol* 13:107–118.
15. Kung L Jr, Demarco S, Siebenson LN, et al. 1997. An evaluation of two management systems for rearing calves fed milk replacer. *J Dairy Sci* 80: 2529–2533.
16. Price EO, Harris JE, Borgwardt RE, et al. 2003. Fenceline of beef calves with their dams at weaning reduces the negative effects of separation on behavior and growth rate. *J Anim Sci* 81:116–121.
17. Haley DB, Bailey DW, Stookey JM. 2005. The effects of weaning beef calves in two stages on their behavior and growth rate. *J Anim Sci* 83:2205–2214.
18. Enríquez DH, Ungerfeld R, Quintans G, et al. 2010. The effects of alternative weaning methods on behaviour in beef calves. *Livest Sci* 128:20–27.

CHAPTER 27

The Downer Cow

Ivette Noami Roman-Muniz
Colorado State University

CONTENTS

DOWNER COW: DEFINITION AND SIGNIFICANCE

Downer or nonambulatory animals represent a welfare issue for livestock agriculture, particularly the dairy cattle industry. In the 1950s, the term "downer" was already being employed by researchers and veterinary practitioners to describe recumbent cattle (Fenwick, 1969). Since then, the condition of recumbency has been the only consistent requirement for classifying cattle as downers. For over six decades, authors have designated varying number of hours of recumbency, degree of alertness, posture, and known or unknown causes of disease as additional requirements to identify cattle as downers (Cox et al., 1986; Correa et al., 1993; Fenwick, 1969; Green et al., 2008; Stojkov et al., 2016; Stull et al., 2007).

Early on, the term downer described cows that remained in sternal recumbency soon after calving, without any obvious illness, but that lacked strength to stand up (Fenwick, 1969). These early reports associated downer cows almost exclusively with milk fever and suggested administration of calcium as the treatment of choice for downers. While the early definitions did not specify a length of recumbency, downer cows have been characterized as recumbent in sternal position for more than 24 hours for no obvious reason by several researchers (Cox et al., 1986; Poulton et al., 2016a, b; Stull et al., 2007). Other authors have proposed using a 12 hour time frame to identify downer cows (Stojkov et al., 2016).

Inconsistent use of the terms downer or nonambulatory by researchers has been problematic as it has resulted in varying conclusions regarding the significance of the problem, the indicated promptness of treatment, and more importantly, the time frame for considering euthanasia as an appropriate course of action to avoid prolonged suffering and to ensure the health and safety of humans and other animals. For example, characterizing downers as cows with no severe systemic disease could have potentially shaped the way dairy producers managed this condition historically. If the perception is that an animal who is alert and in sternal recumbency, but who cannot stand or bear weight, is not afflicted with a systemic disorder, a logical approach would be to provide nursing care while waiting to see if the animal recovers. This is problematic as cows that are alert and able to maintain sternal recumbency could suffer from a variety of metabolic disorders, infectious processes or musculoskeletal injuries that, if untreated, would not improve and very likely worsen. Definitions that specify a shorter time frame and consider a variety of primary causes of prolonged recumbency should aid in early assessment and treatment, and as a result, increase the chances of recovery.

Downer cattle are a serious concern in terms of animal health and welfare, food quality and safety, human safety, consumer perception, and economic losses. Although downer cattle affect both dairy and beef operations, all ages, both sexes and various production stages, the problem is especially significant for adult female dairy cows in early lactation. The National Agricultural Statistical Service reported that 57% of all cattle who became downers were from dairy origin (NASS, 2005). While the percentage of operations with at least one nonambulatory cow during the year varies according to the study, several multistate surveys have revealed that more than three quarters of dairy operations have at least one downer cow within a year (Adams et al., 2015a; Green et al., 2008). In a recent survey, Adams and colleagues reported that 2.6% of U.S. dairy cows became downers during 2013 (Adams et al., 2015a). Some studies estimate that nine out of every 10 downer cattle at slaughter are of dairy origin (Doonan et al., 2003); nonetheless, the beef industry also recognizes the significance of this condition both in economic and animal welfare terms (Smith et al., 2015).

While only 1% of downers occur during transport or at slaughter facilities (Doonan et al., 2003), downer cattle cases have caught consumer's attention. The transport and handling of downers is a welfare issue and images of downer animal mistreatment at slaughter facilities have made the national news more than once and have made animal protection groups and consumers alike question handling practices as well as the safety of beef products.

In addition to eroding consumer trust, downer cattle represent significant economic losses to the dairy and beef industries (Green et al., 2008; NASS, 2005). Downer cattle are responsible for carcass condemnations at slaughter as they represent a food safety and public health concern. Carcasses from downer cattle have greater odds of food-borne pathogen contamination (e.g., *Escherichia coli* and *Salmonella spp.*), and are more likely to have bruising and other defects that compromise meat quality (Stull et al., 2007). Diseases that affect the nervous system of ruminants and that can be transmitted to humans can cause cattle to become recumbent. Rabies and Bovine Spongiform Encephalopathy, or Mad Cow Disease are two examples of infectious causes of downers that could represent a serious risk to the safety of animal handlers at slaughter facilities and to the health of consumers. The ban to the slaughter and processing of nonambulatory cattle (9 CFR 309.3(e)) (FSIS, 2014) was intended to minimize the risk of introducing Mad Cow Disease into the human food supply, while increasing consumer confidence in beef products.

Downer cattle also represent a human safety risk for livestock farm workers. When handling downers (e.g., moving, assessing, treating, providing nursing care, and performing euthanasia), livestock workers are at risk of injury or disease. Close contact with cows has been identified as a factor contributing to work-related injuries on dairies (Roman-Muniz et al., 2006). When asked about occupational safety and health risks, dairy workers identified working with sick animals as an activity with a high risk of injury and illness (Menger-Ogle et al., unpublished).

Only about a third of all downer cows recovers (Adams et al., 2015a; Cox et al., 1986), with the rest being euthanized or dying on the farm. Putting all economic losses and public perception considerations aside, downer cattle represent a significant concern in terms of animal health and welfare. Not managing cattle as to prevent risks associated with prolonged recumbency and later on denying the needed care to minimize further damage and to increase chances of recovery jeopardizes animal welfare and all of the five freedoms as described by the Farm Animal Welfare Council (2009). Preventive strategies and proper management of the downer cow with the goal of preserving her welfare will be the focus of the remainder of this chapter.

RISK FACTORS AND PREVENTION STRATEGIES

Ideally, efforts of livestock producers and workers should focus on preventing new cases of nonambulatory cattle. In order to prevent this condition, it is necessary to be aware of risk factors associated with it. Again, it is important to understand that while the issue of nonambulatory cattle has been highly publicized at slaughter facilities, most downers originate and are managed on the farm. If we are to focus on the prevention of downers, we must focus on best farm management practices that result in fewer occurrences of disease and injury. It is essential to understand that a downer cow can be the result of a multifactorial process and that in many cases determining the contributing factors and managing the affected cow can be a complex task.

Causes of downers can be classified into four general categories: injuries, metabolic imbalances, infectious or toxic diseases (Green et al., 2008). For several decades, cases of prolonged recumbency on dairy farms have been associated with milk fever (hypocalcemia), and other problems around the time of parturition such as dystocia, retained placenta and stillbirths (Correa et al., 1993; Fenwick, 1969). Over-conditioned dairy cows with a variety of metabolic or infectious diseases at the time of parturition have been linked to greater morbidity and mortality in early lactation (Littledike et al., 1981). Dairy owners recognize the challenges of the periparturient period in dairy cows. When surveyed about risks associated with downer cows, producers reported that most cases occur within one day of calving and are associated with dystocia (Cox, 1986). The fact is that after parturition, dairy cows can be compromised with a variety of metabolic conditions, infectious diseases, and lameness and become recumbent (Stojkov et al., 2016).

Greater milk production has also been linked to the occurrence of downer cows on dairy operations. Dairy producers perceive high or average milk-producing cows as the great majority of downers (Cox et al., 1986), and a national survey identified high milk production as a risk factor for downers (Green et al., 2008). Greater milk production per cow brings about challenges that must be met with excellent management practices, including but not limited to nutrition, cow comfort, reproductive management, and disease identification and treatment protocols. Injuries related to facilities design, maintenance, and animal handling are associated with lameness and downers. Concrete flooring has been associated with more lame cows (Adams et al., 2017), and with downer cows (Green et al., 2008). When paired with improper maintenance and incorrect animal handling, concrete flooring can cause injuries that lead to lameness and downers.

Herds with greater number of animals inherently have more opportunity for disease or injury events that can lead to downers. A recent survey identified larger herd size as a risk factor for the occurrence of downer cows (Green et al., 2008). It is worth mentioning that recent studies have also associated larger herds with fewer lame cows (Adams et al., 2017). Lager herds can be managed in such a way as to prevent those disease events that could lead to downers. For example, having sand bedding and having cows on pasture has been associated with fewer lame cows (Adams et al., 2017). Additionally, providing heat abatement and feeding rations formulated by a nutritionist was recently associated with fewer thin cows on dairy farms (Adams et al., 2017). These practices are not exclusive of a specific herd size and can therefore be used to reduce the prevalence of lameness and low body condition, two factors that have been associated with downers. Careful management and closer observations of animals at risk, especially during the periparturient period are key factors in the prevention of downers (Cox et al., 1986; Stull et al., 2007).

Strategies for the prevention and management of downer cows can be classified into broad categories and should include: nutritional management, herd health, cow comfort, calving procedures, facilities, training, and good husbandry (Doonan et al., 2003; Stull et al., 2007). Stull and colleagues (2007) suggest a list of management procedures for the prevention of and treatment of nonambulatory cattle which includes: well-formulated diets that decrease the risk of metabolic disorders associated with prolonged recumbency and that maintain body condition at healthy levels in both young and adult cattle, veterinary care, treatment of downers as emergencies, high-quality nursing care, preventive care aimed at diseases commonly known as primary causes of downer cattle, clean, dry, and comfortable housing, and calving management practices that promote easier calvings and limit opportunities for injury and infections.

As the dairy industry moves toward larger herds with more hired labor, the implementation of adequate standard operating procedures and the training of all dairy employees are of paramount importance if we are to encourage optimal management of individual animals at risk of becoming nonambulatory.

MANAGEMENT TO INCREASES CHANGES OF RECOVERY AND REDUCE SECONDARY DAMAGE

Downer cows should be treated as medical emergencies and their management should have two primary objectives: to correct the primary cause of recumbency and to minimize the chances of secondary recumbency. Achieving these two objectives should result in greater animal welfare, improved prognosis for returning to a healthy and productive status, and fewer losses for dairy producers.

The first step in managing a nonambulatory animal is to perform a basic physical examination. Several dairy personnel should be trained on the basic steps of a physical exam, as depending on just one employee or outside contractor could mean that a downer animal has to wait to be properly assessed if the designated individual is not on the farm when the animal becomes nonambulatory. The herd veterinarian or university extension personnel could serve as resources for training employees on how to conduct a physical examination. Additionally, bilingual (English and Spanish) training programs are available online (Roman-Muniz and Van Metre, 2011) and could be useful for this purpose. Besides any physical findings that can narrow down the primary cause of recumbency, the physical examination should aid in assessing pain severity, prognosis, treatment feasibility, and costs. (Stull et al., 2007). Physical exam findings together with the cow's history should help to make a decision that will result in less suffering for the affected cow and decreased cost for the livestock operation.

While assessing the downer, dairy personnel should evaluate her ability to stand, and decide if further treatment is the best option to preserve her welfare. Researchers have proposed several conditions to choose treatment over euthanasia. Cows that are alert, show no signs of distress, maintain a sternal position, have a normal mentation, are eating and drinking, attempt to stand, and are afflicted with metabolic disorders, infectious diseases, minor musculoskeletal injuries, or dystocia should be treated (Smith et al., 2015; Stojkov et al., 2016). On the other hand, animals experiencing severe uncontrollable pain with a poor prognosis for correcting the primary cause of recumbency should be immediately euthanized. Stojkov and colleagues (2016) propose not treating any animal who is not alert, in lateral recumbency, with fractures, compromised joints, or any major musculoskeletal injuries. Additionally, body condition should be considered as very thin cows (body condition score less than 2.5 on a dairy cattle scale) have lesser chances of recovering (Green et al., 2008). Poulton and colleagues (2016a) advise to euthanize cattle with secondary hip displacement as they commonly relapse after replacement and have a grave prognosis.

Besides identifying the primary cause of recumbency and being able to decide if euthanasia or medical treatment is the best course of action, a physical examination performed promptly after the cow is identified as a downer, should also help prevent further damage associated with prolonged recumbency. Downer cows have greater chances of recovery if treated within 24 hours of becoming recumbent, and unlikely recovery if treatment is begun after 48 hours of recumbency (Stojkov et al., 2016). Six to twelve hours into recumbency, the downer cow will start showing signs of pressure damage to the hind leg located underneath the body. Pressure on the limb will cause damage to the sciatic nerve and necrosis to the muscles of the hind legs (Cox et al., 1982). Additionally, femoral nerve damage is a common finding in downer animals, affecting two thirds of all downer dairy cows (Poulton et al., 2016a).

Secondary damage is a significant cause of animal losses in the form of death or euthanasia on dairy operations. In fact, after being down for more than 24 hours, secondary damage has greater influence on the downer's ability to recover than the primary cause of recumbency. In a study by Poulton and colleagues (2016a), secondary damage was solely responsible for close to three quarters of the cows that didn't recover. The same study revealed that 13% of downer cows that were euthanized or died were lost due to a combination of the primary cause of recumbency and secondary damage. Green and colleagues (2008) reported greater chances of recovery for animals recumbent for less than 24 hours and recommend considering euthanasia for those animals recumbent for longer than 24 hours, based on the welfare issues and chances of recovery associated with prolonged recumbency. Secondary damage must be assessed and in all cows recumbent for a prolonged period (more than 12 hours of recumbency). If present, secondary damage must be addressed in conjunction to treating the primary cause of recumbency.

Animals with a good prognosis and receiving treatment for the primary cause of recumbency should be assessed twice a day for signs of improvement or worsening and should receive the highest quality nursing care, as it determines the downer's ability to successfully recover.

While assessing the downer animal, caretakers should address hydration deficits, inflammation, and infection or metabolic challenges (Stull et al., 2007). Good nursing practices that improve the odds of recovery include providing access to clean water and feed, repositioning frequently, and placing her in an area with clean and deep bedding, or a shaded area in the pasture. While water and food address some hydration and nutritional needs, repositioning frequently (at least every 6 hours, or four times a day) as to alternate the hind leg located underneath the body will help prevent compression damage to muscles and nerves. The surface where the downer cow lies should provide cushion, good traction, in case she attempts to rise, and reduce the opportunity for abrasions, secondary infections, and ischemic muscle necrosis. Additionally, the downer should be provided shelter against inclement weather, as excessive heat, cold, wind, and precipitation can exacerbate signs of illness and limit its ability to recover. If housed in a sick pen, this area must not be crowded, and some suggest that the downer should be housed with not more than two other cows (Stull et al., 2007). Crowded pens could result in further injury to the downer animal and safety risks for caretakers.

It is not uncommon for cows to become recumbent in places other than the sick pen. When a cow falls down in the milking parlor, an alleyway or her home pen, she should be transported to an area where she can be assessed and treated safely. Downers should never be dragged; this inhumane practice results in additional stress and can cause further injury. When transportation is required, it should be done in such a way as to minimize discomfort and prevent further injury, as well as to minimize safety risk for the caretakers. Downer cattle can be rolled into a loader bucket or pulled on a pallet or sled. Either of these methods for transporting downers require properly trained caretakers working as a team.

Besides providing a surface that allows the animal to more safely attempt to rise (deep, clean, bedding with traction), caretakers can assist cows to stand in several ways. Although a majority of dairy owners have reported experience with lifting devices and found them useful (Cox, 1986), there is a lack of literature and guidance as to how to best use them in a way that minimizes further damage and stress (Stojkov et al., 2016). Recommendations on how frequent and the time length per session of lifting cows with hip clamps were not found in published research. Only personnel previously trained should use lifting devices, and they should only be employed with animals who are able to support their own weight. If used incorrectly, devices designed to lift animals can result in further damage to the downer and be a safety risk to both animal and humans. For example, incorrect use of hip clamps has been associated with pain, pressure sores over the hook bones (Van Metre et al., 1996), and laceration of the abdominal muscles (Stojkov et al., 2016). Furthermore, using devices such as slings with one single belly strap can compress abdominal contents and hinder breathing (Stull et al., 2007).

Repeated electric prod use is not appropriate or humane. Although the electric prod can stimulate a cow to rise, she will only rise if she is able to do so. While this may prove useful in an emergency situation, for example, when a cow falls down in the parlor or in a poorly designed chute, using the electric prod in a downer cow repeatedly only adds to stress and discomfort. Facility design, standard operating procedures and appropriate personnel training prior to these events should prove more effective than using the electric prod.

Other methods to help downer cows rise, such as flotation therapy have been found effective, but are also dependent on the chronicity of the recumbency and the quality of nursing care provided to the downer. Flotation therapy is associated with animal stress during the filling and draining of the flotation tank (Stojkov et al., 2016), and should be only be performed by trained and skillful personnel after careful assessment of the animal. If done appropriately and depending on the primary cause of recumbency, flotation therapy can be administered for up to 10 hours per day. This therapy allows the cow to improve circulation to the limbs and strengthen her leg muscles while being supported by water (Van Metre et al., 1996).

Since nursing care influences both the recovery from the initial cause of recumbency and secondary damage, Poulton and colleagues (2016b) describe optimum nursing care as the treatment to address the primary cause of recumbency as well as the secondary conditions resulting from prolonged recumbency. Caretakers should place the downer in a location where they can provide deep soft clean bedding, protect her from adverse weather conditions, assist with lifting when appropriate, reposition every 6 hours, and maintain adequate hygiene. Additionally, caretakers should provide access to good-quality water and feed, udder care for lactating animals, as well as appropriately transport if needed. All these procedures require adequate labor, in terms of number, training and skill. Although many have suggested a timeline for deciding to euthanize, Poulton and colleagues (2016b) argue that if nursing care is adequate and animal welfare is not compromised, the decision to euthanize shouldn't be time dependent. On the other hand, if the downer cow cannot be nursed under appropriate conditions, or if animal welfare is compromised, caretakers should euthanize her. And even when excellent nursing care can be provided, euthanasia should be performed if prognosis for recovery is poor and severe pain is evident (Poulton et al., 2016b).

Despite all the compelling evidence stressing the importance of high-quality nursing care in the successful management of downer cattle, the care of downer cows is often unappreciated and lacking. Although most recently surveyed dairies indicated that they offered shelter to nonambulatory cows, some operations didn't do so for several hours and 9.1% of operations didn't offer downer cows shelter at all. Likewise, water and food was offered on most dairy farms, but 2.6% and 3.4% of operations didn't offer water or food at any time, respectively (Adams et al., 2015a). While the percentages of operations that didn't offer shelter, water or food might seem insignificant, this lack of appropriate nursing care results in compromised animal welfare as well as production losses for dairy operations. Any primary condition, even one as common and easily treated as milk fever could result in treatment failure if appropriate nursing care is not provided. Denying high-quality nursing care in a timely manner to a downer animal compromises its well-being and chances of recovery, and warrants considering euthanasia as the best course of action.

THE IMPORTANCE OF STANDARD OPERATING PROCEDURES, APPROPRIATE TRAINING, AND LABOR MANAGEMENT

Livestock operations that aim to prevent and provide high-quality nursing care to downer animals, should equip personnel with the knowledge and tools to properly identify animals at risk, to effectively assess and treat, and to humanely euthanize downers when needed. Nonetheless, downer cattle management is a topic often overlooked by management on dairy operations. A recent study revealed that written guidelines for handling nonambulatory cattle can be found in less than a quarter of all operations (Adams et al., 2015a). When operations in the same study were categorized by heard size, more dairies with 500 cows or more reported having written guidelines than dairies with fewer than 500 cows. Not having written standard operating procedures results in downers being managed inconsistently, according to the caretaker assigned to the sick pen that day. This, in turn, jeopardizes the chances of recovery and puts animals at risk of unnecessary suffering when treatment or euthanasia decisions are delayed.

Written guidelines or standard operating procedures, should be available, and communicated regularly to all caretakers. Written guidelines should be used to train all new employees and to create a culture that values animal welfare and human safety. To more effectively manage downers, written guidelines should include best practices such as timely assessment of pain, alertness, hydration, and any serious condition that would warrant euthanasia, humanely transporting, properly bedding the pen, providing good-quality water and feed and regularly repositioning the downer every 6 hours. Additionally, there should be guidelines for assessing secondary damage in cows recumbent for more than 12 hours.

Researchers have reported that repeated treatments offered to downer animals, such as flotation therapy, while helpful in many instances, can induce stress and increase costs (Stojkov et al., 2016). Guidelines for repeating treatments or for offering different treatments should be established by management and communicated with caretakers ahead of time. Treatment offered to the downer animal, whether to address the primary cause or the medical sequelae of prolonged recumbency should be based on a physical exam and assessment of the animal's condition. Written guidelines for considering euthanasia when further treatment is not justified or when the downer's condition worsens should be discussed with all personnel, and should stress best practices congruent with the safety of the person performing the euthanasia as well as the humane treatment of the animal to be euthanized.

As previously discussed, training employees on how to perform a basic physical exam is a critical need. Although some primary causes of downers require additional diagnostic procedures or laboratory testing, a basic physical exam should aid in diagnosing some relatively common causes of downer cattle, such as milk fever, uterine or mammary gland infections, and some musculoskeletal or calving injuries. Assessing the downer when first identified, and after treatments will allow for more effective management, more prompt decision making, and less risk of welfare issues. Additionally, being able to identify severe injuries and signs of significant pain should expedite the decision to euthanize. Physical exam of downer animals could be difficult depending on the location of the animal, and any training received by the caretakers should emphasize human and animal safety. With the help of the herd veterinarian, caretakers should learn how to do physical exams consistently and following the same steps. Consistency in the process can help avoid missing important clues, such as changes in rumen motility, mentation, degree of dehydration, and elevated pulse and respiratory rate, both of which could be signs of pain. When examining a dairy cow, it is critical to assess the reproductive tract and the mammary gland, as they are often associated with disease in producing animals. Employees performing the physical examination should carefully assess the musculoskeletal system and inform a supervisor or herd veterinarian if they find any abnormalities as they may warrant euthanizing the cow.

It is worth noting that only one third of dairy producers provide their employees training on management of nonambulatory cattle and just a fifth provide training on euthanasia procedures (Adams et al., 2016b). The need for training is significant if animal well-being is a priority on dairy operations. Training programs that equip caretakers with the skills to maximize the comfort and the chances of recovery of downer animals, as well as the factors to take into account when considering and performing euthanasia show promise in increasing the knowledge of dairy employees regarding welfare-related practices, including euthanasia procedures (Adams et al., 2016). Cultural congruency is important when considering the occupational health and safety of livestock workers (Menger et al., 2016b) and will impact how a multicultural population of livestock workers engages in discussions of animal welfare and euthanasia procedures.

Besides written guidelines and appropriate training for caretakers, providing high-quality nursing care requires to allocate enough labor time to the management of downer animals. Job organization on livestock operations can be a challenge. Time pressures, lack of clarity regarding work responsibility, and issues with work organization have been described by employees of large dairies in the Western U.S. as stressors that may affect work performance (Menger et al., 2016a). If management cannot allocate enough time for employees to tend to downer cows, the caretaker will feel rushed and provide suboptimal care. In addition to reducing the chances of recovery and affecting animal well-being, rushing through the treatment of downer animals could potentially put the caretaker at risk of injury.

WHEN EUTHANASIA IS NEEDED

Euthanasia can be described as the humane termination of an animal's life and it should be performed in such a way that it minimizes or eliminates distress and suffering (Underwood et al., 2013). Euthanasia should be considered in downers with a poor prognosis, when pain is evident and uncontrollable, or when adequate nursing care cannot be provided. Although some researchers argue that with excellent nursing care, the decision to euthanize can be delayed, under most situations, recumbency longer than 24 hours carries a poor prognosis and euthanasia should be considered at this time. All caretakers should focus on diagnosing, treating, and caring for the downer cow in such a way that she can stand up and ambulate within 24 hours. If the best treatment and nursing care is not successful in resolving the primary and secondary issues, then euthanasia should be seriously considered to prevent animal welfare compromise.

In a recent survey, dairy producers reported that just about half of all downer cows were euthanized, with 59% of them being euthanized within 2 days of being identified as nonambulatory (Adams et al., 2015a). The same producers indicated that 17.7% of downers died on the farm. With euthanasia being a common outcome of nonambulatory cattle cases, it is imperative that people performing this procedure are trained to do it effectively and safely. The American Veterinary Medical Association (AVMA) provides guidelines for making a decision regarding euthanasia and the considerations of employing various euthanasia methods in cattle (Underwood et al., 2013).

Herd veterinarians are a resource for farm personnel training on euthanasia procedures, in addition to the development of guidelines for making the decision to euthanize downers. A study by Hoe and Ruegg (2006) revealed that veterinarians were consulted about euthanasia decisions on dairies less than a third of the time. The same study exposed a severe need for training of personnel as producers reported that untrained dairy personnel performed euthanasia 13% of the time, and that gunshot was the most common method. These findings, in addition to the lack of nonambulatory cattle management and euthanasia training reported by Adams and colleagues (2015b), are of concern as the efficacy of euthanasia by gunshot is dependent on the operator's accuracy, and the firearm and bullets chosen by the operator. Without appropriate training, dairy personnel could be inflicting unnecessary suffering to the animal being euthanized and putting their own safety at risk.

Protocols for euthanasia must include who will be performing the procedure, what it entails and how it will be carried out (Smith et al., 2015). Additionally, the owner of the livestock operation or the employee performing euthanasia should have not only the knowledge, but also the equipment to do so correctly. As the goal of euthanasia is to eliminate unnecessary pain and distress by causing immediate loss of consciousness, loss of brain functions, and cardiac and respiratory arrest, personnel performing the procedure must be trained on acceptable methods of euthanasia as well as how to evaluate that the procedure was successful.

If done properly (accurate placement and correct firearm and bullet) euthanasia by gunshot causes enough brain tissue destruction to result in death. Captive bolt guns (penetrating or non-penetrating) are another acceptable method for euthanasia of cattle. When placed correctly, captive bolt guns will cause immediate unconsciousness, but don't always result in death. Because death cannot be assured with just the captive bolt gun, an adjunctive method is necessary to ensure death (Underwood et al., 2013). Adjunctive methods of euthanasia include rapid intravenous injection of potassium chloride, a second gunshot, exsanguination, or pithing. The management team together with the veterinarian should discuss best methods to practice. These discussions should consider animal well-being, facilities on the operation, skill of operators, and the safety of personnel and other animals.

It is absolutely necessary that death is confirmed after the euthanasia is performed and before the disposal of cattle. A combination of signs, including lack of corneal reflex, lack of pulse, lack of heartbeat, and lack of respiratory sounds should be considered to confirm death. The herd veterinarian should teach personnel responsible for euthanizing cattle how to assess these parameters and the human safety measures to be considered while confirming death.

CONCLUSION

As livestock operations become larger, the chances for events that lead to downer animals increase. It is imperative that livestock producers approach this subject proactively by having written guidelines for the prevention of these events and the management of downer animals. Producers should take advantage of available culturally congruent training resources as well as the expertise of the herd veterinarian and university extension personnel as they strive to equip their employees with the knowledge and skills needed to effectively manage downer animals. The goal for the management of downers is to eliminate or reduce animal welfare issues by addressing the main cause of recumbency as well as preventing and addressing secondary damages associated with prolonged recumbency. Prompt identification of the primary problem and excellent quality nursing care are essential to the downer's return to a healthy and productive status. Thorough and frequent assessments of the downer, in conjunction with established treatment protocols will enable caretakers to make the decision to continue treatment or to euthanize when prognosis is poor, suffering is uncontrollable and evident, or needed care is not feasible. By preventing and optimally managing downer cattle, we can improve animal welfare, the health and safety of humans and other animals, the sustainability of the livestock operation, and the public perception of animal agriculture.

Example of high-quality nursing care protocol*. (Adapted from Poulton et al., 2016b; Stojkov et al., 2016; Stull et al., 2007):

- House the downer in a sick pen with no more than two other cows. The area should be sheltered and should protect her against inclement weather.
- Provide soft, deep, clean, and dry bedding that provides traction and minimizes abrasions, sores, and chances for infection.

* Nursing care should be provided only after performing a physical exam and determining that the downer cow should be treated according to farm protocols. The primary cause of recumbency should be addressed while nursing care is provided.

- Ensure that the downer has access to high-quality water and feed at all times. If free choice access is not feasible, water and feed should be provided at least four times a day.
- Reposition the cow as to alternate the hind leg located underneath the body at least four times a day (every 6 hours).
- If needed, move the cow to another location by using a loader bucket or by placing on a pallet. Moving should never cause further injury and discomfort.
- Use lifting devices only when the downer cow can support her own weight after being lifted. A well-trained caretaker should supervise the process and be available to lower her when no longer supporting her own weight.
- Assess the condition of the downer after each treatment (at least four times per day) and inform your supervisor if her condition is improving or deteriorating. This will help with the decision to continue treatment or euthanize.

REFERENCES

Adams AE, Lombard JE, Fossler, CP, Roman-Muniz, IN, Kopral, CA. 2017. Associations between housing and management practices and the prevalence of lameness, hock lesions, and thin cows on U.S. dairy operations. *Journal of Dairy Science.* 100: 2119–2136.

Adams, AE, Ahola JK, Chahine M, Roman-Muniz IN. 2016. Effect of dairy beef quality assurance training on dairy worker knowledge and welfare-related practices. *Journal of Extension.* 54(5).

Adams AE, Lombard JE, Roman-Muniz IN, Fossler CP, Kopral CA. 2015a. Management of nonambulatory dairy cows on US dairy operations. *Abstract in Journal of Animal Science.* 93(s3) *Journal of Dairy Science.* 98, Suppl. 2.

Adams AE, Lombard JE, Roman-Muniz IN, Fossler CP, Kopral CA. 2015b. Management practices that may affect dairy cow welfare on US dairy operations. Abstract in J. Anim. Sci. Vol. 93, Suppl. s3/J. Dairy Sci. Vol. 98, Suppl. 2.

Correa MT, Erb HN, Scarlett JM. 1993. Risk factors for downer cow syndrome. *Journal of Dairy Science.* 76(11): 3460–3463.

Cox VS, McGrath CJ, Jorgensen SE. 1982. The role of pressure damage in pathogenesis of the downer cow syndrome. *American Journal of Veterinary Research.* 43(1): 26–31.

Cox VS, Marsh WE, Steuernagel GR, Fletcher TF, Onapito JS. 1986. Downer cow occurrence in Minnesota dairy herds. *Preventive Veterinary Medicine.* 4(3): 249–260.

Doonan G, Appelt M, Corbin A. 2003. Nonambulatory livestock transport: The need for consensus. *The Canadian Veterinary Journal.* 44(8): 667–672.

Fenwick DC. 1969. The downer cow syndrome. *Australian Veterinary Journal.* 45: 184–188.

FSIS Directive 6100.1 Antimortem Livestock Inspection. 2014. United States Department of Agriculture. www.fsis.usda.gov/wps/wcm/connect/2b2e7adc-961e-4b1d-b593-7dc5a0263504/6100.1.pdf?MOD=AJPERES. Accessed January 24, 2018.

Green AL, Lombard JE, Garber LP, Wagner BA, Hill GW. 2008. Factors associated with occurrence and recovery of nonambulatory dairy cows in the United States. *Journal of Dairy Science.* 91(6): 2275–2283.

Hoe FG, Ruegg PL. 2006. Opinions and practices of Wisconsin dairy producers about biosecurity and animal well-being. *Journal of Dairy Science.* 89(6): 2297–2308.

Littledike ET, Young JW, Beitz DC. 1981. Common metabolic diseases of cattle: Ketosis, milk fever, grass tetany, and downer cow complex. *Journal of Dairy Science.* 64(6): 1465–1482.

Menger LM, Pezzutti F, Tellechea T, Stallones L, Rosecrance J and Roman-Muniz IN. 2016. Perceptions of health and safety among immigrant Latino/ a dairy workers in the U.S. *Frontiers in Public Health.* 4: 106.

Menger LM, Rosecrance J, Stallones L, Roman-Muniz IN. 2016. A guide to the design of occupational safety and health training for immigrant, Latino/a dairy workers. *Frontiers in Public Health.* 4: 282.

NASS. 2005. Non-ambulatory cattle and calves. http://usda.mannlib.cornell.edu/usda/current/nacac/nacac-05-05-2005.pdf Accessed December 13, 2017.Poulton PJ, Vizard AL, Anderson GA, Pyman MF. 2016a. Importance of secondary damage in downer cows. *Australian Veterinary Journal.* 94(5): 138–144.

Poulton PJ, Vizard AL, Anderson GA, Pyman MF. 2016b. High-quality care improves outcome in recumbent dairy cattle. *Australian Veterinary Journal*. 94(6): 173–180.

Roman-Muniz IN, Van Metre DC. 2011. Development of a bilingual tool to train dairy workers in the prevention and management of non-ambulatory cows. *Journal of Extension*. 49(6).

Smith RA, Thomson DU, Lee TL. 2015. Beef quality assurance in feedlots. *Veterinary Clinics of North America Food Animal Practice*. 31(2):269–281.

Stojkov J, Weary DM, von Keyserlingk MAG. 2016. Nonambulatory cows: Duration of recumbency and quality of nursing care affect outcome of flotation therapy. *Journal of Dairy Science*. 99(3): 2076–2085.

Stull CL, Payne MA Berry SL, Reynolds JP. 2007. A review of the causes, prevention, and welfare of nonambulatory cattle. *Journal of the American Veterinary Medical Association*. 231(2): 227–234.

Underwood W, Anthony R, Gwaltney-Brant S, Meyer R. 2013. *AVMA guidelines for the euthanasia of animals: 2013 edition*. Schaumburg, IL: American Veterinary Medical Association.

Van, Metre DC St., Jean G, Vestweber J. 1996. Flotation therapy for downer cows. *Kansas Agricultural Experiment Station Research Reports*. 0(2). doi:10.4148/2378-5977.3245.

Roman-Muniz IN, Van Metre DC, Garry FB, Reynolds SJ, Wailes WR, Keefe TJ. 2006. Training Methods and Association with Worker Injury on Colorado Dairies. *Journal of Agromedicine*. 11(2): 19-26, doi: 10.1300/J096v11n02_05.

CHAPTER 28

Cow Comfort in Intensive and Extensive Dairy Housing Systems

Jesse Robbins
Iowa State University

Alex Beck
Banks Veterinary Service

CONTENTS

When many people think of dairy farming they imagine cows grazing on lush pastures (Cardoso et al., 2015). While grazing-based systems are common in many parts of the world (e.g., Oceania, Chile, Brazil), indoor, zero-grazing systems are much more common in most of the developed world (Barkema et al., 2015). Although there are many benefits associated with the shift to these more intensive systems, it is widely acknowledged they also come with numerous challenges for cattle welfare (Robbins et al., 2016). These include many traditional veterinary and animal science concerns such as high rates of disease, including lameness, mastitis, and other production diseases (Garry, 2004), but also increasingly negative public perceptions associated with a perceived lack of naturalness (Weary et al., 2016).

In response, there is a growing focus on understanding how dairy cattle interact with their engineered environment in hopes of striking a more reasonable balance between the needs of people working on the farm and the needs of the cows. Much of this work has relied on merging the fields of veterinary medicine and ethology with aspects of facility design and management. Behavioral studies often rely on paradigms that include observational and experimental preference and motivation tests (see Fraser and Matthews, 1997) that allow cows to choose aspects of the environment that are important to them. Research focused on health has primarily been led by the veterinary profession (e.g., LeBlanc et al., 2006; Cook and Nordlund, 2009; Barkema et al., 2015) and has played an especially important role in understanding common maladies such as lameness (von Keyserlingk et al., 2009) and mastitis (Barkema et al., 2006).

This chapter will provide a brief overview of the impacts both extensive and intensive dairy systems have on cow welfare. Within the dairy industry and many veterinary and animal science departments, this type of research is often collectively referred to as "cow comfort" research. Unfortunately, this term is vaguely defined so we have elected to simply focus on commonly studied behavioral and health outcomes associated with different systems or specific features of these systems. Where possible we highlight limitations of this work and suggest areas where further research is needed. Given that we are located in North America, much of this chapter will focus on the U.S. and Canadian dairy industries, but we will draw on examples from other parts of the world when appropriate. Lastly, we will focus on the effect housing has on lactating dairy cattle.

BRIEF OVERVIEW OF COMMON DAIRY HOUSING SYSTEMS

Tie Stalls

Most cows housed in North America spend the majority of time indoors in one of several different housing systems. In tie stalls or stanchion barns, cows are tethered in a stall by a piece of rope or chain. Stalls consist of a lying area that is specific to each cow. Although cows have limited ability to move freely and socialize with other animals, tie stalls allow for greater individualized attention than loose housing systems described later. Tie stalls are unique in that cows are typically milked directly in their stalls and have their own individual feed and water stations located at the front of their stall. The frequency and duration of tethering, as well as the opportunities afforded when not tethered, vary considerably. Some farms will keep cows tethered for only a few hours a day and then allow them pasture access when weather conditions permit, whereas others may tether cows almost continuously year-round (Popescu et al., 2013).

Although tie stall housing affords greater individual attention than other housing systems it also faces unique welfare challenges because it entails severe restrictions on the cow's ability to move. Cows tethered for extended periods of time exhibit increased cortisol levels and stereotypical behavior (i.e., tongue rolling) persisting at least 4 months (Redbo, 1992; Redbo, 1993). After being released from prolonged tethering cows also display increased locomotor activity, which some authors see as a likely indication of frustration brought on by immobility (Veissier et al., 2008). Tethering also greatly restricts the cow's ability to engage in many normal social behaviors (e.g., grooming) that cows likely find rewarding (Krohn, 1994).

In the U.S., 39% of farms report using tie stalls or stanchions as their primary form of housing (USDA, 2016). In Canada, approximately 75% of the milk supply comes from cows housed in tie stall housing (Barkema et al., 2015). Tie stalls are much more commonly found in older dairy farms with relatively small herd sizes; thus, they are becoming much less numerous as the dairy industry continues to consolidate into fewer and larger farms (Robbins, et al., 2016).

Free Stalls

More than 40% of all U.S. dairy farms now use free stalls as their main form of housing (USDA, 2016). The growing number of large dairy farms increasingly prefers these systems because unlike most tie stalls, they utilize separate milking parlors that allow for more efficient management of

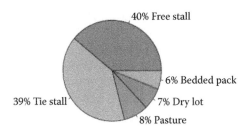

Figure 28.1   US dairy operations by housing type.

(Source: Redrawn from data summarized by the USDA National Animal Health and Monitoring survey completed in 2014 and published in 2016.)

larger herds. In freestall barns cows are housed in groups or pens according to their stage of lactation and/or milk production. As their name suggests, cows are able to move around indoors where they have access to distinct feeding and lying areas connected by concrete alleyways. Lying areas consist of series of stalls separated from one another by partitions. Cows housed in free stalls are typically moved in groups to a milking parlor two to three times each day or, less commonly, provided with access to a milking robot (Jacobs et al., 2012; Figure 28.1).

Other/Mixed Systems

Dry lots and deep-bedded pack barns are two other housing systems gaining traction in the U.S. (Barkema et al., 2015; USDA, 2016). In dry lots (sometimes referred to as open lots) cows are housed in large outdoor dirt lots similar to feedlots used for beef cattle. Dry lots are increasingly popular among large-scale dairies. In 2014, 7% of U.S. dairy operations housed their cows in dry lot systems, yet these farms constituted 17% of farms with ≥500 cows (USDA, 2016). Dry lots are often located in arid climates like the Western U.S. where the risk of excessively muddy conditions is greatly minimized.

Deep-bedded pack barns (sometimes called compost dairy barns) are a relatively novel form of housing now found on 6% of U.S. dairy farms (USDA, 2016). Unlike dry lots, cows housed in deep-bedded pack barns are kept indoors in a barn that usually consists of a concrete feed alley and a large deeply bedded area used for resting. The surface of the bedded area must be regularly aerated in order to compost the accumulation of manure and urine (Janni et al., 2007). Generally speaking, dry lots and deep-bedded pack barns provide a much softer and less-restrictive environment than either tie stalls or free stalls, while still permitting more-intensive management practices than pasture-based systems.

Pasture-Based Systems

Although much more common in other parts of the world, only about 8% of the dairy herds in the U.S. house dairy cattle extensively in pasture-based systems. This figure increases to about 20% when one looks at the percentage of cows that are provided with some access to pasture at some time during their lactation. The prevalence of pasture-based systems varies by region and is heavily dependent on the presence of conditions conducive to high-quality forage growth. The likelihood that farms provide pasture access tends to decrease with increasing herd size. Among farms that reported routinely allowing pasture access, 81% had less than 30 cows (USDA, 2016). Growing consumer demand for organic dairy products, which require pasture access, has contributed greatly to the persistence of pasture-based systems (Stiglbauer et al., 2013).

PROBLEM WITH SYSTEM-LEVEL COMPARISONS

There are many studies comparing housing systems in terms of their effects on cattle welfare. However, this approach is somewhat limited because system-level comparisons make it difficult to identify specific causal factors (Rushen et al., 2008). For example, finding that cows housed in tie stalls tend to have less lameness than those housed in freestall barns (Wells et al., 1993; Sogstad et al., 2005) doesn't allow us to pinpoint the cause of this difference because these systems differ in a variety of ways that could plausibly affect lameness. For this reason, research will often attempt to isolate (statistically or experimentally) specific design features common to different housing systems to see how they impact different welfare indicators. As one might imagine there are strengths and weaknesses associated with each of these approaches, but for the purposes of this chapter we draw upon research using all of them.

LEG HEALTH: LAMENESS AND LEG INJURIES

One of the most commonly studied areas of cow comfort research has been the effect these housing systems have on rates of lameness and leg injuries. Multiple studies have found the percentage of lame dairy cows on farms ranges from 10% to 55% (von Keyserlingk et al., 2012). In addition to imposing significant financial costs in the form of reduced milk production, increased reproductive failure and early culling, the levels of lameness found across multiple studies indicate a significant proportion of dairy cows are frequently in pain (Whay et al., 1998).

The etiology of lameness is multifactorial, but the proximate cause of most cases seems to be a variety of both infectious and noninfectious hoof lesions. Although disagreements about the relative contributions of various environmental factors persist, there is widespread agreement that frequent contact with hard, wet, and unhygienic surfaces plays a major role in the development of the hoof pathologies that lead to lameness (Cook et al., 2004; Solano et al., 2016).

Pasture access is often associated with a reduction in the prevalence of hoof lesions (Fregonesi and Leaver, 2001; Somers et al., 2003; Cramer et al., 2009; Solano et al., 2016) and lameness (Barrientos et al., 2013). One study found zero-grazing farms had 48% more lame cows than those providing pasture access (Haskell et al., 2006). Hernandez-Mendo et al. (2007) found lameness scores improved by more than 20% after cows spent 4 weeks on pasture. Within intensive systems, tie stall farms (Wells et al., 1993; Cook et al., 2003; Sogstad et al., 2005) and compost-bedded farms (Lobeck et al., 2011) both tend to have lower lameness rates than free stall farms. We are not aware of any comparable research addressing how dry lots dairy systems might fare in comparison. Despite the general trend for reduced lameness in pasture-based systems, it is worth noting that some studies have reported positive associations between foot lesions and outdoor access (Haufe et al., 2012). Moreover, lameness rates among grazing herds comparable to those found in well-managed intensive systems have also been noted (Hemsworth et al., 1995; Ranjbar et al., 2016; Bran et al., 2018). Unfortunately, there has been insufficient research exploring risk factors for leg health issues within pasture-based systems.

The benefits of pasture may also extend beyond its contributions to leg health. There is a growing body of evidence that suggests cows are highly motivated to access pasture. Using a weighted gate to measure motivation strength, researchers found cows were willing push as much weight to access pasture as they were to access fresh feed following milking—a time cows are highly motivated to access feed (von Keyserlingk et al., 2017). This effect was largest during the afternoon and evening hours. Although Legrand et al. (2009) reported no overall preference for pasture, they did find evidence that cows had a strong preference for pasture during night time hours when temperature and humidity were at their lowest. In contrast to these studies, Charlton et al. (2011) found cows had a strong preference to remain indoors, regardless of time of day. This contradictory result might be explained by the rather limited exposure of test cows to the outdoor area prior to preference testing. There is clearly a need for much more work elucidating the myriad factors influencing cow preferences for pasture access.

Another common leg health issue are hock lesions. The hock region is especially vulnerable to injury because it has very little protective fatty tissue or muscle. Hock lesions can vary in severity from mild hair loss and inflammation to severe swelling and, in extreme cases, ulceration and joint infection. Mean herd prevalence estimates of hock lesions range from 42% to 73% (Kester et al., 2014). The relationship between hock lesions and lameness is not fully understood, but research has found cows identified as lame in the previous month are more likely to develop subsequent hock lesions (Lim et al., 2013). This suggests hock lesions are more often a consequence of lameness rather than a cause. Once cows become lame they tend to lay down much longer than normal (Ito et al., 2010), which increases the time spent in close contact with abrasive stall surfaces that create hock lesions.

The importance of lying times is a theme found throughout the cow comfort literature. Munksgaard et al. (2005) demonstrated cows will choose to lie down instead of feeding after being deprived of both lying and feeding for extended periods of time (also see Metz, 1985). Cows with restricted lying times also exhibit a greater physiological stress response than cows with an unrestricted ability to lie down (Munksgaard and Simonsen, 1996). Multiple investigations have monitored lying/standing times to assess difference between housing systems. The results are inconsistent with studies finding that lying times increase (Olmos et al., 2009), decrease (Roca-Fernández et al., 2013) or are unchanged (Navarro et al., 2013).

Although a minimum daily lying time of approximately 12 hours/day is often recommended in confinement housing systems (NFACC, 2009), studies have found grazing cattle tend to lie down significantly less than this (Singh et al., 1993). Cows in pasture-based systems appear to have a distinct time budget paradigm where reduced lying times do not necessarily indicate obvious welfare problems. Thus while lying times can be a useful tool for assessment purposes in some circumstances, caution is warranted when drawing inferences about cow welfare between different systems based solely on lying behavior.

The close connection between lying behavior and leg health has led to a great deal of research on the effects of specific stall design features and maintenance practices. Commonly used stall bedding materials include sand, sawdust, straw, dried manure, composted materials, almond hulls and other secondary products. Inorganic materials such as sand are generally preferred because they not only provide excellent cushioning and traction when rising, but also are less conducive to bacterial growth than organic alternatives (Zdanowicz et al., 2004). In an effort to minimize the costs associated with purchasing and maintaining bedding materials, some farms have installed rubber-filled mattresses. Despite their intuitive appeal, several studies have shown that these mattresses often fail to provide a sufficiently comfortable lying surface when used without the addition of deep bedding (Tucker and Weary, 2004; van Gastelen et al., 2011; Barrientos et al., 2013; Chapinal et al., 2013). One study found farms using rubber mattresses had a six fold increase in the number of cows with swollen or ulcerated hocks (Cook et al., 2016). Farms using water-filled mattresses (aka 'waterbeds') have fewer hock injuries compared to those using rubber filled mattresses. However sand-bedded stalls and compost barns still result is fewer hock injuries than both of these mattress surfaces (Fulwider et al., 2007).

Apart from the selection of bedding materials, it is very important that they be dry and sufficiently deep to provide optimal compressibility. Several studies have shown deeper bedding leads to longer lying times (Tucker et al., 2009; Lombard et al., 2010). To avoid concavity, routine maintenance is necessary to ensure bedding materials are evenly distributed across the entire lying surface. Drissler et al. (2005) experimentally manipulated the evenness of bedding material and found lying times increased linearly with bedding evenness (approximately 10 minutes of lying time was lost for every 1-cm decrease in bedding material). The moisture content of bedding material also impacts lying behavior. Cows given dry bedding materials spent approximately 4 hours more per day lying than those with very wet bedding (Fregonesi et al., 2007). Reich et al. (2010) also found a dose-dependent response with lying times gradually increasing across five levels of bedding material dry matter content.

While the selection and maintenance of bedding materials are both very important considerations, they are unable to compensate for improperly sized stalls (Popescu et al., 2013). The average size of dairy cows has increased steadily over the last 30 years (Bouffard et al., 2017), often without concomitant changes in stall designs. This has led to a situation where many farms have stalls that are simply too small for their cows (Zurbrigg et al., 2005; Dippel et al., 2009; Westin et al., 2016; Bouffard et al., 2017). Perching occurs when cows stand with two feet in the stall and two feet in the alley. This behavior is often interpreted as an indication of a cow's reluctance to lie in the stall. Perching behavior has been observed more frequently on farms with stalls that fail to meet standard stall size recommendations (Tucker et al., 2004; Lombard et al., 2010). A more recent study found smaller stall dimensions were associated with increased risk of neck and leg injuries, as well as lameness (Bouffard et al., 2017).

Other aspects of stall design affect stall usage as well. Like neck rails, brisket locators are barriers designed to control the forward limit of the recumbent cow in free stalls. Brisket locators may be constructed from a variety of materials including PVC, wood and concrete. If they protrude too far from the stall surface the cow's ability to normally rise will be impeded and disruptions in normal lying behavior may occur. Tucker et al. (2006) found cows housed in free stalls preferred to lie down, and lied down longer, in stalls that had had their brisket locators completely removed. In a similar fashion, the neck rail functions to delineate the forward limit a cow can comfortably stand within the stall. A large cross-sectional study found stall occupancy significantly increased when the height of the neck rail was raised just 13 cm relative to the bedding surface (Fulwider and Palmer, 2005). However, experimental research failed to detect any effect of neck rail height on lying times (Tucker et al., 2005). This somewhat surprising null result might have been an artifact of the rather limited range of neck rail heights manipulated in the study. The horizontal distance from the neck rail to the stall's rear curb does seem to have an effect on how the stall is used for standing as stall usage increases with the effective length of the stall (Fregonesi et al., 2009). Bernardi et al. (2009) found increasing the distance between the neck rail and the rear curb improved lameness scores. Complete removal of the neck rail leads to substantially longer lying times, however it also leads to dirtier stalls and increased the risk of mastitis (Tucker et al., 2005). This constant tension between providing a comfortable environment for the cow and maintaining a high level of cleanliness is commonly encountered in the cow comfort literature.

MASTITIS

Mastitis, or inflammation of the mammary gland, is the common name used to describe a broad range of symptoms associated with intramammary infections. Mastitis is a painful condition and one of the most common infectious diseases affecting dairy farms (Barkema et al., 2006). Between lost milk production, increased mortality and treatment costs, the average cost of a case of clinical mastitis is estimated to be $179 (Bar et al., 2008). Like most infectious diseases, mastitis is the result of the interplay between host resistance, environmental conditions and pathogen virulence. Over the past 100 years, improved understanding of the range of etiologic agents, programs to control contagious pathogens, and the use of antibiotics and teat sealants at dry-off have contributed to more effective mastitis management protocols (Ruegg, 2017). The development of vaccines against mastitis pathogens, and research into a potential genetic basis for resilience are likely to further improve cow-side resistance (Rupp and Boichard, 2003; Ismail, 2017), but changes in housing and management are currently much more feasible for the average farm.

Cleanliness and hygiene scores are tools commonly used in mastitis prevention programs (Shook et al., 2017). A study of 144 Dutch dairy farms found udder cleanliness scores were positively correlated with cow somatic cell count (SCC; Dohmen et al., 2010). Another study involving 3,554 evaluations from 545 animals on two Brazilian dairies, showed cows categorized as "very clean" consistently had the lowest somatic cell counts, while dirtier cows had the highest scores (Sant'Anna and Paranhos da Costa, 2011).

Exposure to pasture impacts cleanliness scores and subsequent mastitis risk. A study involving more than 1,000 loose-housed cows found pasture access was associated with better cow cleanliness scores (Nielsen et al., 2011). A longitudinal study of dairy cows in the United Kingdom assessed cleanliness scores and found cows tended to be dirtier during times of the year when they were continuously housed indoors compared to when they had pasture access (Ellis et al., 2007). Interestingly, this study also found a correlation between herd cleanliness score and bulk tank somatic cell count (BTSCC), but failed to detect any direct link with clinical mastitis rate.

Other studies have found pasture is associated with an overall reduction in the risk of clinical mastitis (Barkema et al., 1999). One multiyear study comparing dairy cows in both confinement and pasture systems found confinement housed animals experienced 1.8 times more clinical mastitis and eight times the culling rate for mastitis compared to cows on pasture (Washburn et al., 2002). Other research has failed to detect significant SSC differences between pasture-based and continuously housed systems (Arnott et al., 2017).

The use of sand bedding is associated with lower BTSCC, whereas the use of composted manure is associated with higher BTSCC (Wenz et al., 2007). Lower herd SCCs have also been associated with the use of free stalls (Dufour et al., 2011); however, a study of more than 100 Canadian dairy farms failed to detect any differences in the incidence rate of mastitis between tie stalls, free stalls, and pasture systems (Olde Riekerink et al., 2008). These somewhat inconsistent results may be attributable to other unmeasured differences and the sheer complexity of mastitis infections. Future research should consider the interactions between bedding type and other management factors (e.g., bedding frequency, depth, dry matter, lying time, and stall design) as herd level associations likely obscure important details about the pathogenesis of mastitis.

STOCKING DENSITY, GROUPING STRATEGY, AND FEEDING BEHAVIOR

Beyond the constraints imposed by the physical environment, management factors such as group size, stability, and composition can also impact cow comfort. Within the dairy science literature, there have been a number of studies exploring the impact of stocking density on different indices of cow comfort.

An early survey of free stall barns failed to find milk production differences across a wide range of stocking densities ranging from under capacity to >30% overcrowded (Bewley et al., 2001). However, another study involving 47 herds fed identical diets found stall availability and maintenance explained 38% of the variation in milk production (Bach et al., 2008). Subsequent investigations have identified a number of other disadvantages associated with elevated stall-stocking density including: decreased lying times (Telezhenko et al., 2012; Charlton et al., 2014), increased prevalence of severe lameness (King et al., 2016), decreased conception rates (Schefers et al., 2010), and decreased feeding times, along with increased aggressive interactions at the feed bunk (Huzzey et al., 2006). With respect to lying times, reducing the stall-stocking rate from 150% to 100% increased mean lying time by 1.7 hours/day (Fregonesi et al., 2007).

Despite the obvious impact of overstocking, agreement on the level at which the negative impacts occurs remains somewhat elusive. With respect to milk production, it appears overcrowding may need to be relatively severe before production is compromised. For example, negative impacts on lying and ruminating time were not appreciated at a stall stocking density of 113%, but were noted at 131% and 142% (Krawczel et al., 2012). A separate study found notable neuroendocrine and metabolic changes in cattle stocked at 200%, but mild overstocking of 120% did not produce any changes in immune response, milk yield, or reproductive performance (Chebel et al., 2016).

Two separate studies found that cattle in confinement systems display more agonistic behavior than those on pasture (9.5 vs. 1.1 per hour) (Miller and Wood-Gush, 1991), with peaks in aggression occurring around delivery of fresh feed (O'Connell et al., 1989). These studies also noted synchrony of feeding and lying behavior was diminished in confinement systems. Roca-Fernández et al. (2013) reported cattle on pasture had longer feeding times than those in confinement (522 vs. 173 minutes), but also lower milk production (20.1 vs. 27.0 kg/day). A separate study found a similar trend in milk production and suggested cattle on pasture needed disproportionately more time to eat given the generally lower nutritional density of pasture (Navarro et al., 2013). This might also explain the tendency for pastured cattle to have increased standing times and decreased lying times relative to cows housed in intensive systems. Thus, the discrepancies in agonistic behavior found between the extensive and intensive housing systems are likely attributable to extensively reared cattle being preoccupied with grazing and the much lower stocking densities typically found in pasture-based systems.

Differences in feeding and social behavior are also sensitive to stocking rate. Increasing stall stocking density (SSD) from 80% to 100% results in an increase in feed bunk displacements (Lobeck-Luchterhand et al., 2015). Similarly, increased competition for bunk space leads to decreased feeding time, increased idle time spent standing, and increased feed bunk displacements (Huzzey et al., 2006). Proudfoot and colleagues (2009) echoed these results, finding displacements and standing time increased when the ratio of cows to feed bins increased from 1:1 to 2:1. Other authors have reported a linear relationship between displacement behavior and SSD in groups stocked at 100%, 113%, 131%, and 142%. The most overcrowded cows in this study had drastically reduced lying times, as a result of increased competition for stalls (Krawczel et al., 2012).

Stocking density represents only one aspect of pen management's impact on social behavior of cattle. Regrouping cows is a common practice as cows move from one phase of the production cycle to the next. More frequent regrouping leads to increased feed bunk displacements, decreased lying times, and transient losses in milk production (von Keyserlingk et al., 2008). Competitive and agonistic behaviors also occur more frequently among cows subjected to frequent group changes as opposed to those managed in an all-in/all-out model that maintain more stable group structures (Lobeck-Luchterhand et al., 2014). These negative effects may not be borne equally by all cows. Wierenga and Hopster (1990) found the impact on resting time on low rank cows is greater at relatively low stocking densities (1.25 cows per stall), but others have failed to find an effect of social rank (Fregonesi et al., 2007).

Changes in social behavior appear to interact with both stocking density and group composition. For example, at 120% SSD, weekly pen moves increase feed bunk displacements, but do not alter immune responses or several health and production measures, as long as heifers and cows have separate housing (Chebel et al., 2016). In free stall systems, it is very difficult to avoid group changes, especially on smaller farms which may lack excess of pens. In these circumstances decreasing stall-stocking density to 25% or 50% can mitigate increases in competitive behavior and negative effects on lying time, although this may not be realistic on many commercial farms (Talebi et al., 2014).

Several caveats of the research on stocking density and group composition are worth mentioning. First, much of this research has relied on groups of animals (e.g., between 6 and 24 cows) much smaller than those typically found on many commercial farms. It is therefore not clear how translatable these research findings are to farms with larger group sizes. Moreover, the effects of stocking density also apply to pasture-based dairies, yet very little research has investigated the effects stocking rate have on social, feeding, and lying behavior in extensive systems. Unlike intensive systems, forage production thresholds are likely to be the major factor limiting stocking rates in pasture-based systems. More research is needed to understand how stocking density affects extensively raised dairy cattle. Such work would nicely complement existing research programs exploring how management changes can minimize aggressive and competitive social interactions in intensive systems.

Given the negative impacts of overstocking, it seems reasonable to consider why a majority of U.S. dairy farms still maintain a SSD greater than 100% (von Keyserlingk et al., 2012)? We surmise the answer has to do with the fact that overstocking is profitable. Adding more cows dilutes fixed costs over a larger number of production units. While this may reduce individual milk production and negatively impact cattle welfare, it appears to improve overall farm profitability (De Vries et al., 2016). Without attendant changes in market conditions that currently incentivize overstocking, reducing the inclination to overstock will be difficult.

THERMAL STRESS

The modern dairy industry is dominated by continental *Bos taurus* cattle which are adapted to mild European climates. The versatile nature of cattle coupled with technological advances in housing has allowed dairy herds to occupy a diverse array of landscapes from Saudi Arabia to Scandinavia to Brazil. Although the extent and severity vary, the majority of regions where dairy production takes place still experience seasonal struggles maintaining cow comfort in the face of thermal stress. In 2003, it was estimated that the U.S. dairy industry loses $897 million annually due to decreased production, increased mortality and decreased reproduction due to heat stress (St-Pierre et al., 2003). In the next section, we explore the current science addressing the negative effects of thermal stress on cattle and various heat abatement strategies.

Like most mammals, cattle seek to alleviate the discomfort associated with elevated body temperature. Some cattle, particularly *Bos indicus* type animals, have innate mechanisms for heat abatement. These include respiratory rate modifications, changes in hair coat thickness/length, lighter skin pigment, more abundant sweat glands, and increased skin vascularity. Behavioral strategies to mitigate heat stress also occur including decreases in time spent lying and corresponding increases in standing times that help to dissipate heat. As we have seen, these behaviors also increase the risk of hoof lesions and lameness (Cook et al., 2007). There is even some evidence that prolonged heat stress can lead to increased aggression among cattle, especially in competitive feeding situations (Polsky and von Keyserlingk, 2017).

Other responses to heat stress align closely with our own personal experiences. Heat-stressed cattle consume more water (West, 2003) and prefer standing in shade versus lying in unshaded areas, even after being unable to lie down for 12 hours (Schütz et al., 2008). In other cases, cattle modify their behavior in less seemingly rational ways (see review by Polsky and von Keyserlingk, 2017). Dairy cows, for example, prefer shade to sprinklers, despite sprinklers being more effective means of cooling as measured by decreasing respiratory rate (Schütz et al., 2011). It is not yet clear how we ought to address these instances where cow preferences seem to contradict other relevant lines of evidence.

When cattle are unable to adapt and become heat stressed a rapid depression in feed consumption (i.e., dry matter intake) will occur (West, 2003). As cows begin to enter a negative energy balance, milk production will begin to suffer. These changes place sensitive populations (e.g., transition cows) at even greater risk of production diseases like metritis, displaced abomasum and ketosis that typically require immediate veterinary attention (Esposito et al., 2014). Although milk production, like all animal welfare indicators, is by itself an imperfect measure of thermal stress, changes in feeding behavior and intakes are useful indicators of the effectiveness of different heat-stress abatement strategies.

The integration of bovine genetics for high heat tolerance may help to alleviate the severity of heat-stress effects in the long term. However, even well-adapted breeds of cattle will likely experience heat stress in extreme conditions. Currently, management is the most important element in reducing the downstream effects of heat stress. Various engineering solutions have been utilized to mitigate the negative impact of heat stress on modern dairy farms. Shade structures, sprinklers/misters, fans, and/or combinations of all three are the most common methods of heat abatement (Ortiz et al., 2010; Calegari et al., 2012; Anderson et al., 2013). A study by Kendall et al. (2007) demonstrated exposing cows to just 90 minutes of shade and sprinklers in the holding pen prior to milking reduced their respiratory rate by 67%. Such drastic improvements certainly merit greater attention by researchers.

In extensive production systems, cattle spend much of their time outside of controlled barn environments, which may place them at greater risk for heat stress. Shade structures, whether artificial or natural, offer an opportunity for heat-stress mitigation in these more extensive circumstances. Shade use increases with the heat load index and cattle that seek and utilize shade exhibit decreases in respiratory rate, panting score, and lower rectal temperatures (Veissier et al., 2018). Even in temperate climates, cattle with access to shade tend to have increased milk production levels (Van Laer et al., 2015), yet it is still unclear whether this difference is due to changes in feeding behavior, shifting metabolic demands, or some combination.

Confinement systems have the advantage of constant shade, but shade alone may not be sufficient to eliminate all the negative impacts of heat stress (Ortiz et al., 2010). Less common practices, such as air conditioning and conductive stall cooling, offer additional effective means of heat abatement (Collier et al., 2006; Ortiz et al., 2015), but are not yet economically feasible for most farms. Extensive systems can also benefit from engineering solutions. Providing very brief access to shade and sprinklers can result in lower body temperatures lasting up to 4 hours (Kendal et al., 2007). Regardless of production system, producers and their cattle stand to benefit greatly from improved heat abatement strategies.

Although the effects of extreme temperatures on cattle behavior are mostly due to heat stress, cold stress can and does occur as well. These conditions are especially problematic for extensively reared cattle, but also impact cattle housed indoors in insulated barns located in colder climates. One study found that at $-19°C$, cattle in free stall barns had increased feed intakes, but decreased milk production, likely in response to increased metabolic demand for temperature maintenance (Broucek et al., 1991). Even under relatively mild temperatures, prolonged exposure to wind and rain substantially impacts cattle behavior. In these conditions, cows may spend as little as 4 hours lying down (Tucker et al., 2007). Not unlike heat stress, cattle do possess some innate ability to respond to cold stress. In extremely cold conditions, they will deposit larger amounts of subcutaneous and intramuscular fat compared to cattle housed indoors (Mader et al., 1997). Clearly, there is a great need for additional research exploring the effect thermal conditions have on cattle welfare.

CONCLUSIONS

In this chapter, we have discussed how various features of dairy housing systems impact the welfare of lactating dairy cattle. Perhaps unsurprisingly, housing systems that provide cows with soft, clean, dry, cool environments, and the ability to move freely and interact with others, tend to lead to better animal welfare. However, these factors are not the exclusive property of either extensive or intensive systems. Well-managed intensive systems are capable of achieving welfare outcomes comparable to those found in more extensive systems whereas extensive, pasture-based systems are subject to a number of unique welfare challenges (e.g., weather extremes, parasites, nutritional deficiencies, etc.). It is our hope that future research will move beyond system-level comparisons, which necessarily obscure important details, and instead focus on the impact specific features of different housing systems have on animal welfare. If for no other reason, this would lead to more nuanced and informative discussion about the specific strengths and weaknesses of all systems.

Many of the welfare challenges we have discussed involve facility design issues that cannot be easily addressed by later changes in management. Future efforts should encourage greater collaboration between those designing, constructing, and renovating dairy facilities, and animal scientists versed in the relevant scientific literature. Injecting science-based information about the needs of the cows early on in the planning process may help prevent animal welfare issues before they occur and in doing so help avoid the costs associated with expensive housing modifications.

We believe science can and should continue to play a central role in providing valuable information about the impact different housing features have on animal welfare. However, the science of cow comfort should be accompanied by a much greater appreciation of farms as complex, interdependent systems. Failure to recognize this could result in science-based recommendations that improve one dimension of cow comfort (e.g., lameness) at the expense of another (e.g., mastitis). The challenges posed by considering these potential trade-offs are even more formidable when we recognize they extend far beyond animal welfare concerns. Farmers are in the unenviable position of having to constantly balance animal welfare concerns with other, equally urgent concerns revolving around milk quality and safety, environmental stewardship, worker well-being and profitability. More holistic research approaches that better reflect the interconnectedness of the farm as a dynamic system will be needed if we are to foster genuinely sustainable solutions to the challenges facing the dairy industry.

REFERENCES

Anderson, S.D., B.J. Bradford, J.P. Harper, C.B. Tucker, C.Y. Choi, J.D. Allen, L.W. Hall, S. Rungruang, R.J. Collier and J.F. Smith. 2013. Effects of adjustable and stationary fans with misters on core body temperature and lying behavior of lactating dairy cows in a semiarid climate. *J. Dairy Sci.* 96:4748–4750.

Arnott, G., C.P. Ferris and N.E. O'Connell. 2017. Review: Welfare of dairy cows in continuously housed and pasture-based production systems. *Animal.* 11:261–273.

Bach, A., N. Valls, A. Solans and T. Torrent. 2008. Associations between nondietary factors and dairy herd performance. *J. Dairy Sci.* 91:3259–3267.

Bar, D., L.W. Tauer, G. Bennett, R.N. González, J.A. Hertl, Y.H. Schukkne, H.F. Schulte, F.L. Welcome and Y.T. Gröhn. 2008. The cost of generic clinical mastitis in dairy cows as estimated by using dynamic programming. *J. Dairy Sci.* 91:2205–2214.

Barkema, H.W., Y.H. Schukken, T.J.G.M. Lam, M.L. Beiboer, G. Benedictus and A. Brand. 1999. Management practices associated with the incidence rate of clinical mastitis. *J. Dairy Sci.* 82:1643–1654.

Barkema, H.W., Y.H. Schukken and R.N. Zadoks. 2006. Invited review: The role of cow, pathogen, and treatment regimen in the therapeutic success of bovine Staphylococcus aureus mastitis. *J. Dairy Sci.* 89:1877–1895.

Barkema, H.W., M.A.G. von Keyserlingk, J.P. Kastelic, T.J.G.M. Lam, C. Luby, J.P. Roy, S.J. LeBlanc, G.P. Keefe and D.F. Kelton. 2015. *Invited review*: Changes in the dairy industry affecting dairy cattle health and welfare. *J. Dairy Sci.* 98:7426–7445.

Bernardi, F., J. Fregonesi, C. Winckler, D.M. Veira, M.A. von Keyserlingk and D.M. Weary. 2009. The stall-design paradox: Neck rails increase lameness but improve udder and stall hygiene. *J. Dairy Sci.* 92:3074–3080.

Barrientos, A.K., N. Chapinal, D.M. Weary, E. Galo and M.A.G von Keyserlingk. 2013. Herd-level risk factors for hock injuries in freestall-housed dairy cows in the northeastern United States and California. *J. Dairy Sci.* 96:3758–3765.

Bewley, J., R.W. Palmer and D.B Jackson-Smith. 2001. A comparison of free-stall barns used by modernized Wisconsin dairies. *J. Dairy Sci.* 84:528–541.

Bouffard, V., A.M. de Passillé, J. Rushen, E. Vasseur, C.G.R. Nash, D.B. Haley, and D. Pellerin. 2017. Effect of following recommendations for tiestall configuration on neck and leg lesions, lameness, cleanliness, and lying time in dairy cows. *J. Dairy Sci.* 100:2935–2943.

Bran, J.A., R.R. Daros, M.A.G. von Keyserlingk, S.J. LeBlanc, and M.J. Hötzel. 2018. Cow and herd-level factors asscoaited with lameness in small-scale grazing dairy herds in Brazil. *Prev. Vet. Med.* doi:10.1016/j.prevetmed.2018.01.006.

Broucek, J., M. Letkovičová and K. Kovalčuj. 1991. Estimation of cold stress effect on dairy cows. *Int. J. Biometeorol.* 35:29–32.

Calegari, F., L. Calamari and E. Frazzi. 2012. Misting and fan cooling of the rest area in a dairy barn. *Int. J. Biometeorol.* 56:287–295.

Cardoso C.S., M.J. Hötzel, D.M. Weary, J. Robbins and M.A.G. von Keyserlingk. 2015. Imagining the ideal dairy farm. *J. Dairy Sci.* 99:1–9.

Chapinal, N., A.K. Barrientos, M.A.G. von Keyserlingk, E. Galo and D.M. Weary. 2013. Herd-level risk factors for lameness in freestall farms in the northeastern United States and California. *J. Dairy Sci.* 96:318–328.

Charlton, G.L., S.M. Rutter, M. East and L.A. Sinclair. 2011. Effects of providing total mixed rations indoors and on pasture on the behavior of lactating dairy cattle and their preference to be indoors or on pasture. *J. Dairy Sci.* 94:3875–3884.

Charlton, G.L., D.B. Haley, J. Rushen and A.M de Passillé. 2014. Stocking density, milking duration, and lying times of lactating cows on Canadian freestall dairy farms. *J. Dairy Sci.* 9:2694–2700.

Chebel, R.C., P.R.B. Silva, M.I. Endres, M.A. Ballou and K.L Luchterhand. 2016. Social stressors and their effects on immunity and health of periparturient dairy cows. *J. Dairy Sci.* 99:3217–3228.

Collier, R.J., G.E. Dahl and M.J. VanBaale. 2006. Major advances associated with environmental effects on dairy cattle. *J. Dairy Sci.* 89:1244–1253.

Cook, N.B. 2003. Prevalence of lameness among dairy cattle in Wisconsin as a function of housing type and stall surface. *J. Am. Vet. Med. Assoc.* 223:1324–1328.

Cook, N.B., T.B. Bennett and K.V. Nordlund. 2004. Effect of free stall surface on daily activity patterns in dairy cows with relevance to lameness prevalence. *J. Dairy Sci.* 87:2912–2922.

Cook, N.B., R.L. Mentink, T.B. Bennet and K. Burgi. 2007. The effect of heat stress and lameness on time budgets of lactating dairy cows. *J. Dairy Sci.* 90:1674–1682.

Cook, N.B. and K.V. Nordlund. 2009. The influence of the environment on dairy cow behavior, claw health and herd lameness dynamics. *Vet. J.* 179:360–369.

Cook, N.B., J.P Hess, M.R. Foy, T.B. Bennett and R.L. Brotzman. 2016. Management characteristics, lameness, and body injuries of dairy cattle housed in high-performance dairy herds in Wisconsin. *J. Dairy Sci.* 99:5879–5891.

Cramer, G., K.D. Lissemore, C.L. Guard, K.E. Leslie and D.F. Kelton. 2009. Herd-level risk factors for seven different foot lesions in Ontario Holstein cattle housed in tie stalls or free stalls. *J. Dairy Sci.* 92:1404–1411.

De Vries, A., H. Dechassa and H. Hogeveen. 2016. Economic evaluation of stall stocking density of lactating dairy cows. *J. Dairy Sci.* 99:3848–3857.

Dippel, S., M. Dolezal, C. Brenninkmeyer, J. Brinkmann, S. March, U. Knierim and C. Winckler. 2009. Risk factors for lameness in freestall-housed dairy cows across two breeds, farming systems, and countries. *J. Dairy Sci.* 92:5476–5486.

Dohmen, W., F. Neijenhuis and H. Hogeveen. 2010. Relationship between udder health and hygiene on farms with an automatic milking system. *J. Dairy Sci.* 93:4019–4033.

Drissler, M., M. Gaworski, C.B. Tucker, and D.M. Weary. 2005. Freestall maintenance: Effects on lying behavior of dairy cattle. *J. Dairy Sci.* 88:2381–2387.

Dufour, S., A. Fréchette, H.W. Barkema, A. Mussell and D.T. Scholl. 2011. Invited review: Effect of udder health management practices on herd somatic cell count. *J. Dairy Sci.* 94:563–579.

Ellis, K.A., G.T. Innocent, M. Mihm, P. Cripps, W.G. McLean, C.V. Howard and D. Grove-White. 2007. Dairy cow cleanliness and milk quality on organic and conventional farms in the UK. *J. Dairy Res.* 74:302–310.

Esposito, G., P.C. Irons and E.C. Webb. 2014. Interactions between negative energy balance, metabolic diseases, uterine health and immune response in transition dairy cows. *Anim. Reprod. Sci.* 144:60–71.

Fabian, J., R.A. Laven and H.R. Whay. 2014. The prevalence of lameness on New Zealand dairy farms: A comparison of farmer estimate and locomotion scoring. *Vet. J.* 201:31–38.

Fraser, D. and L.R. Matthews. 1997. Preference and motivation testing. In: Appleby, M.C., B.O. Hughes (Eds.), *Animal welfare.* CAB International, Oxon, pp. 159–173.

Fregonesi, J.A., and J.D. Leaver. 2001. Behaviour, performance and health indicators of welfare for dairy cows housed in strawyard or cubicle systems. *Livest. Prod. Sci.* 68:205–216.

Fregonesi, J.A., C.B. Tucker and D.M. Weary. 2007. Overstocking reduces lying time in dairy cows. *J. Dairy Sci.* 90:3349–3354.

Fregonesi, J.A., M.A.G. von Keyserlingk, C.B. Tucker, D.M. Veira, and D.M. Weary. 2009. Neck-rail position in the free stall affects standing behavior and udder and stall cleanliness. *J. Dairy Sci.* 92:1979–1985.

Fulwider, W.K. and R.W. Palmer. 2005. Effects of stall design and rubber alley mats on cow behavior in freestall barns. *Prof. Anim. Sci.* 21:97–106.

Garry, F.B. 2004. Animal well-being in the US dairy industry. In: J. B. Benson, J.B., B.E. B. E. Rollin (Eds.), *The well-being of farm animals.* Blackwell, Ames, IA, pp. 207–240.

Haskell, M.J., L.J. Rennie, V.A. Bowell, M.J. Bell and A.B. Lawrence. 2006. Housing system, milk production, and zero-grazing effects on lameness and leg injury in dairy cows. *J. Dairy Sci.* 89:4259–4266.

Haufe, H.C., L. Gygax, B. Wechsler, M. Stauffacher and K. Friedli. 2012. Influence of floor surface and access to pasture on claw health in dairy cows kept in cubicle housing systems. *Prev. Vet. Med.* 105:85–92.

Hemsworth, P.H., J.L. Barnett, L. Beveridge and L.R. Matthews. 1995. The welfare of extensively managed dairy cattle: A review. *Appl. Anim. Behav. Sci.* 42:161–182.

Hernandez-Mendo, O., M.A.G. von Keyserlingk, D.M. Veira, and D.M. Weary. 2007. Effects of pasture on lameness in dairy cows. *J. Dairy Sci.* 90:1209–1214.

Huzzey, J.M., T.J. DeVries, P. Valois and M.A.G von Keyserlingk. 2006. Stocking density and feed barrier design affect the feeding and social behavior of dairy cattle. *J. Dairy Sci.* 89:126–133.

Ismail, Z.B. 2017. Mastitis vaccines in dairy cows: Recent developments and recommendations of application. *Vet World.* 10:1057–1062.

Ito, K., M.A.G. von Keyserlingk, S.J. LeBlanc and D.M. Weary. 2010. Lying behavior as an indicator of lameness in dairy cows. *J. Dairy Sci.* 93:3553–3560.

Jacobs, J.A. and J.M. Siegford. 2012. Invited review: The impact of automatic milking systems on dairy cow management, behavior, health, and welfare. *J. Dairy Sci.* 95:2227–2247.

Janni, K.A., M.I. Endres, J.K. Reneau, and W.W. Schoper. 2007. Compost dairy barn layout and management recommendations. *Appl. Eng. Agric.* 23:97–102.

Jensen, M.B., and K.L. Proudfoot. 2017. Effect of group size and health status on behavior and feed intake of multiparous dairy cows in early lactation. *J. Dairy Sci.* 100:9759–9768.

Kendall, P.E., G.A. Verkerk, J.R. Webster and C.B. Tucker. 2007. Sprinklers and shade cool cows and reduce insect-avoidance behavior in pasture-based dairy systems. *J. Dairy Sci.* 90:3671–3680.

Kester, E., M. Holzhauer and K. Frankena. 2014. A descriptive review of the prevalence and risk factors of hock lesions in dairy cows. *Vet. J.* 202:222–228.

King, M.T.M, E.A. Pajor, S.J. LeBlanc and T.J. DeVries. 2016. Associations of herd-level housing, management, and lameness prevalence with productivity and cow behavior in herds with automated milking systems. *J. Dairy Sci.* 99:9069–9079.

Krawczel, P.D., L.B. Klaiber, R.E. Butzler, L.M. Klaiber, H.M. Dann, C.S. Mooney and R.J. Grant. 2012. Short-term increases in stocking density affect the lying and social behavior, but not the productivity, of lactating Holstein dairy cows. *J. Dairy Sci.* 95:4298–4308.

Krohn, C.C. 1994. Behaviour of dairy cows kept in extensive (loose housing/pasture) or intensive (tie stall) environments. III. Grooming, exploration and abnormal behaviour. *Appl. Anim. Behav. Sci.* 42:73–86.

LeBlanc, S.J., K.D. Lissemore, D.F. Kelton, T.F. Duffield and K.E. Leslie. 2006. Major advances in disease prevention in dairy cattle. *J. Dairy Sci.* 89:1267–1279.

Legrand, A.L., M.A.G von Keyserlingk and D.M. Weary. 2009. Preference and usage of pasture versus free-stall housing by lactating dairy cattle. *J. Dairy Sci.* 92:3651–3658.

Lim, P.Y., J.N. Huxley, M.J. Green, A.R. Othman, S.L. Potterton and J. Kaler. 2013. An investigation into the association between lameness and hair loss on the hock, in a longitudinal study. *Proceedings of the 17th International Conference on Lameness in Ruminants,* 11–14th August 2013, Bristol, UK, pp. 273–274.

Lobeck, K.M., M.I. Endres, E.M. Shane, S.M. Godden and J. Fetrow. 2011. Animal welfare in cross-ventilated, compost-bedded pack, and naturally ventilated dairy barns in the upper Midwest. *J. Dairy Sci.* 94:5469–5479.

Lobeck-Luchterhand, K.M., P.R.B. Silva, R.C. Chebel and M.I. Endres. 2014. Effect of prepartum grouping strategy on displacements from the feed bunk and feeding behavior of dairy cows. *J. Dairy Sci.* 97:2800–2807.

Lobeck-Luchterhand, K.M., P.R.B. Silva, R.C. Chebel and M.I. Endres. 2015. Effect of stocking density on social, feeding and lying behavior of prepartum dairy animals. *J. Dairy Sci.* 98:240–249.

Lombard, J.E., C.B. Tucker, M.A.G von Keyserlingk, C.A. Kopral and D.M. Weary. 2010. Associations between cow hygiene, hock injuries, and free stall usage on US dairy farms. *J. Dairy Sci.* 93:4668–4676.

Mader, T.L., J.M. Dahlquist and J.B. Gaughan. 1997. Wind protection effects and airflow patterns in outside feedlots. *J. Anim. Sci.* 75:26–36.

Metz, J.H.M. 1985. The reaction of cows to a short-term deprivation of lying. *Appl. Anim. Behav. Sci.* 13:301–307.

Miller, K. and D.G.M. Wood-Gush. 1991. Some effects of housing on the social behavior of dairy cows. *Anim. Prod. Sci.* 53:271–278.

Munksgaard, L. and H.B. Simonsen. 1996. Behavioral and pituitary adrenal-axis responses of dairy cows to social isolation and deprivation of lying down. *J. Anim. Sci.* 74:769–778.

Munksgaard, L., M.B. Jensen, L.J. Pedersen, S.W. Hansen and L. Matthews. 2005. Quantifying behavioural priorities—Effects of time constraints on behaviour of dairy cows, Bos taurus. *Appl. Anim. Behav. Sci.* 92:3–14.

National Farm Animal Care Council. 2009. Code of practice for the care and handling of animals—Dairy cattle. Accessed: January 12, 2018. www.nfacc.ca/codes-of-practice/dairy-cattle/code.

Navarro, G., L.E. Green and N. Tadich. 2013. Effect of lameness and lesion specific causes of lameness on time budgets of dairy cows at pasture and when housed. *Vet. J.* 197:788–793.

Nielsen, B.H., P.T. Thomsen and J.T. Sørensen. 2011. Identifying risk factors for poor hind limb cleanliness in Danish loose-housed dairy cows. *Animal.* 5:1613–1619.

O'Connell, J., P.S. Gille and W. Meaney. 1989. A comparison of dairy cattle behavioural patterns at pasture and during confinement. *Irish J. Agric. Res.* 28:65–72.

Olde Riekerink, R.G.M., H.W. Barkema, D.F. Kelton and D.T. Scholl. 2008. Incidence rate of clinical mastitis on Canadian dairy farms. *J. Dairy Sci.* 91(4):1366–1377.

Olmos, G., L. Boyle, A. Hanlon, J. Patton, J.J. Murphy and J.F. Mee. 2009. Hoof disorders, locomotion ability and lying times of cubicle-housed compared to pasture-based dairy cows. *Livest. Sci.* 125:199–207.

Ortiz, X.A., J.F. Smith, B.J. Bradford, J.P. Harper and A. Oddy. 2010. A comparison of the effects of 2 cattle-cooling systems on dairy cows in a desert environment. *J. Dairy Sci.* 93:4955–4960.

Ortiz, X.A., J.F. Smith, F. Rojano, C.Y. Choi, J. Bruer, T. Steele, N. Schuring, J. Allen and R.J. Collier. 2015. Evaluation of conductive cooling of lactating dairy cows under controlled environmental conditions. *J. Dairy Sci.* 98:1759–1771.

Polsky, L. and M.A.G. von Keyserlingk. 2017. Invited review: Effects of heat stress on dairy cattle welfare. *J. Dairy Sci.* 100:8645–8657.

Popescu, S., C. Borda, E.A. Diugan, M. Spinu, I.S. Groza and C.D. Sandru. 2013. Dairy cows welfare quality in tie-stall housing system with or without access to exercise. *Acta Vet. Scand.* 55:43.

Proudfoot, K.L., D.M. Veira, D.M. Weary and M.A.G. von Keyserlingk. 2009. Competition at the feed bunk changes the feeding, standing and social behavior of transition cows. *J. Dairy Sci.* 92:3116–3123.

Ranjbar, S., A.R. Rabiee, A. Gunn, and J.K. House. 2016. Identifying risk factors associated with lameness in pasture-based dairy herds. *J. Dairy Sci.* 99:7495–7505.

Redbo, I. 1992. The influence of restraint on the occurrence of oral stereotypies in dairy cows. *Appl. Anim. Behav. Sci.* 35:115–123.

Redbo, I. 1993. Stereotypies and cortisol secretion in heifers subjected to tethering. *Appl. Anim. Behav. Sci.* 38:213–225.

Reich, L.J., D.M. Weary, D.M. Veira and M.A.G. von Keyserlingk. 2010. Effects of sawdust bedding dry matter on lying behavior of dairy cows: A dose-dependent response. *J. Dairy Sci.* 93:1561–1565.

Robbins, J.A., M.A.G. von Keyserlingk, D. Fraser and D.M. Weary. 2016. Invited review: Farm size and animal welfare. *J. Anim. Sci.* 94:5439–5455.

Roca-Fernández, A.I., C.P. Ferris and A. González-Rodríguez. 2013. Short communication. Behavioural activites of two dairy cow genotypes (Holstein-Friesian vs. Jersey X Holstein-Friesian) in two milk production systems (grazing vs. confinement). *Span. J. Agric. Res.* 11:120–126.

Ruegg, P.L. 2017. A 100-year review: Mastitis detection, management and prevention. *J. Dairy Sci.* 100:10381–10397.

Rupp, R. and D. Boichard. 2003. Genetics of resistance to mastitis in dairy cattle. *Vet. Res.* 34:671–688.

Rushen J., A.M. de Passillé, M.A.G. von Keyerslingk and D.M. Weary. 2008. *The welfare of cattle*. Springer, Dordecht.

Sant'Anna, A.C. and M.J.R. Paranhos da Costa. 2011. The relationship between dairy cow hygiene and somatic cell count in milk. *J. Dairy Sci.* 94:3835–3844.

Schefers, J.M., K.A. Weigel, C.L Rawson, N.R. Zwald and N.B. Cook. 2010. Management practices associated with conception rate and service rate of lactating Holstein cows in large, commercial dairy herds. *J. Dairy Sci.* 93:1459–1467.

Schütz, K.E., N.R. Cox and L.R. Matthews. 2008. How important is shade to dairy cattle? Choice between shade or lying following different levels of lying deprivation. *Appl. Anim. Behav. Sci.* 114:307–318.

Schütz, K.E., A.R. Rogers, Y.A. Pouloun, N.R. Cox and C.B. Tucker. 2010. The amount of shade influences the behavior and physiology of dairy cattle. *J. Dairy Sci.* 93:125–133.

Schütz, K.E., A.R. Rogers, N.R. Cox, J.R. Webster and C.B. Tucker. 2011. Dairy cattle prefer shade over sprinklers: Effects on behavior and physiology. *J. Dairy Sci.* 94:273–283.

Shook, G.E., R.L. Bamber Kirk, F.L. Welcome, Y.H. Schukken and P.L. Ruegg. 2017. Relationship between intramammary infection prevalence and somatic cell score in commercial dairy herds. *J. Dairy Sci.* 100:9691–9701.

Singh, S.S., W.R. Ward, K. Lautenbach, J.W. Hughes and R.D. Murray. 1993. Behaviour of first lactation and adult dairy cows while housed and at pasture and its relationship with sole lesions. *Vet. Rec.* 133:469–474.

Sogstad, A.M., T. Fjeldaas, O. Osteras, and K. Plym Forshell. 2005. Prevalence of claw lesions in Norwegian dairy cattle housed in tie stalls and free stalls. *Prev. Vet. Med.* 70:191–209.

Solano, L., H.W. Barkema, E.A. Pajor, S. Mason, S.J. LeBlanc, C.G.R. Nash, D.B. Haley, D. Pellerin, J. Rushen, A.M. de Passillé, E. Vasseur and K. Orsel. 2016. Associations between lying behavior and lameness in Canadian Holstein-Friesian cows housed in freestall barns. *J. Dairy Sci.* 99:2086–2101.

Somers, J.G.C.J., K. Frankena, E.N. Noordhuizen-Stassen and J.H.M. Metz. 2003. Prevalence of claw disorders in Dutch dairy cows exposed to several floor systems. *J. Dairy Sci.* 86:2082–2093.

St-Pierre, N.R., B. Cobanov and G. Schnitkey. 2003. Economic losses from heat stress by U.S. livestock industries. *J. Dairy Sci.* 86:E52–E77.

Stiglbauer, K.E., K.M. Cicconi-Hogan, R. Richert, Y.H. Schukken, P.L. Ruegg and M. Gamroth. 2013. Assessment of herd management on organic and conventional dairy farms in the United States. *J. Dairy Sci.* 96:1290–1300.

Talebi, A., M.A.G. von Keyserlingk, E. Telezhenko and D.M. Weary. 2014. Reduced stocking density mitigates the negative effects of regrouping in dairy cattle. *J. Dairy Sci.* 97:1358–1363.

Telezhenko, E., M.A.G. von Keyserlingk, A. Talebi and D.M. Weary. 2012. Effect of pen size, group size and stocking density on activity in freestall-housed dairy cows. *J. Dairy Sci.* 95:3064–3069.

Tucker, C.B., D.M. Weary and D. Fraser. 2003. Effects of three types of free-stall surfaces on preferences and stall usage by dairy cows. *J. Dairy Sci.* 86:521–529.

Tucker, C.B. and D.M. Weary. 2004. Bedding on geotextile mattresses: How much is needed to improve cow comfort? *J. Dairy Sci.* 87:2889–2895.

Tucker, C.B., D.M. Weary and D. Fraser. 2005. Influence of neck-rail placement on free-stall preference, use, and cleanliness. *J. Dairy Sci.* 88:2730–2737.

Tucker, C.B., G. Zdanowicz and D.M. Weary. 2006. Brisket boards reduce freestall use. *J. Dairy Sci.* 89:2603–2607.

Tucker, C.B., A.R. Rogers, G.A. Verkerk, P.E. Kendall, J.R. Webster and L.R. Matthews. 2007. Effects of shelter and body condition on the behavior and physiology of dairy cattle in winter. *Appl. Anim. Behav. Sci.* 105(2007):1–13.

Tucker, C.B., D.M. Weary, M.A.G. von Keyserlingk, K.A. Beauchemin. 2009. Cow comfort in tie-stalls: Increased depth of shavings or straw bedding increases lying time. *J. Dairy Sci.* 92:2684–2690.

USDA. 2016. Dairy 2014, Dairy Cattle Management Practices in the United States, 2014 USDA–APHIS–VS–CEAH–NAHMS. Fort Collins, CO #692.0216. www.aphis.usda.gov/animal_health/nahms/dairy/downloads/dairy14/Dairy14_dr_PartI.pdf.

Van Gastelen, S., B. Westerlaan, D.J. Houwers and F.J.C.M. van Eerdenburg. 2011. A study on cow comfort and risk for lameness and mastitis in relation to different types of bedding materials. *J. Dairy Sci.* 94:4878–4888.

Van Laer, E., F.A. Tuyttens, B. Ampe, B. Sonck, C.P. Moons and L. Vandaele. 2015. Effect of summer conditions and shade on the production and metabolism of Holstein dairy cows on pasture in temperate climate. *Animal*. 9:1547–1558.

Veissier, I., S. Andanson, H. Dubroeucq and D. Pomiès. 2008. The motivation of cows to walk as thwarted by tethering. *J. Anim. Sci.* 86:2723–2729.

Veissier, I., E. Van Laer., R. Palme, C.P.H. Moons, B. Ampe, B. Sonck. S. Andanson and F.A.M. Tuyttens. 2018. Heat stress in cows at pasture and benefit of shade in a temperate climate region. *Int. J. Biometeorol.* 4:585–595.

von Keyserlingk, M.A.G., D. Olenick and D.M. Weary. 2008. Acute behavioral effects of regrouping dairy cows. *J. Dairy Sci.* 91:1011–1016.

von Keyserlingk, M.A.G., J. Rushen, A.M. de Passillé and D.M. Weary. 2009. Invited review: The welfare of dairy cattle—Key concepts and the role of science. *J. Dairy Sci.* 92:4101–4111.

von Keyserlingk, M.A.G., A. Barrientos, K. Ito, E. Galo and D.M. Weary. 2012. Benchmarking cow comfort on North American freestall dairies: Lameness, leg injuries, lying time, facility design and management for high-producing Holstein dairy cows. *J. Dairy Sci.* 95:7399–7408.

von Keyserlingk, M.A.G., A.A. Cestari, B. Franks, J.A. Fregonesi, D.M. Weary. 2017. Dairy cows value access to pasture as highly as fresh feed. *Sci. Rep.* 7:44953.

Washburn, S.P., S.L. White, J.T. Green Jr. and G.A. Benson. 2002. Reproduction, mastitis, and body condition of seasonally calved Holstein and Jersey cows in confinement or pasture systems. *J. Dairy Sci.* 85:105–111.

Weary, D.M., B.A. Ventura and M.A.G. Von Keyserlingk. 2016. Societal views and animal welfare science: understanding why the modified cage may fail and other stories. *Animal*. 10:309–317.

Wells S.J., A.M. Trent, W.E. Marsh, and R.A. Robinson. 1993. Prevalence and severity of lameness in lactating dairy cows in a sample of Minnesota and Wisconsin dairy herds. *J. Am. Vet. Med. Assoc.* 202:78–82.

Wenz, J.R., S.M. Jensen, J.E. Lombard, B.A. Wagner and R.P. Dinsmore. 2007. Herd management practices and their association with bulk tank somatic cell count on United States dairy operations. *J. Dairy Sci.* 90:3652–3659.

West, K.W. 2003. Effects of heat-stress on production in dairy cattle. *J. Dairy Sci.* 86:2131–2144.

Westin, R., A. Vaughan, A.M. de Passillé, T.J. DeVries, E.A. Pajor, D. Pellerin, J.M. Siegford, E. Vasseur and J. Rushen. 2016. Lying times of lactating cows on dairy farms with automatic milking systems and the relation to lameness, leg lesions, and body condition score. *J. Dairy Sci.* 99:551–561.

Whay, H.R., A.E. Waterman, A.J.F. Webster and J.K. O'brien. 1998. The influence of lesion type on the duration of hyperalgesia associated with hindlimb lameness in dairy cattle. *Vet. J.* 156:23–29.

Wierenga, H.K. and H. Hopster. 1990. The significance of cubicles for the behaviour of dairy cows. *Appl. Anim. Behav. Sci.* 26:309–337.

Zdanowicz, M., J.A. Shelford, C.B. Tucker, D.M. Weary and M.A.G. von Keyserlingk. 2004. Bacterial populations on teat ends of dairy cows housed in free stalls and bedded with either sand or sawdust. *J. Dairy Sci.* 87:1694–1701.

Zurbrigg, K., D. Kelton, N. Anderson, S. Millman. 2005. Stall dimensions and the prevalence of lameness, injury, and cleanliness on 317 tie-stall dairy farms in Ontario. *Can. Vet. J.* 46:902.

Index

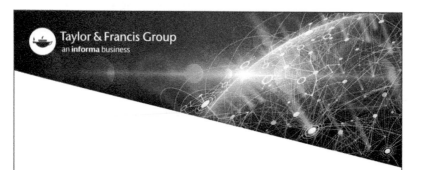

Taylor & Francis Group
an **informa** business

Taylor & Francis eBooks

www.taylorfrancis.com

A single destination for eBooks from Taylor & Francis
with increased functionality and an improved user
experience to meet the needs of our customers.

90,000+ eBooks of award-winning academic content in
Humanities, Social Science, Science, Technology, Engineering,
and Medical written by a global network of editors and authors.

TAYLOR & FRANCIS EBOOKS OFFERS:

A streamlined
experience for
our library
customers

A single point
of discovery
for all of our
eBook content

Improved
search and
discovery of
content at both
book and
chapter level

REQUEST A FREE TRIAL
support@taylorfrancis.com

 Routledge
Taylor & Francis Group

 CRC Press
Taylor & Francis Group

Printed and bound by CPI Group (UK) Ltd, Croydon, CR0 4YY

01/11/2024

01782598-0003